HUMAN-MACHINE INTERACTIVE SYSTEMS

AF740194

LANGUAGES AND INFORMATION SYSTEMS

Series Editor: Shi-Kuo Chang
University of Pittsburgh
Pittsburgh, Pennsylvania
and Knowledge Systems Institute
Skokie, Illinois

*Series was formerly entitled
MANAGEMENT AND INFORMATION SYSTEMS

HUMAN-MACHINE INTERACTIVE SYSTEMS
Edited by Allen Klinger

LANGUAGES FOR AUTOMATION
Edited by Shi-Kuo Chang

MANAGEMENT AND OFFICE INFORMATION SYSTEMS
Edited by Shi-Kuo Chang

VISUAL LANGUAGES
Edited by Shi-Kuo Chang, Tadao Ichikawa, and Panos A. Ligomenides

VISUAL LANGUAGES AND APPLICATIONS
Edited by Tadao Ichikawa, Erland Jungert, and Robert R. Korfhage

VISUAL LANGUAGES AND VISUAL PROGRAMMING
Edited by Shi-Kuo Chang

A Continuation Order Plan is available for this series. A continuation order will bring delivery of each new volume immediately upon publication. Volumes are billed only upon actual shipment. For further information please contact the publisher.

HUMAN–MACHINE INTERACTIVE SYSTEMS

Edited by

Allen Klinger

University of California at Los Angeles
Los Angeles, California

PLENUM PRESS • NEW YORK AND LONDON

QA
76.9
.H85
.H8

Library of Congress Cataloging-in-Publication Data

Human-machine interactive systems / edited by Allen Klinger.
p. cm. -- (Languages and information systems)
Includes bibliographical references and index.
ISBN 0-306-43758-9
1. Human-computer interaction. 2. Man-machine systems.
I. Klinger, A. (Allen), 1937- . II. Series.
QA76.9.H85H89 1991
005.1--dc20 91-18966
CIP

CARDIFF
6 MAR 1995

ISBN 0-306-43758-9

© 1991 Plenum Press, New York
A Division of Plenum Publishing Corporation
233 Spring Street, New York, N.Y. 10013

All rights reserved

No part of this book may be reproduced, stored in a retrieval system, or transmitted in any form or by any means, electronic, mechanical, photocopying, microfilming, recording, or otherwise, without written permission from the Publisher

Printed in the United States of America

CONTRIBUTORS

RUSSELL J. ABBOTT • *Department of Mathematics and Computer Science, California State University, Los Angeles, California 90032*

SANJAY BHASIN • *Image Processing Center, Drexel University, Philadelphia, Pennsylvania 19104*

JOHN BRADLEY • *Computer and Information Science Department, University of Pennsylvania, Philadelphia, Pennsylvania 19104*

ALFONSO CARDENAS • *Computer Science Department, University of California, Los Angeles, California 90024-1596*

BERNARD DIMSDALE • *IBM Scientific Center, Santa Monica, California 90404*

SUSAN F. EHRLICH • *User Interface Department, Wang Laboratories, Inc., Lowell, Massachusetts 01851*

FILIP FUMA • *Computer and Information Science Department, University of Pennsylvania, Philadelphia, Pennsylvania 19104*

ALESSANDRO GIACALONE • *Department of Computer Science, State University of New York at Stony Brook, Stony Brook, New York 11794-4400*

ALFRED INSELBERG • *IBM Scientific Center, Santa Monica, California 90404 and Department of Computer Science, University of Southern California, Los Angeles, California 90089-0782*

KEITH S. JOSEPH • *Communications Applications Research, Pacific Bell, San Ramon, California 94583*

ARIE KAUFMAN • *Department of Computer Science, State University of New York at Stony Brook, Stony Brook, New York 11794-4400*

ALLEN KLINGER • *Computer Science Department, University of California, Los Angeles, California 90024-1596*

ROBERT R. KORFHAGE • *Department of Information Science, University of Pittsburgh, Pittsburgh, Pennsylvania 15260*

DEEPA KRISHNAN • *Department of Information Science, Faculty of Science, The University of Tokyo, 7-3-1, Hongo, Bunkyo-ku, Tokyo 113, Japan*

TOSIYASU L. KUNII • *Department of Information Science, Faculty of Science, The University of Tokyo, 7-3-1, Hongo, Bunkyo-ku, Tokyo 113, Japan*

STEPHEN R. LEVINE • *User Interface Department, Wang Laboratories Inc., Lowell, Massachusetts 01851*

JODY PAUL • *Social Policy Department, The Rand Corporation, Santa Monica, California 90406*

ARTURO PIZANO • *Computer Science Department, University of California, Los Angeles, California 90024-1596*

J. W. ROACH • *Department of Computer Science, Virginia Polytechnic Institute and State University, Blacksburg, Virginia 24061*

YUKARI SHIROTA • *Software Research Center, Software Division, RICOH Company, Ltd., Tomin-Nissei-Kasugacho-Building, 1-1-17, Koishikawa, Bunkyo-ku, Tokyo 112, Japan*

OLEH J. TRETIAK • *Image Processing Center, Drexel University, Philadelphia, Pennsylvania 19104*

RONI YAGEL • *Department of Computer Science, State University of New York at Stony Brook, Stony Brook, New York 11794-4400*

PREFACE

Many hardware devices present either results or alternatives selected by computers to users. A few are video display terminals (VDTs), touch-tone telephones, and computer-generated speech systems. In part this book concerns the impact and implications of such tools. Alternatively this is an attempt to provide material for researchers, students, and managers concerned with computer interfaces. The subject of computer interfaces is at one level a technical subarea sharing common interests with the broad disciplines of computer science, psychology, and bioengineering. However, it is also a topic thrust to the forefront of interest of a wide variety of individuals who confront one of the most striking technological changes that has occurred in human history—the introduction of contact with computing devices as an essential component of many kinds of ordinary transactions. Point of entry sales, travel and entertainment reservations, and library information, are commonly conducted today by interaction with digital calculating devices that did not exist in the recent past.

The papers in this book present several concerns arising from the widespread use of computing. One involves the future implications of further advances of this technology. This is a twofold issue: (a) the potential consequences of changing the basic way that information is managed in areas ranging from design, engineering, and management/planning to information access, education, and clerical function; and (b) improvements that could be instituted from further development of the special characteristics of display techniques, technologies, and algorithms. The latter provides another way to describe the chapters that appear here.

The prevalence of information-presenting hardware in society today is a consequence of computer programs that turn digital computers into useful symbol-handling tools. The utility comes from two features not immediately apparent when computers are viewed as calculating machines: high speed, and large volume of accessible data. The authors of the chapters presented here address several ways that both features have been put to use; they also reflect on the consequences of existing or likely new technology involving massive increase in the amount of storage, hence the size of data sets that are computer-accessible.

The contributions fall into two main groups. One concerns such broad themes as *knowledge*, *information*, and *communication*, as well as recent special emphasis in computing on techniques involving *menus*, *icons*, and *objects*. The other involves extending the capability of the computer to images, and senses, items normally the province of people such as *visualization*, *depth*, and *touch*. The first group represents data and applications currently well underway in terms of adaption to the computer, but also includes recent development of techniques to support interactive and personal computing. The second group extends computing power to design, planning, and complex data. Hence there are two parts to the organization of chapters: part one is called "Accessing Knowledge," part two, "Display and Design."

In both areas new computer-based products require *learning*. Knowledge of proper keystroke sequences or technical areas of applied mathematics, computer software, or hardware systems: prefix and postfix expressions; stack, tree, and array concepts; and active and secondary memory are relatively technical for the current workplace. The rapid pace of technological change and lack of standardization leads to needs that are being met by individuals such as those who contribute to this volume. Many of the chapters in this book address learning issues, and concern steps toward more useful methods to handle tasks using computer technology.

The book began as an attempt to communicate with a few graduate students in the UCLA Computer Science Department. Many thanks are due them, and my colleagues, for their helpful assistance to me that enabled me to gain some competence in dealing with computer interfaces. Thanks are also due to the organizers of the Third International Conference on Human–Computer Interaction, that took place in Boston, September 1989, particulary to Michael J. Smith, University of Wisconsin, the Program Chairman, whose electronic mail helped me to the process of organizing two sessions at that meeting. Six of the chapters were presented in those sessions in earlier form and one other was included in the session proposal. Valued assistance by Susanna Reyn aided in preparing manuscript copies of several of the papers. Numerous individuals contributed technical expertise to further the process of evolving the contributions as they are now in this book; a partial list of technical reviewers appears following the Table of Contents. Apologies are offered to those who have been inadvertently omitted, and grateful thanks to all.

The fourteen chapters in this book, like the broad society as a whole, are only beginning the process of coming to terms with the implications of the new reality, that powerful information-handling tools are widely available, and that many kinds of individuals deal with computer interfaces. Some are scientists and engineers. Others are unfamiliar with science and mathematics, and often resent computer requirements. We all, the authors, reviewers, and other contributors to this work, have begun to address the wide use of computers in design or commerce, but there is much more to be concerned with as we put greater and greater information-handling power in the hands of society. The hope here is that this work will contribute to our understanding of the power and potential of the new computer technology.

ALLEN KLINGER

Los Angeles, California

CONTENTS

PART I
ACCESSING KNOWLEDGE

PART II
DISPLAY AND DESIGN

I

ACCESSING KNOWLEDGE

THE FREESTYLE SYSTEM
A DESIGN PERSPECTIVE

STEPHEN R. LEVINE AND SUSAN F. EHRLICH

1. Introduction

The increase in speed, memory, and display characteristics of personal computers creates a new opportunity to support a larger set of fundamental types of human communication. In order to communicate with groups of people in the past, we have used standard alphabets that are efficiently reproduced using typing, printing, word processing, and copiers. People have shared these typed messages across distance using sophisticated means of transporting both paper and electronic mail. While we have grown to accept our dependence on the written word to reach large numbers of people quickly, we have also sought ways to use technology to capture more fundamental types of human communication.

The telephone has allowed us to speak across distance, while the television has allowed us to capture most dimensions of communication—including speech, pointing, and gesture—in one-to-many communication. We have continued to look for two-way and multiway communication media that can capture these multidimensional fundamentals. Video conferencing and videophones have been tried with some limited success.[1, 2]

People can now *point* to things on computer screens with free-roaming cursors, capture *voice* in compressed computer files, as well as capture *hand-drawn input* into drawing packages, typically with a mouse. In face-to-face meetings, people rarely use these types of communication separately. There is a need, then, to simultaneously capture them as they are *used in concert* and share them via electronic mail.

The Freestyle system was designed to capture synchronized pointing, speaking, drawing, and writing with electronic versions of familiar tools: paper, pencil, and voice handset/speaker-phone (Fig. 1). People simultaneously use combinations of real world versions of these tools to

STEPHEN R. LEVINE and SUSAN F. EHRLICH • *User Interface Department,* Wang Laboratories, Inc., Lowell, Massachusetts 01851.

FIGURE 1. Components of the Freestyle system including pencil, tablet, and voice handset. © Wang Laboratories Inc., 1989.

communicate with each other in the initial stages of almost all work. As a project solidifies, a formal record is often kept in a typed form that can easily be reproduced and distributed. But up to this point, the computer has not supported the capture and distribution of synchronized verbal and handwritten communication. Nor has it supported such synchronized discussions *about* formal documents.

This new type of electronic mail is made possible partially because of the increasing ability of high-speed graphic interfaces to provide rich, immediate, multidimensional *feedback* that is similar to that in the real world. (See discussion of the data glove[3] and the alternate reality kit[4] below.) Even as display capabilities of personal computers have been improving, so has the ability to interconnect them. With the general introduction of networking into the work force, there has been an upsurge in the use of computer-based mail systems. The Freestyle project was in the position to take advantage of existing networks.

The design of this system was based on a combination of original ideas and discovery of needs through laboratory and field research. It was partially driven by a theory of users' communication needs and by a particular approach to graphics design based on real-world mimicry. This chapter will outline how various contributions to the design were interwoven as it evolved over time from an initial prototype through multiple versions that were tested in business environments. (See also Francik and Akagi,[5] Perkins *et al*,[6] and Hsiao and Levine[7] for details of the design process.)

2. *User Interface Design*

Freestyle was designed to be a voice and handwriting based communication system that sits on top of the current world of personal computing. It leaves intact the DOS operating system and PC software used for calculation, typing, or record keeping (e.g., spreadsheets, word processing, database, etc.)

2.1. *Image Annotation*

The starting point was to enable the user to capture any visible screen shown on a standard PC (including a completely blank screen, or an image that was previously scanned into the PC) and convert it into an "electronic piece of paper." Once the screen was frozen, it could then be pointed at and written on using a tablet with a pencil-shaped stylus (Francik and Akagi[5]).

The system also provided the ability to record speech via a handset/speaker-phone even as the user is writing or pointing with the stylus (Hsiao and Levine[7]), thus supporting four fundamental means for synchronized

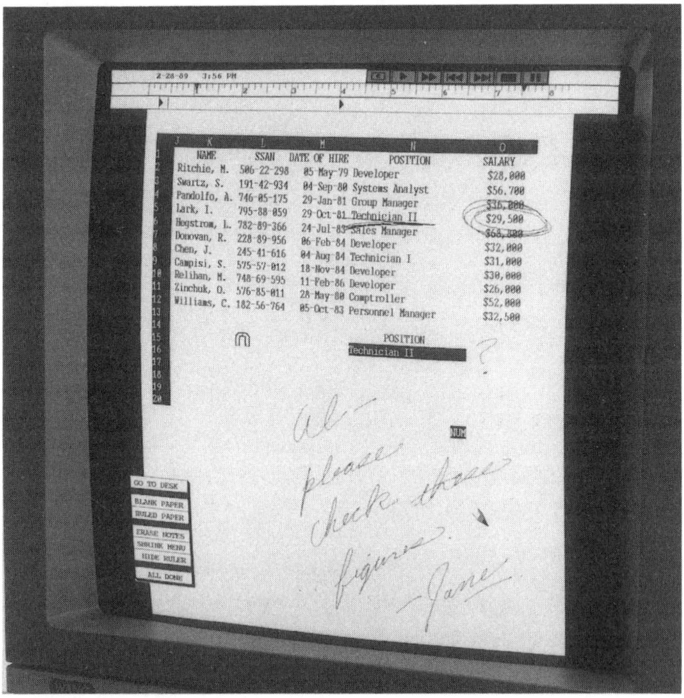

FIGURE 2. The Freestyle annotation environment. © Wang Laboratories Inc., 1989.

communication from the real world: pointing, speaking, drawing, and writing. The ability to type anywhere on the screen image was also provided. The combined inputs produce a dynamic multimedia message that can be mailed to others via a minicomputer network and played back. (See Fig. 2.)

The stylus and voice handset are familiar communication tools. The system capitalized on this familiarity. In order for the user to write on the screen, he or she only needs to touch the pencil to the tablet and begin writing. (No mode switch or menu navigation is required.) This mimics direct writing on a familiar piece of paper. The stylus was designed to have a pink eraser that works like a regular eraser. This direct mimicry of components of familiar tools was intended to lead people to use them without any training.

2.2. *Graphic Desktop*

The system also provides a familiar desklike environment for keeping track of the various pieces of electronic paper that are created by the user (Fig. 3). This environment was intended to be independent from the user

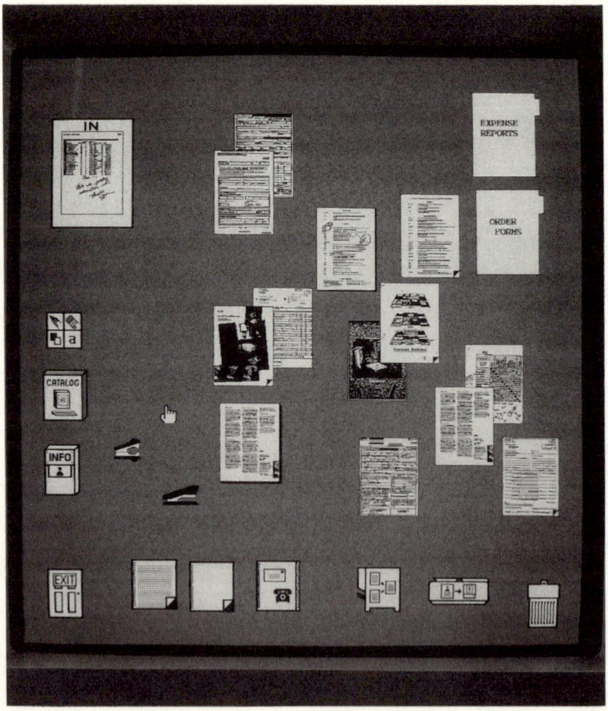

FIGURE 3. The Freestyle desktop. © Wang Laboratories Inc., 1989.

interface of the file structure of the computer (i.e., the DOS operating system).

The annotated "electronic paper" images are stored by the user in miniature form on the desk (which is accessed via a pop-up menu during annotation). The miniatures (or "stamps") first appear in a graphical "inbox." Then they can be moved around the desk with the stylus using a drag motion. A down–up motion of the pen opens selected items to full size—ready for additional annotation.

The Freestyle desktop is an environment for storing and manipulating images used primarily as electronic mail. This contrasts with desktop systems like the Apple MacIntosh that support management of editable character-based files and programs.

Recent research has shown the power of providing immediate graphic responses to fine motor input. For example, the data glove[3] allows finely articulated three-dimensional control of graphic objects on a screen via electrical feedback from the relative positions of the fingers in the glove. Another example is the alternate reality kit,[4] where graphic elements respond to mouse movements according to physical laws of nature (e.g., gravity). Commercially available desktop computers have also begun to allow users to manipulate objects with natural motion feedback (e.g., movement of icons on computer "desktops"). This allows users to predict and instantly confirm system behavior.

Real-world desks are covered with paper and folders. People maintain mental models of the spatial locations of papers and categories of papers on their desks and in their offices.[8] People overlap paper, they create piles, they put things in folders, and they put things of critical importance in places of prominence. Our goal was to provide ways to directly and flexibly mimic these activities.

For example, the miniature pieces of paper can be overlapped and laid on top of each other on haphazard or neat piles as the user wants. The movement of the elements is exactly as fast or slow as the movement of the stylus. The graphic detail of the image is stable during movement and follows rules of object constancy[9, 10] (e.g., hidden objects can be found again "behind" occluding objects). This creates an impression of delicate control over concrete objects. It was important that the stamps maintain as much of the detail of the larger images as possible to allow direct recognition without requiring labeling.

The desk also provides the icons used for organizing the stamps (e.g., stapler, staple remover, folders, trash can), and the icons used for managing communication (e.g., printer, scanner, phone book, individual "mail slots"). Individual pieces of paper can be "stapled together" by moving a stapler icon to a pile. This allows the user to move a pile of pages as a unit. A staple remover icon allows the users to separate the pages. The images can also be inserted into labeled folders.

2.3. Controlling Communication

The annotated images are sent around within an electronic mail network. Complete analogies to real-world experience could not be efficiently used for functions like mailing and printing, so we resorted to the use of icons on the desktop for these and related functions. Rather than taking an electronic piece of paper and putting it in an envelope and handwriting an address, the user drags the piece of paper to a phone book icon, which results in the display of an alphabetized listing from which names can be selected with the stylus. In additon, users can set up individual mail slot icons for specific people or groups that they communicate with frequently. They can also convert an image into a voiceless FAX image to be sent out over a telephone line. Because only a few operations are required to mail or FAX items in this environment, it is more efficient than the real world of interoffice paper mail.

2.4. Metaphor and Efficiency: A Delicate Balance

One of the challenges was maintaining the balance between real-worl mimicry and the need for time saving though less familiar functions. The design team's goal was to minimize learning time for the screen capture and annotation functions and provide responsive feedback, but without sacrificing efficiency. For example, the eraser could be used on small areas, but there is also a much more efficient menu pick for erasing all annotations made in that session (Fig. 4). descriptive verbal labels were needed for such menu functions.

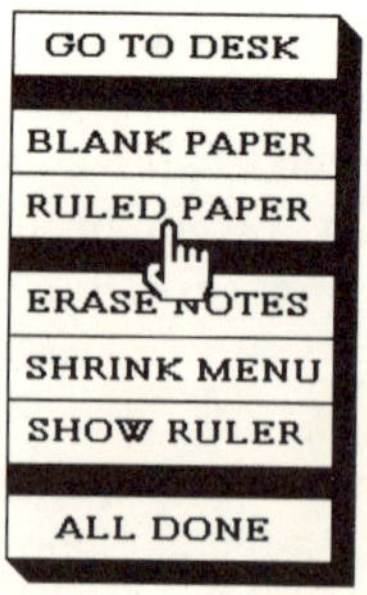

FIGURE 4. The annotation menu. © Wang Laboratories Inc., 1989.

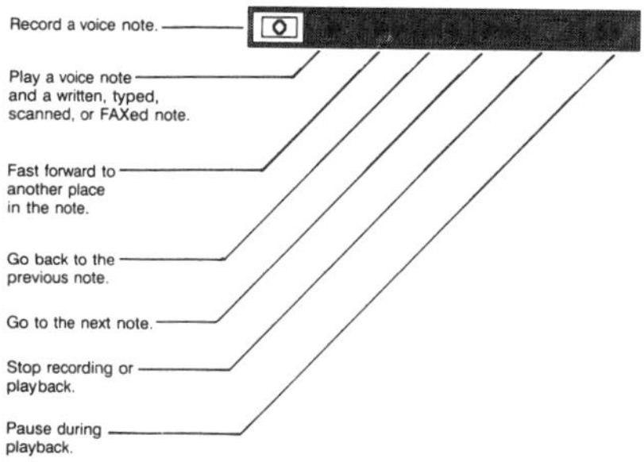

FIGURE 5. Voice playback controls. © Wang Laboratories Inc., 1989.

Also, during live conversation there are very complex social rules for stopping people speaking that were clearly beyond the powers of the Freestyle user interface (e.g., polite interruptions, a raised eyebrow). The system needed a "second best" analogy for controlling the playback of speech. The analogy that was selected was the playback controls on a tape deck or a compact disk machine (Fig. 5). These controls are not as familiar as pen, paper, and handsets.

Users are given the sensation of having successfully changed control states by several means. For example, scanned images in the system are larger than either the half page or full page monitor image and therefore need to be scrolled to be viewed in their entirety. To guide moving the image to reveal the bottom part, it was not possible to directly mimic grabbing the edge of the paper. A substitute mechanism was needed with feedback that was as natural as possible. In this case, the user moves the stylus to the edge of the screen (tablet) until a hand appears on the edge that looks like it is grabbing the image. The user thus knows that she or he is in position to touch the tablet and drag the image up and down within the screen.

A related problem was supporting the user in transitions between multiple environments. Because the Freestyle system captured any type of screen (anything from a local Lotus spreadsheet to a remote application run on a mainframe) the Freestyle system had to coexist with a multitude of "computer cultures" including the DOS file listing and Wang's regular electronic mail system. Freestyle did not attempt to provide a consistent application environment, but rather coexisted with the preexisting environments. A clean, universal mechanism was needed for switching between these environments, making it obvious when the user was in frozen annotation mode versus live application mode.

The most salient feature for distinguishing frozen annotation mode from live application mode was reversing light-on-dark screens to dark-on-light. This added to the sense of familiarity because dark writing typically appears on light paper. The context switch was also made obvious by providing a single pop-up menu and graphic pen activated controls for voice, typing, and navigation that always appear during annotation sessions.

A menu pick in the annotation environment and an icon in the desktop environment return the user to the application he or she was previously using (in the state it was left in upon touching the tablet). The number of operations was kept small within both environments. This helped minimize the conflicts in culture with "live" applications. All functions in the annotator are controllable with the pen and the handset/speaker-phone, thus requiring no memorization of keyboard commands that might conflict with other environments. In both the annotation and desk environments *only two pen motions* are available for controlling function access:

- Touch and lift on a graphic (icon, menu, stamp, button) to select a function or perform an action
- Touch and drag to relocate graphics and scroll

A cursor in the shape of a pencil point is displayed on the screen when the pen is moved around just above the tablet. Writing occurs, of course, when the user touches down on the tablet.

Extensive testing of the system with novice users provided input into iterative versions of the design of the system and showed that the concepts of pointing with the pen, writing, function selection with the pen, speaking while writing, and movement of graphic elements with the pen required minimal training. (See Perkins *et al.*[(6)])

3. *System Use*

While aspects of the system attempted to mimic real-world behavior, the overriding design goal of the product was to enhance communication in business environments.[(11)] This goal was supported by iterative evaluation of prototypes used for real business communication. The evaluation team started with interviews with potential users in a variety of business environments. As prototype versions of the product emerged and were placed into internal test applications, the users were interviewed about the match between the system features and their communication needs. Finally, the researchers followed the product out to its first customer test sites.

The field research was focused on understanding and molding two unique aspects of the system design: the ability to capture screens as the backgrounds of the messages, and dynamic multimedia annotation. As the interviews progressed, we also focused on the design of the graphical desk for supporting filing and retrieval as we shall describe below.

3.1. Captured Screens as Background

The background images in Freestyle can be created from familiar images in the office environment via program screen capture, scanning, or original hand drawing. The first internal business group to use Freestyle scanned in paper travel authorization forms for signature processing. Also, during design, the designers used the prototype system to send rough hand-drawn sketches of icons back and forth to each other with voice explanations of the tradeoffs between the variants in the icons.

We carried out preparatory field interviews that showed that in engineering and graphics departments, graphic materials guide the progress of work, suggesting that capture of such graphics could be an important use of Freestyle. These graphics can be complex and go through repeated revision cycles. Rough sketches are used during preliminary stages of coordination. Other materials can include CAD drawings, hand drawings, photographs, and page layouts. Revisions are required during the various review cycles or if there are any last minute changes in the product designs. These preliminary interviews anticipated the role that screen capture would play in the system.

The evaluation team later observed real use of various types of message backgrounds in the customer test sites. They included scanned images of forms, invoices, FAX's, handwritten messages, and newspaper clippings. They also included application screens such as database query results (with superimposed handwritten comments), spreadsheet screens, *and* electronic mail screens with handwritten comments.

3.2. Dynamic Multimedia Annotations

In Freestyle, simultaneous written and voice annotations are played back together while the receiver watches the sender point to parts of the image, write, and speak, almost as if the user is watching a video tape at her or his terminal. The business need for annotation was assessed by interviewing people in the field about their current annotation and commenting activities.

The preliminary field research showed that engineering groups use a variety of hand markings during their daily work. Some provide *focus of attention.* The prototype engineer has to locate the particular corner of a design drawing, mark it, and clearly state the problem: "Does this require a bend of a fold?" "When this grille is formed on the lathe, the lip tends to elongate—we suggest the design be changed along these dimensions." A number of the problems must be described at *multiple levels*. Pointing directs the listener's attention to particular parts within the drawing, while verbal comments explain the consequences of the desired action.

As users described how they negotiate design changes, *sequential pointing* within a diagram provides critical order information for problem resolution. This predicted a need for Freestyle's live playback capabilities.

The first internal prototype pilot was set up to support signature cycles for scanned-in forms. Signatures and handwritten/voice annotations were also used at the customer test sites. People used handwriting and voice to ask each other questions and give each other answers about the status of invoices, or the reasons for changes in database or spreadsheet information. The users also communicated with each other about problems in spatial layout. After considerable time using the system, users began to produce

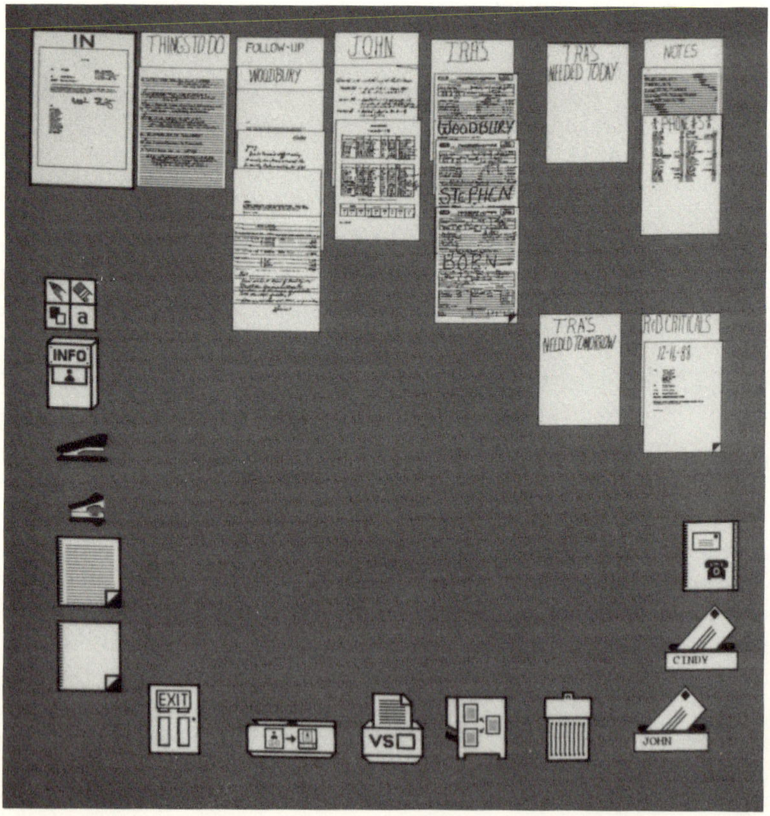

FIGURE 6. Secretary's desktop. © Wang Laboratories Inc., 1989.

"multimedia" greeting cards to, for example, congratulate a new parent on the arrival of a child with combined drawn and spoken imagery (e.g., a baby crying).

3.3. The Graphical Desk

The fluidity of the desktop allows it to be used as a specialized kind of indexing system supporting specific types of filing and retrieval. The basic functions that are available are (1) free, instantaneous motion of any object on the desktop; (2) the ability to overlap "stamps"; (3) the ability to create "piles"; (4) the ability to create hierarchically stapled piles; and (5) the ability to place things in folders.

These basic capabilities created indexing and organizing opportunities as illustrated in Figs. 6 and 7 below. These photographs were taken from the actual working desk of an executive user of a prototype version of the system

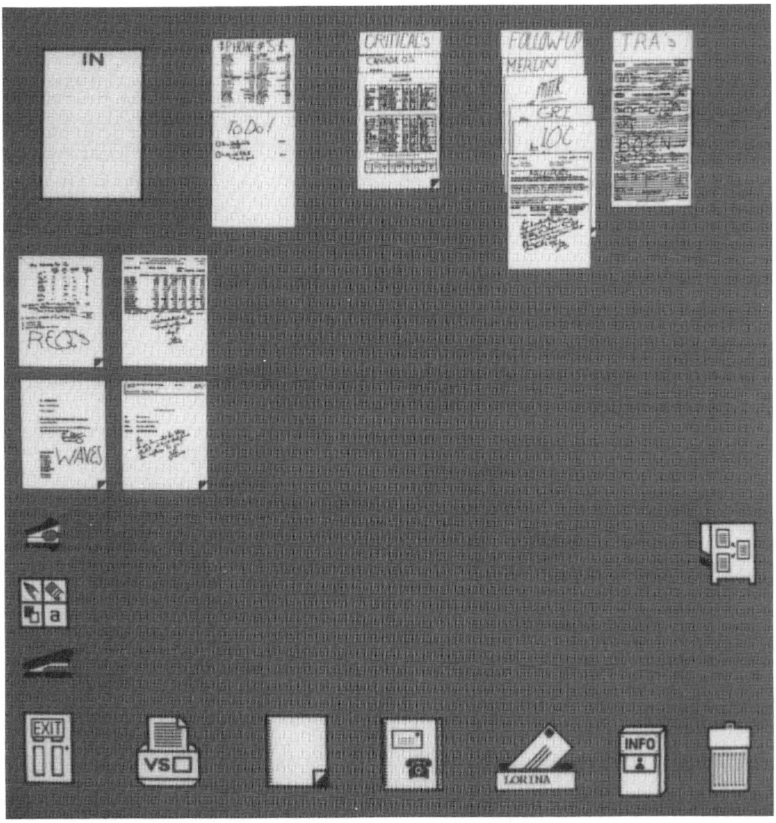

FIGURE 7. Executive's desktop. © Wang Laboratories Inc., 1989.

and his secretary after several months of experience using the system for ongoing communication (folders were not yet available). The structure of the desk evolved substantially over time, with the users themselves creating new uses of the basic desk elements. A key use was approving travel request forms for a large customer service organization (see below for description of application terms used on the desktop).

Labeling Files. Perhaps the most salient organizing tools on these users desks are the labels. Labels are created for piles by writing large leters on blank electronic pieces of paper (obtained by selecting an icon) and then overlapping image stamps on top of those blank pages. Labels are also applied to the images themselves (e.g., **PHONE**) to support easy recognition.

Grouping Files. The images can be grouped into categories that are related to specific aspects of business as shown here. The nature of the categories is flexible in that it is possible to have superordinate categories (e.g., TRA: Travel Request Approval forms) and *at the same time* view separate but related subordinate piles (TRAs today versus TRAs tomorrow). Space is reserved for time-critical communications down at the side of the executive's desk. The creation of these nonstandard hierarchies is achieved directly through labeling, physical movement, and sorting.

The visual images themselves are manipulated during sorting, not file names as in traditional hierarchical filing systems. Notice that the ability to overlap stamps is used to create a view of the approximate number of images within a pile (equal to the approximate amount of pending work) while still conserving space. A subtle clue of the size of a pile of stacked stamps is given by the variable size of a "shadow."

Regrouping Files. Regrouping of files requires no keyboard commands or file renaming but rather is done directly through motion of the miniature images. This allows a relatively high rate of change of file groupings. For example, travel request forms are moved from one pile to another on a daily basis. This provides the user with an easy way of maintaining a dynamic record of the progress of work as it goes through different stages.

Regrouping of piles is, of course, very similar to regrouping of paper piles in the real world. The miniaturization of the stamps provides a "birds's eye view" of the organization of the piles that has similar advantages to miniature models used in other environments (e.g., tactical models of military operations). The Freestyle desk is superior to spreading real paper on a conference table or the floor both because it does not compete for real space and because much smaller physical movements are required to reorganize the pages.

Locating Images. Users locate particular miniature images with a combination of visual cues in the miniatures themselves and the image's relative position on the desk (e.g., in a pile). Just as in the familiar paper world of the desktop, the images can be distinguished and selected on the basis of distinctive overall pattern or gestalt (see distinguishing characteristics of spreadsheets, electronic mail, and memos with distribution lists in Fig. 7 above). Retrieval of a particular image is very fast if the whole image or part of the image is visible on the screen. The desk does suffer from similar problems found on real desks as the volume of images increases. Things do get lost under other things, but they are found again by visual sorting.

Storage versus Throughput. The lack of file names (that can be listed in an index and searched) puts restrictions on the use of the desk for long-term filing and retrieval. Named filing was provided via a special "file cabinet" icon to help serve the need of long-term storage. But it is important to emphasize the need for "informal" visual pile structure for handling communication flow of *messages that often last only for short duration* and do not merit the investment of additional typed data entry. The Freestyle desk actively presents temporary but short shelf-life information for easy use.

Sharing Organizational Structure. One unexpected group effect was observed in the organization of desktop structure by co-workers who were cooperating on the same procedures. Even though there is no desktop organizational software shared between the users, desktop organization was shared. This included shared categories ("TRAs," "Criticals"), shared spatial organization of categories (TRAs on right side of desk), and shared images used on a repeated basis (a list of phone numbers). Slight *differences* in organization between users revealed differences in functional roles of the collaborators. For example, the secretary needed to maintain finer-level distinctions of TRAs to guide the detailed flow of communication.

The desks of other users started to reveal organizational structure. Piles are created representing *projects* as well as *people* responsible for certain activities, independent of whether the named people are accessible on the Freestyle network. We believe desktop organization may be able to support and help reveal the hidden communication structures of organizations that are sometimes obscured by official organizational charts.

3.4. Anticipating the Relationship to Other Communication Media

The design of this system had certain characteristics that position it with respect to other communication media. The role that this system will play in business communication will be influenced by the media that will be used in concert with it (meetings, telephones, paper and pencil, electronic mail, voice mail, and fascimile).[8] This section contrasts Freestyle's features with those of other media.

Meetings. Like meetings, the Freestyle multimode annotation allows sharing of paper (in electronic form), writing, voice comments, and pointing by people within the same minicomputer network. But meetings have major advantages over Freestyle, particulary in the area of turn-taking.[12] Body language, silent gesture, and spoken interruptions guide who talks next and therefore the course of the discussion. They also provide immediate interpretations of what is being said (e.g., confirmations versus rejection). With asynchronous mail turn-taking is only carried out in an extended time frame. Yet, the ability to quickly mail the multimedia annotated images around the network to people who are *geographically distributed* provides some of the advantagese of face-to-face meetings.

The fact that original Freestyle messages are played back at a different time by the receiver means that the individuals who are communicating with each other do not have to schedule some meetings in the first place. Users also can avoid traveling long distances in order to communicate informally. Our preliminary field work in a variety of departments inside Wang revealed that meeting management and scheduling were the primary causes of communication breakdowns, particulary in the graphics-dependent applications, (See also Ref. 13.)

Telephone. While telephone communication is usually very fast, there are often serious delays in completing calls because of the lack of availability of the called party—telephone tag.[14] If the topic of conversation is visual in nature (see earlier description of interpretation of an engineering drawing), then there can be serious communication breakdowns as two- and three-dimensional visual information has to be translated into a one-dimensional string of words. There is no way to point to part of a drawing in context unless both parties hold separate copies, and even there it is difficult to establish long distance frames of reference (see Bly[15]).

While there are no live conversations with Freestyle, voice comments can be sent as quickly as electronic mail messages and the juxtaposition of the voice with the image means a major increase in the amount and quality of information that can be communicated. The advantage of any type of asynchronous messaging is that all parties do not have to be at their workstations at a particular time. Telephones, of course, have the major advantage of being ubiquitous.

Paper and Pencil. The Freestyle system is both a consumer and creater of paper via scanning and printing and is therefore not really an alternative. It does provide a fast channel for moving the images on paper that does not require human sorting and physical portage. The captured computer screens that are later printed can be used in other contexts (e.g., live meetings) and can be distributed to those without Freestyle systems. The miniature images on the desktop have some characteristics in common with paper (ability to make piles, put things in folders). Though the Freestyle pencil is very similar to a regular pencil, the user must adapt to writing on a tablet rather than directly on the screen. This requires some initial adjustment.

Electronic Mail. Character-based electronic mail systems are the most common computer-based communication systems. They have been shown to have a very large impact, unifying the collaborative efforts of geographically dispersed workers. (See, for example, Eveland and Bikson[16].) But such systems have the same restrictions that telephone conversations have in that they cannot support communication that is dependent on visual information in the form of diagrams, drawings, photographs, and pictures.

Electronic mail systems have the major advantage of maintaining a machine-readable message content (as opposed to a frozen background image with machine-uninterpretable voice and hand-drawn annotations). Character-based electronic mail systems can themselves analyze the content of messages to determine actions that should be taken (e.g., Malone *et al.*[17]).

Voice Mail. Voice mail allows voice capture like Freestyle and has the same powerful forwarding and distribution capabilities, but it does not allow juxtaposition of voice with image and writing. Voice mail has found an important communication niche for people who are on the move (e.g., sales personnel) and need to communicate critical update information quickly (e.g. inventory levels) and as an immediately available message-taker that does perfect "transcription" when people are away from their phones.[11, 14] Voice mail has the same advantage as the telephone: access from any telephone in the international network.

Video Conferencing. Video conferencing captures a broader range of fundamental types of communication than Freestyle (e.g., body language, facial expressions, hand gestures). Yet attempts to support visual communication with video conferencing have not been as successfull as had been anticipated.[1, 2] Video conferencing often requires that all of the communicators assemble at the same time for meetings. Such conferencing has not therefore provided the time-independent communication that people have come to depend on with electronic mail. Communication with Freestyle also contrasts with common videoconferencing implementations that focus attention on the speakers' faces and bodies. The images of Freestyle direct the user's attention to the objects of work as opposed to the participants in the meeting. Because messages are both created and played back privately, they avoid feelings of being intruded upon.

A recent study by Bly[15] demonstrated that videoconferencing (in which the cameras *did* focus on the visual objects of work) was superior to telephone conversations about the same visual work. On the other hand, video conferencing was less satisfactory than face-to-face meetings in that study. In face-to-face meetings, participants were able to point to the visual materials being *used by the other person.* Freestyle messages do capture pointing but in asynchronously delivered messages.

Facsimile. Facsimile, like Freestyle, allows fast transmission of "electronic paper" but without the attached synchronized voice. The Freestyle system itself allows images to be FAXed directly from the workstation, avoiding the steps of printing out and scanning the image into a FAX machine. Voice comments are lost in these images, however.

It seems clear that each of the listed means of communication will be chosen on the basis of the specific communication needs of particular situations. Meetings will continue to be favored because of their high "bandwidth." With increasing geographic distance, users will make choices between character-based mail (to allow computer evaluation of content as well as such things as joint editing and programming), voice mail (to allow access from any telephone and as a "message-taker"), and Freestyle.

It is too early to tell all the possible uses of multimedia messaging, but as more systems become available, this type of system may be chosen for complex multichannel communication with commenting and review cycles and for *informal* communication of any type (e.g., planning, early design, etc.). The familiarity of the modes of communication (e.g., voice, writing) also make this type of system immediately attractive to non-computer-users/nontypists (e.g., some executives), creating a long-awaited sense of both trust of the system and opportunity for self expression.

4. Emerging Applications

We are still in the earliest stages of identifying where the greatest payoff will be for this and future systems of its type. The system provides a bridge between traditional paper-based work and the use of computer programs. Both sets of human activities will continue to guide the development of new features as well as the integration requirements of the two worlds. Our field interviews helped reveal some of the new opportunities that such systems will have for supporting cooperative work.

4.1. Preservation of Context

Currently we take for granted the fact that long-distance communication, under time pressure, is only so effective. We have learned to live with the relatively small bandwidth of telephone communication. With the emergence of personal facsimile systems, businesses have recognized the efficiency of sending full images that do not require verbal translation. While image-based mail greatly increases the number of bits that must flow through our communication channels, it actually decreases the noise in the system by providing the context people need for *accurate interpretation* of information—"a picture is worth a thousand words." The addition of voice and handwritten annotation further reduces the noise by focusing attention on the relevant information.

One of the users interviewed told the story of a one-year negotiation between two companies on different sides of the world concerning the design of a commodity product. Miscommunications kept occurring until finally all parties decided they needed to have a conference call. But the engineering problem was essentially visual and involved the interpretation of a diagram (specific orientation of parts of the product). Fast transmission of engineering drawings with appropriate interpretation via annotation would have allowed the negotiation to be completed much more quickly.

Similar problems in interpretation were reported by a carpenter who had to build according to specification from both architects and general contractors. Miscommunications about the interpretations of the drawings often cost him weeks of effort and delays. Again, fast transmission of complex drawings with superimposed interpretation promises to have a major impact for such applications.

4.2. Planning and Iteration

Intimately related to the need for context information is the need to support rapid informal and often ad hoc communications that is the mainstay of any kind of project initiation. Design of products, organizational structures, page layouts, and buildings evolve very rapidly in the early stages, often requiring input from multiple contributors with unique expertise. The quality of collaboration is directly dependent on the ease of communication. Often the less formal, the better. Stylus and voice annotation can play roles that are similar to the white board used in a conference room. The shelf-life of an idea may be relatively short but with such informal communication mechanisms within easy reach of geographically distributed contributors, good ideas can be recognized as such more quickly by the right set of people.

4.3. An Emerging Challenge for Computer Science

There has been hope throughout the industry that users could control increasingly complex and powerful computer systems with minimal training. Much energy has been put into creating artificially intelligent machines that can understand natural language and spoken voice, can produce voice, and can make complex decisions. The Freestyle project sidestepped the need to copy the logic of users in code or to teach the computer to think. Freestyle does not attempt to understand. Instead the system captures greater and greater levels of detail to be presented to others for human interpretation.

There emerges then, a kind of debate for the future: Will our investment in computer systems development have the fastest and strongest impact by *copying* human intelligence in code, or will the larger impact come from providing more and more sophisticated high bandwidth *capture* of human intelligence for use by others and providing increasing responsiveness to human movement? The anwer to this question may depend on the extent to which both types of computer support can be interwoven to support both predictable aspects of human problem solving and the creative, ad hoc insights that help to drive innovation.

ACKNOWLEDGMENTS

We would like to thank Ellen Francik, Ron Perkins, and Victoria Moseley-Dials for valuable comments. We would also like to acknowledge the contributions of the many members of the design and evaluation teams to the ideas described here. We especially want to acknowledge the contributions provided by members of the development team, Alex Harui, Mike Schirpke, Charles Paress, Ken Knowlton, Bruce Brown, Karen Donoghue, Chuck Hsiao, Donna Ajgaonkar, Dave Wilson, Mary Ystueta, Mary Jane Boyd, Steven Teng, Steve Boylan, Pat Martin, Michael Kudla, and Betty Lorenson; the internal evaluation team, Ellen Francik, Ron Perkins, Louis Blatt, Barbara Kelly, and Kenichi Akagi; and finally those who carried out the field-based research, Ellen Francik, Donna Cooper, Constance Perin, Sharon Dodson, and Amy Bucklin.

References

1. C. Egido, Videoconferencing as a technology to support group work: A review of its failure, Proceedings of the Conference on Computer-Supported Cooperative Work, pp. 13–24, Portland, September 26–28, 1988.
2. R. Johansen and C. Bullen, Thinking ahead: What to expect from teleconferencing, in I. Greif, Ed., *Computer-Supported Cooperative Work: A Book of Readings*, pp. 185–198, Morgan Kaufman Press, San Mateo, 1988.
3. T.G. Zimmerman, J. Lanier, C. Blanchard, C. Bryson, and Y. Harvill, A hand gesture interface device, Proceedings of the CHI '87 Human Factors in Computing Systems and Graphics Interface, pp. 189–192, Toronto, April 5–9, 1987.
4. R.B. Smith, The alternate Reality Kit: An animated environment for creating interactive simulations, Proceeding of 1986 IEEE Computer Society Workshop on Visual Languages, pp. 99–106, June, 1986.
5. E. Francik and K. Akagi, Designing a computer pencil and tablet for handwriting, Proceedings of the Human Factors Society 33rd Annual Meeting, Denver, October 16–20, 1989.

6. R.D. Perkins, L.A. Blatt, D. Workman, and S.F. Ehrlich, Iterative tutorial design in the product development cycle, Proceedings of the Human Factors Society 33rd Annual Meeting, Denver, October 16–20, 1989.
7. C. Hsiao and S.F. Levine, Voice annotations on Wang Freestyle System, Official Proceedings of Speech Tech '89, pp. 365–367, New York City, May 2–4, 1989.
8. T. Malone, How do people organize their desks? Implications for the design of office information systems, *ACM Trans. Office Inf. Syst.* **1**, 99–112, 1983.
9. J. Piaget, *The Origins of Intelligence in Children*, International University Press, New York, 1952.
10. J. Piaget, *The Construction of Reality in the Child*, Basic Books, New York, 1954.
11. S.F. Ehrlich, Strategies for encouraging successful adoption of office communication systems, *ACM Trans. Office Inf. Syst.* **5**, 340–349, 1987.
12. J.C. Tang and L.J. Leifer, A framework for understanding the work space activity of design teams, Proceedings of the Conference on Computer-Supported Cooperative Work, pp. 244–249, Portland, September 26–29, 1988.
13. G. Mintzberg, *The Nature of Managerial Work*, Harper and Row, New York, 1973.
14. S.F. Ehrlich and E.A. Akiba, Successful implementation of voice store and forward technology: A case study, *ACM SIGOA Bull.*, 19–26, 1985.
15. S.A. Bly, A use of drawing surfaces in different collaborative settings, Proceedings of the Conference on Computer-Supported Cooperative Work, pp. 250–256, Portland, September 26–28, 1988.
16. J.D. Eveland and T.K. Bikson, Work group structures and computer support: A field experiment, Proceedings of the Conference on Computer-supported Cooperative Work, pp. 324–343, Portland, September 26–28, 1988.
17. T. Malone, K.R. Grant, and F.A. Turbak, The information Lens: An intelligent system for information sharing in organizations, Proceedings of the CHI '86 Human Factors in Computing Systems, pp. 1–8, Boston, April 13–17, 1986.

ACTION PROPAGATION MODEL FOR USER INTERFACE PROGRAMS

YUKARI SHIROTA AND TOSIYASU L. KUNII

1. Introduction

In this chapter, we introduce a user interface specification method and define a mathematical model to represent action dependencies that occur on the user interfaces. In spreadsheets, cascades from initiated actions occur and dependent cells and graphical displays are automatically changed. Process constraint retention capability is facilitated in several intelligent CAD systems: Thinglab,[(2)] Juno,[(4)] Animus,[(3)] and GRAFLOG,[(5)] where user-defined relationships among cells and graphical elements including icons are maintained following value changes. We call such automatic recalculation facilities collectively "action propagation." The specification method we proposed in Refs. 6 and 7 allows nonprogrammers to specify such action propagations. The purpose of the chapter is to define a general model of the action propagations to analyze and optimize the dependencies of the generated code.

The next section gives an overview of the specification language. Section 3 defines the graph-based action propagation model named Action Propagation Graph (APG). This section also indicates the role of this model in optimizing code.

2. Software Specification

This section gives an overview of the user interface named Enhanced Menu-Based Software (EMBS) and its specification language. The EMBS

YUKARI SHIROTA • Software Research Center, Software Division, RICOH Company, Ltd., Tomin-Nissei-Kasugacho-Building, 1-1-17, Koishikawa, Bunkyo-ku, Tokyo 112, Japan. TOSIYASU L. KUNII • Department of Information Science, Faculty of Science, The University of Tokyo, 7-3-1, Hongo, Bunkyo-ku, Tokyo 113, Japan. This chapter is an extended version of "Automatic Generator for Enhanced Menu-Based Software—Program-Specification-by-Examples," by Y. Shiorta and T.L. Kunii in *Proc. HCI International '89*, Elsevier, Amsterdam, 1989.

serves a user interface for application programs and consists of "menu screens" and the links between them. A menu screen is an image displayed on a terminal screen, including the following three image elements: (1) cell image: an area by "["and"]"; (2) button image: a boxed area; and (3) background image: all images except cells and buttons. The EMBS action invocation is triggered by clicking a button or by changing the value of a cell. In general, the action propagation may cascade widely to other menu screens; the invoked actions retrieve data from cells on other menu screens and from the underlying database management system to perform calculations. A cyclic propagation is prohibited: The descriptive power of the EMBS is discussed in Section 3.4.

Figure 1 illustrates an EMBS for a restaurant accounting system, with the capability for a waitress to input orders (e.g., "coffee 2") that are automatically used to change dependent data fields—"Amount", "Tax", "Total"—in an existing bill record. (The bill is the primary menu screen shown in Fig. 1, while the order is a secondary one.)

In order to define his/her EMBS, the user creates a specification file which consists of two definition parts: a *screen definition* part and an *action definition* part. In the screen definition, the user draws an image of a menu screen, placing example values in cells and buttons. The example values are used as the identifiers in the action definition part that follows. The action

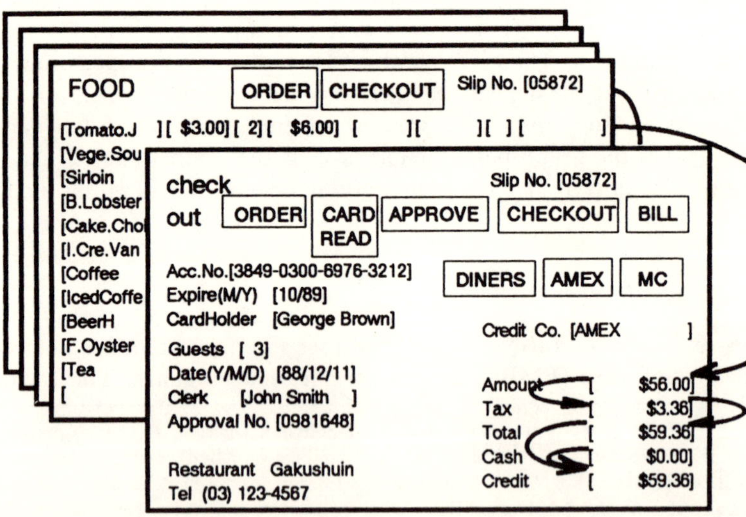

FIGURE 1. The Enhanced Menu-Based Software (EMBS) architecture and the action propagation example.

BNF grammar for an action statement:

```
< action statement >
  ——▶< trigger list > "—<" < action > "—>" < object list >
  ——▶< trigger list > "—<" < action > "—"
  ——▶< expr > "—>" < object >
< trigger >      ——▶ < button >
                 ——▶ < cell >
< object >       ——▶ < cell >
```

Examples:

```
DINERS AMEX MC —< select() >—> VISA
8208-0186 VISA --< find(card_member) >--> John-Brown 02/90
( $150.00 + $30.00 ) / 3 —> $60.00
```

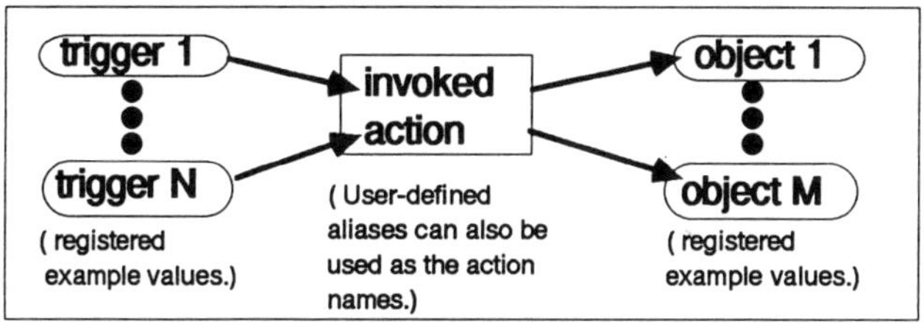

FIGURE 2. The BNF grammar for the action statement with examples.

is specified as action statements in the action definition part. Figure 2 shows the BNF grammar for the action statements with examples. One action statement consists of a *trigger* list, an invoked *action function*, and an *object* list. The action functions are either user-defined subroutines or mathematical operators such as "+" and "/". Action statements are categorized into the following three types: (1) action statements with objects; (2) action statements with an empty object list; (3) arithmetic expressions.

The intermediate functions performed by an automatic generator to translate the user-defined specifications into code are: (1) screen image analysis; (2) action syntactical parsing; and (3) propagation optimization.

3. *Action Propagation Graph*

In the section, we define and explain a graph-based model named Action Propagation Graph to represent the propagation of actions.

3.1. Definition

DEFINITION: *Action Propagation Graph (APG)*. The APG is a directed arc-labeled graph over V. A tuple $G=(F, P, V)$ is defined to be the APG if and only if the arc set P is contained in the Cartesian product of the node set F twice with the arc-label set V, i.e., $P \subseteq F \times V \times F$.

The nodes and the arc labels of the APGs correspond, respectively, to the action functions and the variables (cells and buttons) in the EMBS action propagations. An arc of the APG (f, v, f') corresponds to the fact that the action function f will have effect on the cell v, and consequently, the cell v will activate the action function f' as the trigger cell.

The APG is the internal form for the user interface automatic generator to represent the action propagation. In the case of the automatic EMBS generator, the user-defined action statements are translated to the APG representation. Figure 3 illustrates some APG representations of the action statements, where rectangles represent nodes, and circles arc labels. The action in the EMBS is invoked by a button click or by a cell value change. We call the users' direct button clicks and cell value changes *users' direct input functions* collectively, and represent them by u. Whenever a cell value is changed, the *cell value display function* is invoked, so that the users can view the change. The cell value display functions are contained in the set of leaf nodes in the APG. We represent a member of the set by d.

To simplify the descriptions, we define *source nodes* and *sink nodes* of an arc as follows: For arc (f, v, f'), the node f is called the source node of the arc, if and only if f is not a users' direct input funtion, and f' is called the sink node of the arc, if and only if f' is not a cell value display function. Let so(v) and si(v) represent the source node and the sink node of v within the arc.

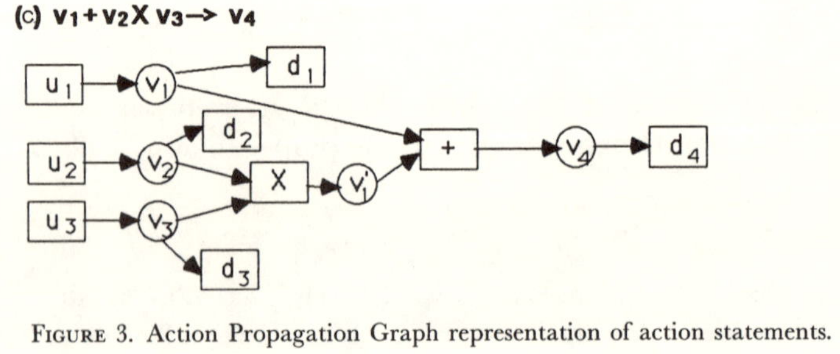

FIGURE 3. Action Propagation Graph representation of action statements.

3.2. Composition of Two Action Propagation Graphs

The EMBS generator reads in action statements sequentially from the specification file. Then the action statements will be translated to the APGs. Let the APGs corresponding to the action statements be $G_i = (F_i, P_i, V_i)$ $(i = 1, 2, ...)$, and let the APG corresponding to the whole action specifications be $G = (F, P, V)$. The generator assembles $G_i (i = 1, 2, ...)$ into G. In order to describe this process, we define the composition of two APGs. Let the symbol "$\circ$" represent the composition operator. We can define $G_1 \circ G_2$, provided that any object cell of the action statement corresponding to G_1, is a member of the trigger list of the action statement corresponding to G_2, or provided that any object cell of the action statement corresponding to G_2, is a member of the trigger list of the action statement corresponding to G_1. Otherwise, the union of G_1 and G_2 is defined as $G_1 \circ G_2$.

By analogy of the triggers and the objects in the action statements specified for the EMBS, we generalize the trigger and object concepts for the APG as follows:

Definition: *Triggers and Objects.* Let $G = (F, P, V)$ be the APG. U and D are, respectively, the set of users' direct input functions and the set of cell value display functions.

T: a set of triggers, $T = \{v | v \in V \text{ and } f \in U \text{ for } {}^{\forall}(f, v, f') \in P\}$

and

O: a set of objects, $T = \{v | v \in V \text{ and } f' \in D \text{ for } {}^{\forall}(f, v, f') \in P\}$

A trigger $t \in T$ is defined as an arc label that does not have any source node. And an object $o \in O$ is defined as an arc label that does not have any sink node. Let T_i and O_i be the trigger set of G_i and the object set of G_i, for $i = 1, 2, ...$, respectively.

Figure 4 presents an example of the composition of two APGs. In this figure, the triggers of A_1 are v_1 and v_8, and the objects of A_1 are v_2 and v_3. In the $A_1 \circ A_2$, the triggers are v_1, v_4, v_5, and v_8, and the objects are v_2 and v_6.

Here we define $G_1 \circ G_2$, if $O_1 \cap T_2 \neq \emptyset$ or $T_1 \cap O_2 \neq \emptyset$, by the algorithm shown in Fig. 5, where G_1 and G_2 are APGs corresponding to one or several action statements. In the APG expressing an action statement, there is one and only one node that is neither a users' direct input function nor a cell value display function; the node corresponds to the action function in the action statement. Therefore, when $v_1 \in O_1$, v_1 has only one source node in G_1, and when $v_2 \in T_2$, v_2 has at least one sink node in G_2. Consequently, for each v in $(O_1 \cap T_2)$, both so(v) in G_1 and si(v) in G_2 always exist. Moreover, in the APG corresponding to one action statement, there is no

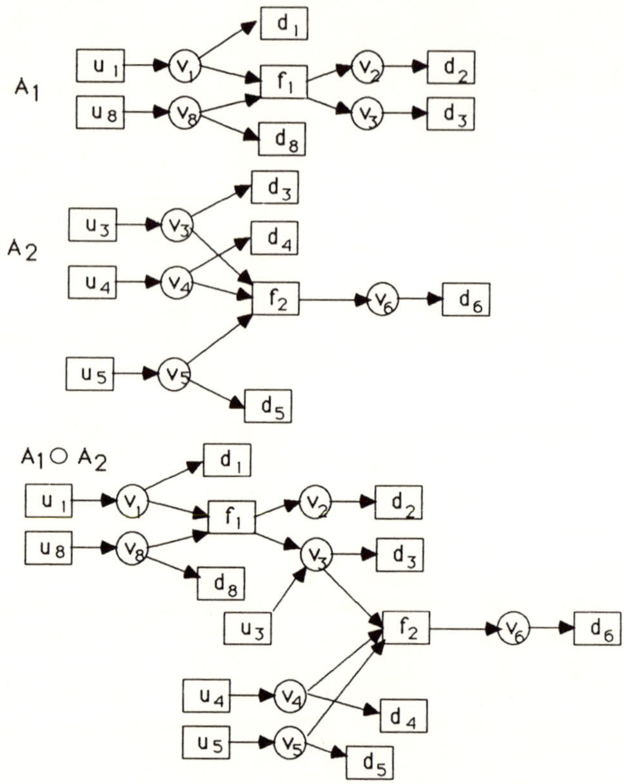

FIGURE 4. Composition of two action propagation graphs.

```
F=F1∪F2 ;
P=P1∪P2 ;
V=V1∪V2 ;
if (O1∩T2≠∅) {
        for each v in (O1∩T2) {
                fi = so(v) in G1 ;
                for each fj which is si(v) in G2 {
                        P=P ∪ (fi, v, fj) ;
                }
        }
}
if (T1 ∩ O2 ≠∅) {
        for each v in (T1 ∩O2) {
                fj = so(v) in G2 ;
                for each fi which is si(v) in G1 {
                        P=P ∪ (fj, v, fi) ;
                }
        }
}
```

FIGURE 5. Algorithm for two Action Propagation Graph composition.

```
G=G_1 ;
for (i=2; i ≤ n ; i++) {
      G = G ○ G_i ;
}
```

FIGURE 6. Aggregation algorithm.

arc label v that is both a trigger and an object; the intersection of $O_1 \cap T_2$ and $T_1 \cap O_2$ is always the null set.

To make the aggregated G, the algorithm shown in Fig. 6 is performed. (Let n be the number of action statements in the specification file.) When composing APGs, we must verify the "acyclicity" and the "uniqueness of the source node" for the consistency of the resultant APG:

1. Acyclicity. The APG must be acyclic to avoid infinite loops in executing the EMBS. Therefore, we check whether the resultant APG is acyclic, every time the composition occurs. For this check, we use the depth-first search algorithm[7]; after adding (f, v, f') to P, we start the depth-first search from the node f. During the search, if f is found again , it means there exists a back path to f, i.e., the APG has a cycle, then an error message will be issued. If the search is completed to the end, it will turn out that the APG is acyclic. In Fig. 7, the result of the composition contains a path back to f_1.

Here we estimate the running time spent on the acyclic check. Assume the resultant APG has e arcs and $n \leqslant e$ nodes. The time spent on the depth-first search of an entire graph is $O(e)$, because no node is visited more than once. In this check, we traverse the sub-APG of which the top node is f. Therefore, the running time is proportional to the number of the nodes in the sub-APG. As the average case, we assume the APG be tree-structured, and also that the sub-APG hardly diverges. Provided that the APG has no cycle, the running time to traverse the sub-APG is $O(\log n)$ approximately.

2. Uniqueness of the source node. The source node must be unique for each arc label, if it exits. This restriction prohibits the redundant definition of a cell. If there are two source nodes for the same cell in the APG, the cell definition will be redundant, because the source node of the cell specifies the procedure for obtaining the cell value. Therefore, we have to make sure $f_i \in U$ for $(f_i, v, f_i') \in P$, before each arc composition (f_i, v, f_j) *to* P. If another source node of v has been found in the symbol table, the arc composition is rejected, and a diagnostic message is issued.

We perform this check by using additional reference bits for each piece of cell information in the symbol table. Initially, all the reference bits are cleared to 0, and if a cell is an object in the action statement, then the reference bit will be set to 1. In the course of the APG compositions, before adding each arc (f_i, v, f_j) to P, we inspect the reference bit of v. If it is 0,

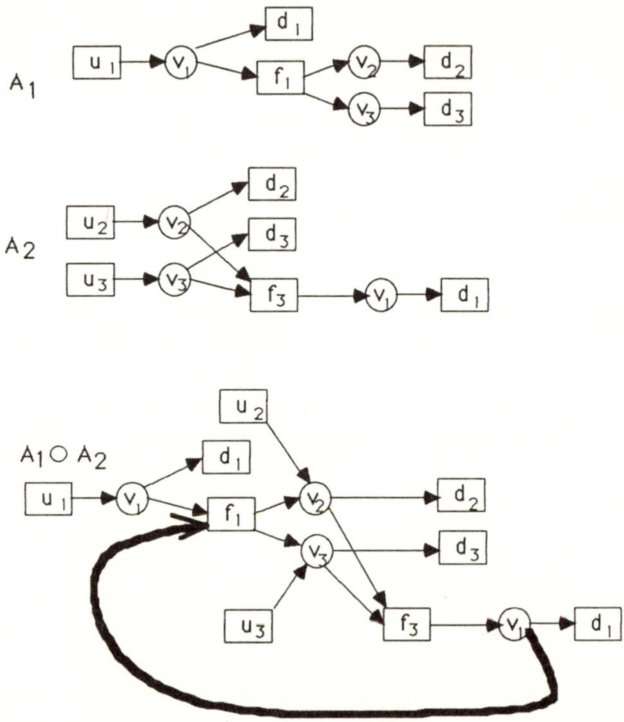

FIGURE 7. Erroneous composition result with a path back to f_1.

we proceed to add this arc to P, and set the reference bit. If the reference bit is 1, however, this arc addition will be rejected. In the EMBS system, operations on the symbol table have been implemented by hashing. We can, therefore, access the symbol table including this check in a constant time per operation on the average.

3.3. *Optimization*

In the section, we present methods for increasing the efficiency of the action propagation. The optimization of the action propagation code is very important in a variety of applications. The more complicated or wider the action propagation is, the later the turnaround comes back. One of our purposes is to speed up the action propagations occurred in scientific applications including complex and iterative mathematical calculations.

3.3.1. Topological Sorting. When we end-user directly accesses a users' direct input function u, the sub-APG with the initial node u will be extracted from the whole APG, and then the sub-APG will be executed. Here, let us consider the execution ordering of the function nodes in the action propagation subgraph. Take the APG shown in Fig. 8, for example. There, assume that the functions are executed in the preordered traversal sequence, i.e., ($u_9, f_0, f_1, d_2, +, d_8, d_1, f_4, +, d_8, d_6, d_4, f_5, +, d_8, d_7, d_5, d_9$). Then we can get the correct result; however, the shared arc $(+, v_8, d_8)$ will be executed three times. The most efficient order in executing the functions is a linear order, where node m always appears before node n, provided that there exists a directed arc $m \rightarrow n$. For example, ($u_9, f_0, d_9, f_1, d_1, f_4, d_4, f_5, d_5, d_2, d_6, d_7, +, d_8$) is one of the linear orders; in general, a linear order is not always determined uniquely.

A topological sort that can be accomplished by using the depth-first search algorithm can give a linear ordering for nodes of a direct acyclic graph (DAG).[1] By using the topological sort, let us calculate the linear ordering with respect to the function nodes. The algorithm is presented in

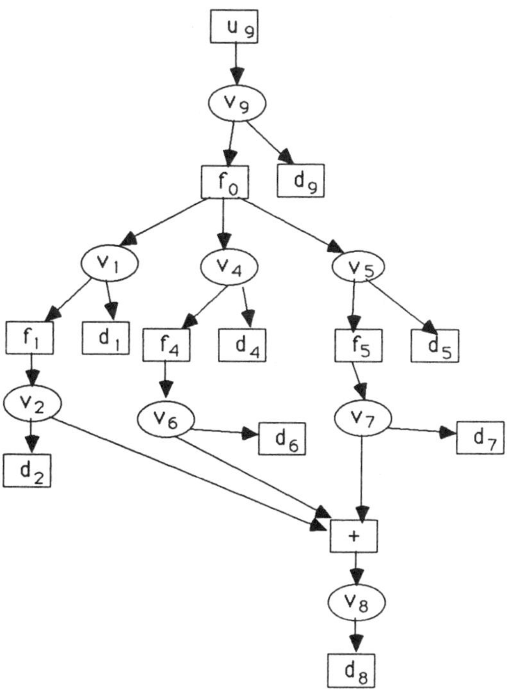

FIGURE 8. Topological sorting. The topological sort assigns a linear ordering to the nodes of the APG such as $d_9, f_0, d_5, f_5, d_7, d_4, f_4, d_6, d_1, f_1, +, d_8, d_2$. The most efficient way to executing the action functions is to follow the linear order, because in that way, the action can propagate correctly and without redundancies.

```
Tsort(n)
        struct node n;
{       struct node w;

        push(n);
        n.mark = 1; /* visited mark */
        for (each node w adjacent to n) { /* directed arc w →n */
              if (w.mark == 0)
                    Tsort(w);
        }
        w = pop();
        print w;
}
```

FIGURE 9. Algorithm to calculate the linear ordering of action functions.

Fig. 9. If the number of arcs in the target APG is n, "Tsort" in the list is invoked $O(n)$ times on the average. As a result of the algorithm, the reverse topological order of the nodes is printed. We must, therefore, execute the function nodes in the reverse order.

The topological order calculated for each u is embedded in the program code, and later in the execution of the program, the action functions will be invoked following the order. The preceding use of topological sorting is easily extended to cases where an APG has several initial nodes.

3.3.2. Delayed Action Propagation. In executing the EMBS, the action propagation may cascade widely. For example, suppose that the cursor x, y coordinate values are changed by a mouse and every time the user moves the mouse, the wide action propagation will occur. Then, the invisible parts of the action propagation go in vain except for the last change. We therefore delay the execution of the invisible parts until the system control transfers to the menu screen, i.e., until the menu screen is displayed.

Figure 10 illustrates a restaurant order action propagation with the initial node u, where the following three arcs represented by bold arrows, (u, #5, what_is), (u, #5, →), and (total, $142.50, →), are shared and allow decomposition into three subordinate action graphs whose arcs follow:

⟨menu-screen 0⟩:
{(u', $142.50, →), (→, #$150.00, cal_tax), (cal_tax, $15.00, +),
(+, $165.00, −), (−, $65.00, d)};

⟨menu-screen 1⟩:
{(u, #5@2, what_is), (what_is, #coffee, retrieve),
(retrieve, #$1.50, ×), ($u$, #5@2, →), (→, #5, ×),
(×, #$7.50, total)}; and

⟨menu-screen 2⟩:
{(u, #5, d)}.

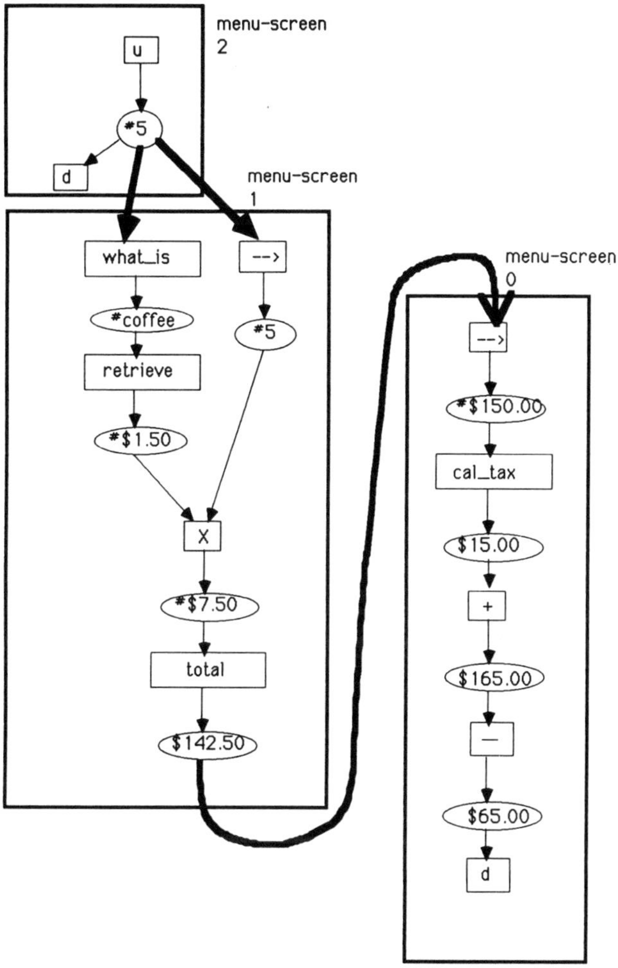

FIGURE 10. Restaurant order action propagation. The original rooted APG devides into the three subrooted APGs corresponding to the different menu screens. Example values in circles are cell identifiers defined in the specifications.

Here "u'" is the users' direct input function for the cell named "$142.50," and the digit that follows "@" represents the menu screen number.

The APG division algorithm is presented in Fig. 11; if a path $f_1 v_1 f_2 v_2 \ldots$ where v_1 and v_2 belong to the different menu screens is found, the arc (f_1, v_2, f_2) becomes a cut point among the menu screens. The function mnum( ) maps any v into the menu-screen number that the v belongs to. The function back_arc( ) is used to check that the sub-APG whose top is f_2 does not contain any arc back to the former APG.

```
DivideAPG(f1)
        struct node f1;
{       struct node f2, f3;
        struct label v1, v2;

        for (each arc(f1, v1, f2)∈ P) {
              for (each arc (f2, v2, f3)∈ P) {
                 if (mnum(v1) ≠ mnum(v2)) {
                    if (back_arc(mnum(v1), f3)==0) { /* if there is no back arc */
                       P = P − (f1, v1, f2);
                       Cutpoint = Cutpoint ∪ (f1, v1, f2);
                    }
                 }
                 DivideAPG(f2);
              }
        }
}
back_arc(menu_num, n)
        struct node n;
        int       menu_num;
{
        struct node w;
        struct label v;
        int   rval;

        n.mark = visited;
        for (each arc (n, v, w)∈P) {
              if (w.mark==unvisited) {
                    if (mnum(v)==menu_num) return(1); /* found the back arc */
                    rval = back_arc(menu_num, w);
                    if (rval==1) return(1);
              }
        }
        return(0);
}
```

FIGURE 11. Algorithm to divide an action propagation graph. The algorithm divides an APG $G=(F, P, V)$ into several sub-APGs which are independent of each other in the sense that they do not have any effect on the other menu screens. Initially, the set of cut points named "Cutpoint" is cleared to 0.

In executing the program, the action propagation is buffered at the cut point. Later, when the transition to the menu screen occurs, the system will invoke the users' direct input function for the hung-up cell, which causes the same effect as when the user directly changes the value of the cell. If the buffered menu screen contains several cut points, the cut points will be executed in the order of the hung-up time.

3.4. Power

This section concerns the descriptive power of APGs. Since they are directed acyclic in nature, interation over action functions is not contained, while an action function can involve loops. To make flow-of-control structures as simple as possible for nonprogrammers, we have layered the programming complexity into (1) EMBS specification language for the nonprogrammers, and (2) the conventional language implementation of action functions by programmers.

In the EMBS, most actions are in the fixed forms. This was made possible through many years of actual experience to accumulate the knowledge necessary to simplify job procedures as a set of fixed forms. Even if the procedures include complex control flows (e.g., some of them may execute iteration until some variables converge), they can easily be specified as one action function. In addition, the users can specify the flow-of-control structures by the action function if needed. Thus, we believe the descriptive power of the APGs is enough to meet the users' requirements.

4. Conclusions

In this paper, we described Enhanced Menu-Based Software (EMBS) specification language and also a graph-based model named Action Propagation Graph (APG) to represent the action propagation in the user interface software. The user-defined specifications are translated into the Action Propagation Graphical representation, which is the internal data structure of the automatic EMBS generator. In the process of the program code generation, Action Propagation Graph can be used as a helpful tool to analyze and optimize the dependencies of the data fields.

References

1. A.V. Aho, J.E. Hopcroft, and J.D. Ullman, Chap. 6, Directed graphs, in *Data Structures and Algorithms*, Addison-Wesley, Reading, Massachusetts, 1982, pp. 198–229.
2. A. Borning, The programming language aspects of Thinglab: A constraint-oriented simulation laboratory, *ACM Trans. Programming Lang. Syst.* **3**(4), 353–387, 1981.
3. R.A. Duisberg, Animated graphical interfaces using temporal constraints, Human Factors in Computing Systems: Proc. SIGCHI '86, Boston, April 13–17, 1986, pp. 131–136.
4. G. Nelson, Juno, A constraint-based graphical system, Proc. Computer Graphics: SIGGRAPH '85 Conf., San Francisco, Vol. 19, No. 3, July 22–26, 1985, pp. 235–243.
5. L.A. Pineda and N. Chater, GRAFLOG: Programming with interactive graphics and PROLOG, *Proc. Computer Graphics International '88, Geneva*, Springer-Verlag, Heidelberg, 1988.
6. Y. Shirota, Y. Shirai, and T.L. Kunii, Sophisticated form-oriented database interface for nonprogrammers, *Visual Database Systems: PROC. the IFIP TC 2/WG 2.6 Working Conference on Visual Database Systems*, T.L. Kunii, Ed., Tokyo, April 3–7, North-Holland, Amsterdam, 1989, pp. 127–155.
7. Y. Shirota and T.L. Kunii, Specification and automatic generation of intelligent graphical interfaces, Proc. VL89 IEEE Workshop on Visual Languages, Rome, October 4–6, 1989, pp. 7–12.

3

A FULLY INTEGRATED NATURAL LANGUAGE KNOWLEDGE BASE MANAGEMENT SYSTEM

RUSSELL J. ABBOTT

1. Introduction

Many systems have dealt with one or another aspect of natural language and knowledge base management. There have been natural language query and analysis systems (see Refs. 13, 16–18, and for a survey Ref. 14), deductive database systems (see Ref. 9 for a survey), rule-based expert systems (see Ref. 10 for a survey), knowledge representation systems (see Ref. 12 for a survey), and logic programming systems.[11] NLP,[15] a computer system for Natural Language Processing, puts together in one package many of the features found in these more specialized systems.

Because it is often easier to get an intuitive feeling for a system by working with it than by reading a description of it, we first present an annotated example interaction with NLP before discussing it in detail. User inputs follow the "$\rangle$" prompt. Everything else is NLP output.

"Lema," for "lexicon/schema entry," is used to define new terms. The notations are: "proper" for proper noun, "vt" for transitive verb, and no definition for count noun. Count nouns are declared as types.[2] If no explicit supertype is declared, new types are situated directly below "thing" in the type hierarchy:

$\rangle$lema: Mary - proper; car; like - vt; own - vt.
user fact 1: "car" is a type.
user rule 1: If $\langle A \rangle$ is a car
then $\langle A \rangle$ is a thing.

RUSSELL J. ABBOTT • Department of Mathematics and Computer Science, California State University, Los Angeles, California 90032.

Next is our first real rule. NLP translates the rule into an internal form and echoes it back to the user.

⟩Mary likes anyone who owns a car.
user rule 2: If ⟨*A*⟩ owns ⟨*B*⟩; and
A⟩ is a person; and
B⟩ is a car
then Mary likes ⟨*A*⟩.

We can use the rule intensionally by turning on "intensional_reasoning," a system flag.

⟩turn on intensional_reasoning.
Your request: turn intensional_reasoning on.
intensional_reasoning is now on.
Done.

(To save space, from here we do not show NLP's confirmation messages.) We can now ask a question thas uses this rule. Note that the trivial intensional answer ("any ⟨*P*⟩ that Mary likes") is not given.

⟩who does Mary like
You have asked for values of ⟨*X*⟩ for which:
Mary likes ⟨*X*⟩.
Assuming that there is some ⟨*P*⟩ that satisfies:
⟨*P*⟩ is a car:
any ⟨*Q*⟩ that satisfies: ⟨*Q*⟩ is a person; and
⟨*Q*⟩ owns ⟨*P*⟩.
Assuming that there is some ⟨*P*⟩ that satisfies:
⟨*P*⟩ is a car:
any ⟨*Q*⟩ that satisfies: ⟨*Q*⟩ is a man; and
⟨*Q*⟩ owns ⟨*P*⟩.
Assuming that there is some ⟨*P*⟩ that satisfies:
⟨*P*⟩ is a car:
any ⟨*Q*⟩ that satisfies: ⟨*Q*⟩ is a woman; and
⟨*Q*⟩ owns ⟨*P*⟩.

The second and third answers derive from built-in rules to the effect that man and woman are subtypes of person.

Explanations are available. First NLP interacts (not shown) with the user to determine which answer is to be explained. (We selected the "any man who owns a car" answer.) Then it provides the explanation:

```
    >why.
You have asked: why.
assumption: ⟨Q⟩ owns ⟨P⟩
assumption: ⟨Q⟩ is a man
sys rule 15: If ⟨A⟩ is a man
             then ⟨A⟩ is a person.
CONCLUSION 1: ⟨Q⟩ is a person
    (from: sys rule 15, assumption)
assumption: ⟨P⟩ is a car
user rule 2: If ⟨A⟩ owns ⟨B⟩; and
             ⟨A⟩ is a person; and
             ⟨B⟩ is a car
             then Mary likes ⟨A⟩.
CONCLUSION 2: Mary likes ⟨Q⟩
    (from: user rule 2, assumption, conclusion 1, assumption.)
```

With a single input we can extend the chain of reasoning by adding two additional rules. Note that the new rules require a skolem car that depends on the variable for "every man":

```
    >every man owns a car.
user rule 3: If ⟨A⟩ is a man
             then some car is a car.
user rule 4: If ⟨A⟩ is a man
             then ⟨A⟩ owns some car.
```

Now when we ask "who does Mary like?" we get

```
any ⟨P⟩ that satisfies: ⟨P⟩ is a man.
```

We can now turn off intensional reasoning and create an extensional man for Mary:

```
    >turn off intensional_reasoning.
Your request: turn intensional_reasoning off.
    >lema: John - proper.
    >John is a man.
user fact 2: John is a man.
```

When we again ask whom Mary likes, "Checking descriptions..." marks NLP's effort to ensure against references (in extensional queries only) to unknown entities such as "a female president of the United States":

```
    ⟩who does Mary like.
You have asked for values of ⟨X⟩ for which:
    Mary likes ⟨X⟩.
Checking descriptions ...
Descriptions ok.
John.
```

This ends the introductory interaction. In the remainder of this report we discuss in turn NLP's knowledge management system (Section 2); NLP's natural language processor (Section 3); semantic analysis in NLP (Section 4); NLP's inference engine (Section 5); and lessons learned (Section 6).

2. *Knowledge Management System*

At its core NLP is a knowledge base and an inference engine. We refer to this combination as NLP's Knowledge Management System (KMS). Everything else, while important, is peripheral. After a brief introduction, this section discusses the forms of knowledge maintained in the knowledge base.

2.1. *KMS Basics*

KMS is a modification of the standard Prolog[7] database and inference mechanism. It is best to think of NLP's inference engine as a conclusion-drawing machine and of NLP's knowledge base as instructions to the inference engine for drawing conclusions. The knowledge base is a collection of facts and rules. A fact, $\langle$*Conclusion*$\rangle$, may be understood as an instruction to the inference engine: $\langle$Conclusion$\rangle$ may be concluded. A rule, *If* $\langle$*Conditions*$\rangle$ *then* $\langle$*Conclusion*$\rangle$, may be understood as an instruction to the inference engine: $\langle$Conclusion$\rangle$ may be concluded if $\langle$*Conditions*$\rangle$ can be concluded. These are Prolog unit clauses and nonunit clauses, respectively. We use the term "clause" to refer to facts and rules together.

Queries (or in Prolog terminology "goals") are transformed into potential conclusions that the inference engine is asked to conclude. Both queries and rules may have logical variables. For example, if the knowledge base includes the fact *man*(*John*) (John is a man), then the query *man*(X) (Is there an X who is a man), can be concluded with X instantiated to John. Similarly if the knowledge base included the rule *person*(X) *if* *man*(X) then the query *person*(X) can also be concluded with X instantiated to John.

2.2. *Primitive Metapredicates*

One can imagine that all knowledge is encoded in terms of a set of metapredicates joined by a fixed set of operators. The most basic metapredicates are the relation metapredicates. The relation metapredicates may be written *relation*(*Relation*, Arg_1 ..., Arg_n), which may be read: The relation *Relation* holds among the arguments, Arg_1, ..., Arg_n. There is one relation metapredicate for each arity.

In general the *Relation* depends on the user's domain, although there are some *Relations* predefined in NLP. The following relation types are defined:

1. *relation*(*Entity*, *Variable*). *Variable* is the entity *Entity*. For example, relation(John, X) for X is John.
   In all that follows, when we use a constant in place of logical variable one should read the constant as standing for an additional implicit goal predicating the constant of a variable in that position. For example, *relation*(p, ..., *John*) should be read *relation*(p, ..., X) *if relation* (*John*, X).
2. *relation*(*Category*, *Entity*). *Entity* is a *Category*. For example: John is a man.
3. *relation*(*Role*, *Filler*, *Possessor*). *Filler* is *Possessor's Role*. For example, Bill is John's father.
4. *relation*(*Preposition*, *Entity*, *Entity*). The *Preposition* relation holds between the two Entities. For example, John is in New York.
5. *relation*(*Subcategorizing_Adjective*, *Entity*, *Entity*). The *Subcategorizing_Adjective* relation holds between the two Entities. For example, John is responsible for Mary.
6. *relation*(*Verb*, *Entity*, ...). The *Verb* relation holds among the Entities. For example, John sees Mary. The number of arguments depends on the verb. Intransitive verbs have one argument; transitive verbs have two; ditransitive verbs have three.

For simplicity we often write *relation*(*Relation*, Arg_1, ..., Arg_n) as *Relation* (Arg_1, ..., Arg_n).

2.3. *Modal Operators*

NLP supports modalities in a very limited way. Any of the following modalities may be predicated of any predicate: *can*, *may*, *might*, *should*. We notate this feature: *Modal*(*Predicate*). For example, *can*(*see*(*John*, *Mary*)) may be read: John can see Mary. Allowing null_modal as the null modal we can say that all relations are modal relations—that is, *null_*modal(P) for P.

While modalities may be expressed, there is nothing built into NLP's inference mechanism regarding modalities. That is, even if the knowledge base contains *must(see(John, Mary))* the inference engine will not necessarily conclude *see(John, Mary)* should that goal be launched. Users may add rules to link modal conclusions to nonmodal conclusions. That is, a user could add the rule: *see(John, Mary) if must (see(John, Mary))*.

2.4. The Occurs Predication

NLP supports a limited form of higher-order predication called the "occurs predication." Occurs predications occur when one is asserting that a conclusion occurs in a certain way. For example, the assertion *John kissed Mary on the ear* not only lets one conclude the John kissed Mary, but also that he kissed her on the ear. If one had added, "in the car" then one could also conclude that John kissed Mary in the car. And if one had also added, "softly" then one could conclude the John kissed Mary softly.

The general form of occurs predication is *occurs(Predicate, Attributes)*, where *Predicate* is any modal (or primitive) predication and *Attributes* is a set of attribute–value pairs. For example,

occur(kiss(John, Mary), { no_attribute: in(the car),
no_attribute: on(the ear),
no_attribute: softly}).

predicates *in(the car)*, *on(the ear)*, and *softly* of *kiss(John, Mary)*. (The attribution no_attribute will be explained shortly. The actual objects for "the car" and "the ear" are skolems.)

Conclusions involving occurs predications are drawn according to the following rule.

occurs(P, Advs) if occurs(P, Advs_1) and subset(Advs, Advs_1).

That is, if Advs is a subset of Advs_1.

It follows from this that if the collection of adverbial phrases is empty the occurs predication is simply the first argument. That is, *occurs*$(\langle P\rangle, \{\ \}) = \langle P\rangle$. This holds for both goals and conclusions. Therefore if we allow *Attributes* to be the empty set we can say that all predications seen so far may be expressed as occurs predications.

The actual *Attributes* and their values vary depending on the *Predicate* of which occurs is predicated.

1. *Adverbials.* Adverbials are allowed on relations derived from verbs, as in the example above. In this case, the attribute is always *no_attribute* and the value is derived from either a prepositional phrase or

an adverb. The adverbial represents a place, a manner, an instrumentiality, etc.

2. *Restrictive Properties.* Restrictive occurs predications are derived from what are known as a restrictive adjectives. These are adjectives, e.g., mortal, that predicate a property directly of an entity. Restrictive occurs predications are occurs predications of individual predications.

   Only one attribute–value combination appears in restrictive occurs predications, although there may be many occurs predications for the same individual (predication). For example, *John is male* is *occurs(John(X), {gender:male})*. The attribute itself is optional and may be either "no_attribute" or a named attribute, as in this case. The value is required and is either an adjective (in this case *male*) or a numeric value, e.g., an age.
3. *Scalar Properties (syncategorematic adjectives).* Scalar properties are occurs predications to *Category* and *Role* relations. These occurs predications are derived from scalar adjectives. As with Restrictive occurs predications, only one attribute–value combination appears in these occurs predications, although there may be many occurs predications for the same predication.
   a. *occurs(⟨Category Predication⟩, {Attribute})*. For example, "John is a tall midget" is *occurs(midget(John), {height: tall})*
   b. *occurs(⟨Role Predication⟩, {Attribute})*. For example, "John is a good father" is *occurs(father(John, X), {no_attribute: good}) if father(John, X)*

2.5. *Negation*

NLP supports a form of negation. We notate this feature *false(P)*, which may be read *P is false*. *P* may be any relation predicate, any modal, any occurs predication, or any negated predication.

While providing the ability to express negative information, NLP does not reason from contrapositives. Given, for example, *mortal(X) if man(X)* and *false(mortal(Superman))*, NLP will not conclude *false(man(Superman))*. Users may add such contrapositive rules themselves, though.

2.6. *Quantified Predication Structures*

NLP provides a number of quantified predication structures. These may be used for expressing goals but not for expressing conclusions. The general form is *quantified_predication(Quantifier, Variable, Description, Goal)* where both *Description* and *Goal* are goals. The following *Quantifiers* are supported.

1. *the. Variable* is the one and only entity that satisfies *Description.* It also satisfies *Goal.* If there are multiple *X*'s that satisfy *Description* when this goal is launched, an interaction with the user is begun to determine which one is intended. If there are no such *X*'s or if the *X* that satisfies *Description* does not satisfy *Goal* the "the" quantified predication goal fails.
2. *cardinality*(*Lower, Upper*). Between *Lower* and *Upper* instances of *Variable* that satisfy *Description* also satisfy *Goal.*

2.7. *Operators*

The operators *and*, *or*, and *whenever* may be used to join goals together to form more complex goals.

1. *and.* The goal formed by the *and* operator succeeds if the two goals that *and* joins both succeed with the same instantiation of variables.
2. *or.* The goal formed by the *or* operator succeeds if either of the two goals that *or* joins succeeds.
3. *whenever.* The goal formed by the *whenever* operator *whenever*(*Antecedent, Consequent*) succeeds if for every instantiation of variables resulting when *Antecedent* succeeds, *Consequent* succeeds with the same instantiation.

2.8. *Summary of Knowledge Structures*

In summary, NLP provides the following structures for expressing knowledge. The following is to be read as an abstract syntax for NLP's knowledge structures and not as the syntax for a specific language. We call this knowledge representation system Silog. Below are abstract syntax specifications for NLP knowledge structures.

⟨Clause⟩	→⟨Conclusion⟩ | ⟨Rule⟩
⟨Rule⟩	→⟨Conclusion⟩ if ⟨Goal⟩
⟨Goal⟩	→⟨Goal⟩ and ⟨Goal⟩ | ⟨Goal⟩ or ⟨Goal⟩ | whenever(⟨Goal⟩, ⟨Goal⟩) | ⟨Prolog Predicate⟩ | ⟨Quantified Predication⟩
⟨Quantified Prediction⟩	→the(⟨Variable⟩, ⟨Description⟩, ⟨Goal⟩) | ⟨Conclusion⟩ | cardinality(⟨Variable⟩, ⟨Description⟩, ⟨Goal⟩, ⟨Number⟩, ⟨Number⟩)
⟨Conclusion⟩	→false(⟨Conclusion⟩) | ⟨Occurs Predication⟩ | ⟨Prolog Predicate⟩
⟨Occurs Predication⟩	→occurs(⟨Modal Predication⟩, ⟨Attributes⟩) | ⟨Modal Predication⟩
⟨Modal Predication⟩	→⟨Modal⟩(⟨Relation Meta Predication⟩) | ⟨Relation Meta Predication⟩
⟨Modal⟩	→can | may | might | must | should
⟨Rel. Meta-Pred⟩	Entity(Variable)) | Category(Entity) | Role(Entity, Entity) | Relation (⟨Entity⟩,

2.9. *Accessing External Databases and Prolog*

As shown above, one of the options for ⟨Goal⟩ is an arbitrary Prolog goal. This lets users call their own Prolog programs, as in *John is happy if Mary is happy and married(John, Mary)*. Since Silogic's Knowledge WorkBench[15] provides access from prolog to external databases, data in existing databases is directly available.

Also, since ⟨Prolog Goal⟩ is one of the options for ⟨Conclusion⟩, it is possible for user programs to access information in the KMS knowledge base, as in *happy*(⟨*A*⟩) *if* ⟨*A*⟩ *is happy*, which provides prolog access to all happy individuals in the knowledge base.

3. *Natural Language Processing*

The grammar implemented by the NLP parser assumes a form of Government Binding theory as developed in Refs. 5 and 6. Although specific details of the implementation may differ in form from the theory, the spirit of the theory has been maintained. (See Ref. 3 for a discussion of the relationship between current linguistic theory and traditional computational approaches to natural language processing.)

3.1. *X-Bar Theory*

Chomsky[4] argued that there is a structural parallelism between ⟨Sentence⟩s and ⟨Noun Phrase⟩s. Specifically, let the variable X range over grammatical categories (Noun, Verb, Adjective, Preposition, etc.) and let n range over "bar levels," i.e., levels of derivation in phrase structure rules. Then every phrase structure rule must be of the form

$$X(n) \rightarrow \ldots X(n\text{-}1) \ldots$$

This constraint guarantees that every phrase in the hierarchical representation of a sentence must have a head, an element that is syntactically "central" to the rest of the phrase. If we stipulate that 2 is the highest bar level, i.e., the "phrase" level, and we call level 1 the "(single) bar" level, then we can define the phrase structure rules for English as

$$\langle XP \rangle \rightarrow \langle \text{Specifier} \rangle \ldots \langle \text{Specifier} \rangle \langle X\text{-Bar} \rangle$$
$$\langle X\_\text{Bar} \rangle \rightarrow \langle X \rangle \langle \text{Complement} \rangle \ldots \langle \text{Complement} \rangle$$

where typical specifiers are adjective phrases for nouns, and typical complements are objects for verbs and prepositions.

A word is said to "subcategorize" for its complements, and word subcategorization structures may be stored in the lexicon. For example, *responsible* subcategorizes for a single complement, a prepositional phrase headed by *for*, as in *John is responsible [for the pizza]*. Similarly *put* subcategorizes for two complements, a noun phrase and a prepositional phrase, as in *John put [the book] [on the table]*. Combining *X*-Bar theory with lexical specification of subcategorization information, the number and structure of the phrase structure rules is simplified dramatically. Currently NLP covers a significant portion of English with fewer than four dozen rules, and many of these deal with very special cases such as the need to convert between *what* and *which*.

3.2. The Extended Projection Principle

The "projection principle" requires that the lexical subcategorization requirements of a head must be satisfied at every level of representation. This principle means that if an item subcategorizes for an NP, then an NP must be present. Consider a transitive verb like "hit," which requires a single noun phrase complement. Given this subcategorization, *Who did John hit?* would appear not to be well formed since no noun phrase follows "hit." The solution is to assume an empty noun phrase (denoted here by [np e] yielding *Who did John hit [np e]?* The empty complement is then bound to *who*. The projection principle along with (a) the requirement that every empty category have an antecedent and (b) constraints on where the antecedents of empty categories may appear are combined in the parser to prevent NLP from accepting certain nongrammatical inputs such as, **John hit [np e]* and **who did pictures of [np e] annoy John*.

3.3. Parse Tree Nodes

A parse tree node reflects the state of the parse of part of the original input. There are three major types of nodes: *x* level, *x*-bar level, and *xp* level nodes. Nodes are represented by Prolog terms of the form:

⟨*x*-level⟩(Cat--Subcategory Requirements, FS--Tree)

The functor ⟨*x* level⟩ is one of *x*, *x* bar, or *xp*; *Cat* is the node's grammatical category, for example: noun (common/3^s) for a third person singular common noun; *Subcategory_Requirements* lists the subcategories for the node's complements; *FS* is the node's functional structure expressed in a predicate calculuslike notation; and *Tree* is the subtree for the node. Coded into the representation of the tree is the "gap status" of the node, i.e., whether the node includes an empty category, an antecedent of an empty category, or neither. In addition, both the syntactic category of the empty category (either the trace or the antecedent) and the semantic variable associated with the trace are included in the tree. In effect, the two first elements of nodes are input elements projected from the node's head; the second two are output elements reflecting the structure of the entire node.

3.4. The NLP Parser

The NLP parse engine is a bottom-up, shift-reduce, context sensitive, rewriting system. In addition, it has the ability to call itself recursively to implement top-down parsing. The parser operates in three passes: NP Pass, S Pass, and F Pass.

1. The *NP Pass* builds noun phrases and prepositional phrases. It proceeds in a single, bottom-up, nonrecursive pass over the input nodes.
2. The *S Pass* combines noun phrases, verbs, prepositional phrases, and other sentence elements into sentences. It too is bottom-up except when the parser calls itself recursively to parse the complements of x-level nodes before attaching them. For example, to parse *John hit the man who kissed Mary* the relative clause *who kissed Mary* must be parsed before *the man who kissed Mary* can be attached as a complement to *hit*.
3. In most cases the *F Pass* does nothing. But when the user responds to an NLP question by entering a noun phrase, e.g., "the man who owns the dog," the F Pass completes the parse of these noun phrases. This is needed since noun adjuncts, which may be relative clauses as in this case, are not attached during the NP Pass. Normally they are attached when the noun that they modify is attached (to something) as a complement. But when the noun phrase appears by itself, they must be attached separately.

4. Semantic Analysis

For the most part the derivation of Silog from a parse tree follows straightforwardly. In general, every noun phrase in an input sentence gives rise to a quantified predication:

quantified_predication(Quantifier, Variable, Description, Predication) where the ⟨Quantifier⟩ depends on the determiner of the noun phrase; the ⟨Description⟩ is derived in a natural way from the remainder of the noun phase, i.e., from whatever adjective specifiers and nominal adjuncts the noun phrase contains; and the ⟨Predication⟩ depends on the rest of the sentence in which the noun phrase appears. We do not go into detail regarding the intuitive transformations, but here is a simple example. *Every rich man likes some pretty woman* becomes *all*(*M*, (*man*(*M*) *and rich*(*M*)), *exists*(*W*, (*woman*(*W*) *and pretty*(*W*)), *like*(*M*, *W*))).

Below we discuss some of the problems encountered in semantic analysis. Among the problems not discussed are scope ambiguities of negation (e.g., what is the scope of *not* in *John is not a happy man* and *John did not read a book*) and the interplay between transformations that map the entire parse tree and others that map substructures.

4.1. Copulas

Copulas cause difficulties if the complement of the copula is a noun phrase, e.g., John is a man. In this case, a distinction is made between referential and predicative noun phrase complements. Indefinite and possessive noun phrases are considered predicative. All others are considered referential. A noun phrase is indefinite if (1) it is not a proper noun and (2) either (a) it has no determiner or (b) its determiner is "a" or "an."

Given these definitions, NLP's heuristic semantic rules are as follows:

1. If a copula sentence includes one referential and one predicative noun phrase, independent of which is the subject and which the complement, then the sentence is interpreted as predicating the predicative noun phrase of the referential noun phrase. For example, *John is a man* and *A man is John* are interpreted identically as *man*(*John*).
2. If a copula sentence includes two referential noun phrases, then the referents of the two noun phrases are equated. For example, *John is Bill* is interpreted as *John*(*Bill*) and *Bill*(*John*).
3. If a copula sentence includes two predicative noun phrases, then the "more predicative" noun phrase is predicated of the less predicative noun phrase, or, if they are equally predicative, then the complement is predicated of the subject. With respect to predicative strength, possessive noun phrases are considered less predicative than indefinite noun phrases. If the less predicative noun phrase is a possessive, it is understood as a condition rather than as a reference. (See the discussion of determiners and quantifiers below.) Here are some examples, *John's father is a man* and *A man is John's father* are both interpreted as the assertion: *man*(*X*) *if father*(*X*, *John*). Also,

A man is a fool is interpreted as *exists*(*M*, *man*(*M*), *fool*(*M*)), and *John's father is Mary's uncle* is interpreted as *uncle*(*X*, *Mary*) *if father*(*X*, *John*).

4.2. Scoping

In general, sentences are scoped from left to right. Thus *Every man loves some woman* yields *every*(*M*, *man*(*M*), *exists*(*W*, *woman*(*W*), *love*(*M*, *W*))) whereas *Some woman is loved by every man* yields *exists*(*W*, *woman*(*W*), *every*(*M*, *man*(*M*), *love*(*M*, *W*))).

An exception is that prepositional phrase nominal adjuncts take scope over the noun phrases that they modify. For example, *Every member of some committee is a fool* is *exists*(*C*, *committee*(*C*), *every*(*M*, *member*(*M*, *C*), *fool*(*M*))).

4.3. Coordination

Coordinations are fully distributed. This is fine for *Janet and Mel are happy*, which is interpreted as *happy*(*Janet*) and *happy*(*Mel*), but not so fine for *Janet and Mel are married*, which is given a parallel interpretation anyway.

4.4. Noun Phrases

Noun phrases with identical syntactic structures may map into different Silog structures depending on the head noun. For example, see Table 1.

4.5. Disjunctions

Disjunctions in conclusions are mapped into negations, i.e., *P* or *Q* is mapped into the two rules: *P if false*(*Q*) and *Q if false*(*P*). This is not wholly satisfactory. Given *If* ⟨*X*⟩ *is a person then* ⟨*X*⟩ *is a man or* ⟨*X*⟩ *is a woman* and *John is a person*, the response to "Is John a man?" (or to "Is John a woman?") will be "I don't know." because neither *John is not a man* nor *John is not a woman* can be proved.

TABLE 1
Noun Phases with Identical Syntax, but Differing Semantic Structures

Example input	Head noun property	Mapping
John's dog is happy.	*Dog* is a type.	Happy(*X*) if have(John, *X*) & dog(*X*).
John's father is happy.	*Father* is a role.	Happy(*X*) if father(*X*, John).
John's mood is mellow.	*Mood* is an attribute—*mellow* is a value of *John's mood.*	Occurs(John(*X*), {mood: mellow}).
John's mood is volatile.	*Mood* is an attribute—*Volatile* is a property of *John's mood.*	Can't handle.

TABLE 2
Implicit Quantifiers

Free	A *free* quantifier is generated whenever NLP encounters a variable, e.g., ⟨*X*⟩.
Free possessive	A *free* possessive quantifier is generated whenever NLP encounters a variable followed by a possessive indicator, e.g., ⟨*X*⟩'s.
⟨No quantifier⟩	Proper nouns refer directly to individual entities and imply no further range.
Possessive	A possessive quantifier is generated whenever NLP encounters a possessive structure such as *'s* or *whose.*

4.6. Determiners and Quantifiers

The mapping of determiners onto quantifiers is one of the more difficult aspects of NLP. The tables above and below summarize the interpretations. In some cases the tables list quantifiers that do not occur directly in English. Table 2 summarizes the situations that give rise to these implicit quantifiers.

In some cases quantifiers in facts and rule conclusions are interpreted differently from the way that they are interpreted in queries and rule conditions. Table 3 shows the interpretations of quantifiers in facts and rule conclusions. Other quantifiers are mapped onto those above (Table 4).

Table 5 shows the interpretation of quantifiers in queries and rule conditions. Other quantifiers map onto the ones above (Table 6).

TABLE 3
Basic Quantifier Interpretations for Facts and Rule Conclusions

Quantifier	Interpretation	Example
Every	Applies to every entity that satisfies the given description.	Every man is mortal
⟨No quantifier⟩	The indicated entity.	John is mortal
Some	NLP creates un unnamed entity that satisfies both the given description and the additional conditions of the sentence.	Some man is mortal
The ⟨singular noun⟩	*The* ⟨singular noun⟩ is always treated intensionally. This is the interpretation that one would normally expect in sentences such as *The president of the United States lives in Washington D.C.* The interpretation is: if there is a unique entity that satisfies the given description then the rest of the sentence is true of that entity.	The man is mortal

TABLE 4
Quantifier Mappings to Basic Quantifiers for Facts and Rule Conclusions

Quantifier	Same as	Examples
A, one, one of the	*Some*	A man is mortal. One man is mortal. One of the men is mortal.
All, any, each, free, free possessive	*Every*	All men are mortal. ⟨*X*⟩ is mortal.
The ⟨plural noun⟩	*Each*	The men are mortal.
Possessive	*The*	John's dog is mortal.
⟨Quantifier⟩ of the	⟨*Quantifier*⟩	All of the men are mortal.

TABLE 5
Basic Quantifier Interpretations for Queries and Rule Conditions

Quantifier	Interpretation	Example
Each	A separate answer is given for each entity that satisfies the given description.	Is each man mortal?
Every	Requires that all entities satisfying the indicated description also satisfy the indicated condition.	Is every man mortal? John is mortal if all men are mortal.
How many	Requests a count of the number of entities that satisfy both the indicated description as well as the other conditions in the question.	How many men are mortal?
⟨No quantifier⟩	Refers to the indicated entity only.	Is John mortal?
Some	Asks whether there is an entity satisfying the given description that also satisfies the question conditions.	Is some man mortal?
The ⟨singular noun⟩	Taken as requiring that a unique entity satisfy the given description and then taken as referring to that entity. Evaluated anew whenever the query or assertion condition is evaluated.	Is the man mortal? John is happy if the woman whom he likes is happy.
⟨Number⟩	Asks whether there are exactly ⟨number⟩ entities that satisfy both the given description and the question conditions.	Are five dogs mortal?
Which	Asks to be told which entity that satisfies the given description also satisfies the question condition.	Which man is mortal?
Whose	Which person or thing's	Whose dog is mortal?

TABLE 6
Quantifier Mappings to Basic Quantifiers for Queries and Rule Conditions

Quantifier	Same as	Example
A, any, free, free possessive, one, one of the	Some	Is a man mortal?
All	Every	Are all men mortal?
The ⟨plural noun⟩	Each	Are the men mortal?
Possessive	The	Is John's dog mortal?
⟨Quantifier⟩ of the	⟨Quantifier⟩	Are any of the men mortal?

5. *The NLP Inference Engine*

This section discusses the features of NLP's inference engine that distinguish it from pure Prolog.

5.1. *Mixed Initiative*

NLP provides a mixed initiative capability. If the user response to an NLP question does not match the form expected, NLP processes the input in the normal top-level way and then returns to its internally generated interaction.

5.2. *Knowledge Base Consistency*

Whenever new information is to be asserted to the knowledge base, NLP checks to ensure that the new information does not violate the validity constraints currently in force. There are four user-settable levels of validity checking:

1. Make no checks; accept everything.
2. Do not accept duplicate information.
3. Do not accept information that itself is contradicted in the knowledge base. In this case, not all possible contradictions are examined, only contradictions of new information. At this level too, NLP checks for functional dependency violations caused directly by new information. By default, attributes are understood to be single valued, although they may be set to be multivalued. This mechanism catches the old barber paradox. As the following example shows, once a contradiction is detected NLP gives the user the option of withdrawing the new input or of looking at the proof of the contradiction. Using NLP's mixed initiative capability, the user can then remove other information, in which case, if the

contradiction is no longer in effect, NLP will accept the new information.

>Anyone $\langle X\rangle$ who doesn't shave $\langle X\rangle$ is shaved by the barber.
user rule 14: [Not repeated here.]
>The barber does not shave anyone $\langle X\rangle$ who shaves $\langle X\rangle$.
user rule 15: [See the explanation below.]
user fact 20: Figaro is a barber.
>Figaro shaves Figaro.
Your input statement leads to a contradiction. Type "why" to see the contradiction. (In the explanation your input will be labeled "temporary.")
You may remove one or more of the clauses that leads to the contradiction, or I can ignore your input statement.
Should I ignore your input statement?
[y, why] (default y) >why.
user fact 20: Figaro is a barber.
temporary: Figaro shaves Figaro.
user rule 13: If $\langle A\rangle$ is a barber
then $\langle A\rangle$ is a person.
CONCLUSION 1: Figaro is a person
(from: user rule 13, user fact 20.)
user rule 15: If $\langle A\rangle$ is the barber; and
$\langle B\rangle$ shaves $\langle B\rangle$; and
$\langle B\rangle$ is a person
then $\langle A\rangle$ does not shave $\langle B\rangle$.
CONCLUSION 2: Figaro does not shave Figaro
(from: user rule 15, user fact 20, temporary, conclusion 1.)

Attempting to enter *Figaro does not shave Figaro* also leads to a contradiction.

4. Do not accept any information that causes a user specified validity constraint to be violated. Users may express validity constraints with inputs of the form $\langle S\rangle$ *is invalid*, where $\langle S\rangle$ is any sentence acceptable to NLP. Since any Prolog goal is acceptable as a sentence, the user can define arbitrary consistency checks.

5.3. *Negation*

As indicated above, it is possible to assert negative information explicitly. In standard Prolog, negation is implemented by what is called "negation by failure," which is essentially a closed world assumption, i.e., that the inability to prove something licenses the conclusion of its negation.

Through flag settings, NLP allows both the closed world and open world assumptions, the latter taken to mean that the inability to prove something means only that it is unknown whether or not it is true.

In NLP, P and *false*(P) are separate and independent goals having nothing to do with each other. Since NLP does not include proof by contradiction, even if P and *false*(P) were both provable, NLP would not be able to "prove anything." In addition, NLP's consistency checking mechanism is capable of preventing P and *false*(P) from being provable simultaneously. This precludes any need to check *false*(P) when P is needed in a proof.

Negation is most visible when responding to yes–no questions. If NLP is incapable of answering a yes–no question in the affirmative, it attempts to prove the negation of the yes–no goal. If that succeeds it responds "No" to the question. As indicated above, if the closed world assumption is in effect, the negation of a goal is assumed if the goal fails; otherwise the negation must be proved explicitly.

5.4. *Intensional Reasoning*

As the introductory example showed, NLP provides an intensional reasoning capability. When the intensional reasoning flag is on, NLP makes assumptions about logical variables that allow it to prove goals. These assumptions are then shown to the user. NLP includes a feature that permits the user to limit the assumptions that NLP will make. With the flag "full_intensionality" on, any assumption is acceptable; otherwise only assumptions specifically authorized by the user are made. The user authorizes an assumption by referring to a hypothetical entity. For example, *Some hypothetical entity is a man* allows NLP to let *man*(X) succeed by assumption.

When finding intensional answers, NLP does not ensure that the answers are internally consistent. For example, assume *$\langle X \rangle$ is a hermaphrodite if $\langle X \rangle$ is a man and $\langle X \rangle$ is a woman* is in the knowledge base. Then if NLP is asked "Who is hermaphrodite?" it will answer *any $\langle P \rangle$ that satisfies: $\langle P \rangle$ is a man and $\langle P \rangle$ is a woman.* This answer will be produced even if there are validity constraints or rules such as *$\langle X \rangle$ is a man and $\langle X \rangle$ is a woman is invalid*, *If $\langle X \rangle$ is a man then $\langle X \rangle$ is not a woman*, or *If $\langle X \rangle$ is a woman then $\langle X \rangle$ is not a man.*

5.5. *Whenever Structures*

A *whenever* structure is a type of if–then statement: If (or *whenever*) some condition holds, then some conclusion also holds. For example, consider this query: "Does every rich woman like some handsome man?" For NLP to answer this query affirmatively, it must ensure that there are *no* rich women who do not like some handsome man. (In answering such queries, NLP first attempts to prove the result intensionally. Only if that fails does it make an exhaustive search of the knowledge base looking for a counterexample.)

In queries of this sort there is always the possibility that the antecedent will not succeed at all; that is, that there are no rich women in the knowledge base. In such cases, the anwer should still be "Yes." But, for NLP to say "Yes" without further explanation would be confusing. In cases such as these, NLP produces its answer followed by a footnotelike message to the effect that the antecedent to a *whenever* structure was null:

>Does every rich woman like some handsome man?
You have asked whether: (if $\langle X \rangle$ is a woman; and
$\langle X \rangle$ is rich
then
$\langle Y \rangle$ is a man; and
$\langle Y \rangle$ is handsome; and
$\langle X \rangle$ likes $\langle Y \rangle$).

Yes.

Note: it cannot be shown that
$\langle P \rangle$ is a woman; and
$\langle P \rangle$ is rich

5.6. Protecting Against Loops

It is easy for users to enter loops into a knowledge base. A simple loop can be generated from the inputs:

>John is happy if Mary is happy.
>Mary is happy if John is happy.

The question, "Is John happy?" if executed naively, would lead to a loop. NLP protects against loops of this sort (loops caused by left recursion in the proof tree) by keeping track of all outstanding rule heads, i.e., goals and the clauses with which they are currently unified, including an identification of the clause itself. If a new unification is structurally identical (the atoms match and the variables are distributed identically) to an outstanding one, the clause body is failed rather then attempted. This mechanism was discovered independently but is essentially the same as reported in Ref. 8.

Loop trapping permits NLP to offer the user the ability to state two-way definitions such as

>Man: a male person.

5.7. Checking for Vacuous Descriptions

NLP checks for vacuous descriptions in extensional questions. Suppose that there are no known men in the knowledge base:

⟩Is any man happy.

You have asked whether: $\langle X \rangle$ is a man; and

$\langle X \rangle$ is happy.

Checking descriptions ...

No $\langle A \rangle$ is known to satisfy the description:

$\langle A \rangle$ is a man

5.8. Default Reasoning

NLP provides a simple default reasoning capability:

⟩If $\langle X \rangle$ is a person then $\langle X \rangle$ is a man unless $\langle X \rangle$ is not a man.

⟩"man" is disjoint from "woman".

⟩John is a person.

⟩Is John a man?

Yes.

Since "unless" generates a Prolog "not," i.e., failure rather than negation, NLP runs into problems when the attempted proof of the unknown clause itself runs into a default cause. Given $\langle A \rangle$ *unless* $\langle B \rangle$ and $\langle B \rangle$ *unless* $\langle C \rangle$ the goal $\langle A \rangle$ will fail since $\langle B \rangle$ will succeed by default. This yields the following unintuitive possibility:

⟩$\langle X \rangle$ is a man unless $\langle X \rangle$ is a woman.

⟩Is George a man?

Yes.

⟩$\langle X \rangle$ is a woman unless $\langle X \rangle$ is a man.

⟩Is George a man?

I don't know.

⟩Is George a woman?

I don't know.

5.9. Consultation Mode

NLP allows the knowledge engineer to enter rules that will cause NLP to interact with the user. There are two forms:

1. "the user says so." A rule of the form ⟨*Conclusion*⟩ *if the user says so* is taken by NLP as an instruction to ask the user about *Conclusion.* NLP will generate an appropriate question. If *Conclusion* contains logical variables, the user will be prompted for these too:

> ⟩⟨*X*⟩'s age is ⟨*Y*⟩ if the user says so.
> ⟩⟨*X*⟩ is an adult if ⟨*X*⟩'s age is greater than 17.
> ⟩Is Mary an adult?

You have asked whether: Mary is an adult.
Are there values of ⟨*Y*⟩ for which Mary's age is ⟨*Y*⟩?
    [yes., no., d. (don't know)] ⟩yes.
For which ⟨*Y*⟩ is it the case that Mary's age is ⟨*Y*⟩
    ⟩18.
Yes.

2. Knowledge engineer supplied questions. A rule of the form ⟨*Conclusion*⟩ *if the user says* ⟨*Response*⟩ *to* ⟨*Query*⟩ lets the knowledge engineer specify the question NLP will ask.

> ⟩⟨*X*⟩'s age is ⟨*Y*⟩ if the user says ⟨*Y*⟩ to "How old is ⟨*X*⟩?".

In both cases it is possible to tell NLP whether to consult the user before or after attempting to prove the goal from information in the knowledge base. It is also possible to tell NLP how long to keep such user-supplied information: the duration of the query, the duration of the session, indefinitely.

5.10. Explanations

As shown earlier, when the user asks for an explanation, NLP linearizes the proof tree and displays it. The user may ask "Why" in response to an NLP answer or may ask "Why ⟨yes–no question⟩" directly.

5.11. Tabulate

NLP offers a simple report generation capability.

> ⟩John likes Mary.
> ⟩Frederick likes Mary.
> ⟩tabulate table(⟨*X*⟩, ⟨*Y*⟩) such that ⟨*Y*⟩ likes ⟨*X*⟩.

Mary	John
Mary	Frederick

The goal following the "such that" can be anything that NLP can parse. The variables in it may be displayed in any order in the table.

5.12. Summary

NLP offers in an integrated package many of the features found only in systems that focus on a single capability. While, as always, compromises were necessary and some goals were not achieved, NLP demonstrates that it is possible to build a full featured, domain-independent, natural language knowledge base system.

6. Lessons Learned

In this section we review the lessons learned as a result of developing the NLP system.

6.1. English as a Programming Language

Natural Language Processing appears to offer the prospect of unprogrammed interaction with computers. Our experience has shown, however, that the best that can now be done is to implement what one might call *English as a programming language.* This is in contrast to what one may naively expect of a natural language system: a conversational partner.

Through informal experimentation, we discovered a very important distinction between how one is able to interact with other people and how one is forced to interact with a computer. The distinction concerns the direction and degree of empathy flow, where empathy means putting oneself in another's place. When interacting with people, a speaker can generally assume that listeners have the ability, at least to some extent, to put themselves in the speaker's place and to understand a communication as it was intended. When interacting with a computer system, such an assumption is completely unwarranted.

When interacting with a computer one must (consciously or unconsciously) (a) build a conceptual model of the system with which one is interacting and (b) frame one's interaction to manipulate that model to achieve the desired effect. In other words, one must put oneself in the computer's place, which is exactly what a programmer does when writing a program.

Thus, even if one could communicate with a computer in completely free natural language, one would still have to be aware of how one's statements were being interpreted and the state of the system doing interpreting. One would have to approach the system the way programmers and users now learn to approach computer systems: the system is a device and one must learn how to manipulate it through the use of whatever interface it makes available. Even if that interface is a natural language, the manipulative stance cannot be avoided.

For the foreseeable future, it appears that no natural language system will provide an interface comparable in naturalness and informality to that which one experiences when interacting with a person. This is because for the foreseeable future we will not be able to build a computer system that in any meaningful sense can put itself in its user's place; the user will always be forced to put himself or herself in the system's place, i.e., to program the system. (For a more detailed discussion of these issues see Ref. 1.)

A second lesson we learned is that a natural language capability, even were it fully empathic, still would not be sufficient. At the very best, a computer system that could interact in natural language would offer the same kind of interface that one has when talking over the telephone. For most of the tasks that one wants to do with computers, such an interface is not adequate.

Does this mean that natural language processing is not worth pursuing? Not at all. Natural language frequently provides efficiency of expression, especially in noun phrases. It appears that the real leverage offered by natural language processing is as the basis for a knowledge programming language. For fully functional interactions, what is needed is a combination of a visual, point-and-click interface along with a natural language capability: one wants to be able to talk about what one is looking at and manipulating.

6.2. *The Fate of the NLP System*

NLP was developed as the primary product of a start-up company. Unfortunately, the company failed. Interestingly, its failure was not due to technical weaknesses in the product. The company failed either because the wrong product was developed or because the marketing arm of the company did not know how to sell the product that was developed. This claim can be made with certainty because no customers or potential customers spent enough time with the system to discover its weaknesses.

It appears that despite the apparent attractiveness of natural language processing, the actual uses of such a system are less obvious than one might suppose. The reasons for this are those discussed in the previous section: natural language, when processed by a computer is just another programming language. It is worth noting, though, that when the company failed, a Japanese company bought the rights to the system.

References

1. R.J. Abbott, Knowledge abstraction, *CACM* **30**(8), 664–671, 1987.
2. R.J. Abbott, Program design by informal English descriptions, *CACM* **26**(11), 882–894, 1983.
3. R.C. Berwick, Transformational grammar and artificial intelligence: A contemporary view, *Cog. Brain Theory* **6**(4), 383–416, 1983.
4. N. Chomsky, Remarks on nominalization, in *Studies on Semantics in Generative Grammar*, Mouton, The Hague, 1970.
5. N. Chomsky, *Lectures on Government and Binding*, Foris Publications, 1981.
6. N. Chomsky, *Some Concepts and Consequences in the Theory of Government and Binding*, MIT Press, Cambridge, Massachusetts, 1982.
7. W.F. Clocksin and C.S. Mellish, *Programming in Prolog*, Springer-Verlag, New York, 1982.
8. M.A. Covington, A Further Note on Looping in Prolog, ACMC Research Report 01-006, Advanced Computational Methods Center, University of Georgia, 1984.
9. H. Gallaire, J. Minker, and J.-M. Nicolas, Logic and databases: A deductive approach, *ACM Surv.* **16**(2), 153–185, 1984.
10. F. Hayes-Roth, Rule-based systems, *CACM* **28**(9), 921–932, 1985.
11. R. Kowalski, *Logic for Problem Solving*, North-Holland, Amsterdam, 1979.
12. G. McCalla and N. Cercone (Eds.), Approaches to knowledge representation, Special issue of *Computer EEE* **16**(10), 1983.
13. M.C. McCord, Semantic Interpretation in the Epistle System, Proceedings of the Second International Logic Programming Conference, Uppsala, Sweden, July, 1984.
14. E. Rich, Natural language interfaces, *IEEE Comput.* **17**(9), 39, 50, 1984.
15. Silogic, Inc., The Knowledge WorkBench User's Manual and The Knowledge Workbench Reference Manual, Silogic, Los Angeles, 1986.
16. A. Walker and A. Porto, KBO1, A Knowledge Based Garden Store Assistant, Report RJ 3928 (44528), IBM San Jose Research Laboratory, June, 1983.
17. H.D. David Warren and F.C.N. Pereira, An Efficient Easily Adaptable System for Interpreting Natural Language Queries, *Am. J. Comput. Ling.* **8**(3-4), 110–122, 1982.
18. E. Wehrli, A Government-Binding Parser of French, Working paper No. 48, Institut pour les Etudes Semantique et Cognitives, University de Geneve, 1984.

COMMUNICATING WITH PICTORIAL DATABASES

ARTURO PIZANO, ALFONSO CARDENAS, AND ALLEN KLINGER

1. Introduction

This chapter presents a model and a language for specifying spatial queries: predicates involving the position on objects in a database picture. For example, "hospitals located *within* ten miles of a given school," "cities *adjacent* to Los Angeles," or "streets that *intersect* Highway 10 *in* downtown LA." Most existing pictorial query languages are text oriented, e.g., GRAIN,[2] IDMS, PSQL,[11] MIM,[13] and PROBE.[10] Others such as QPE[3] and PICQUERY,[9] follow the query-by-example approach (insertion of example values in tables), but do not make use of the visual properties of pictorial data. The language presented here differs from these methods by using pictures for specifying queries. These pictures, termed *query pictures*, visually depict conditions to be satisfied by the results of a query. For instance, Fig. 1a sketches a street intersection in which the condition "automobiles and people simultaneously entering a crosswalk" is satisfied. We enable new interaction mechanisms built upon the use of visual structures to manipulate spatially related objects. The work is related to *visual programming languages* as defined in Ref. 4. The following discussion addresses the problem of translating query pictures into first-order predicate calculus expressions describing queries. These expressions may be used by a query-processing mechanism to identify database pictures satisfying the desired conditions.

Query pictures resemble actual database occurrences. Our approach is similar to languages proposed in Refs. 1 and 14, since we also describe queries through visual examples. However, pictures in our case contain arbitrary entities and relationships, as opposed to specific sets of geometric shapes and spatial associations. We also recognize queries that involve non-pictorial data and provide tools for their description.

ARTURO PIZANO, ALFONSO CARDENAS, and ALLEN KLINGER • Computer Science Department, University of California, Los Angeles, California 90024-1596.

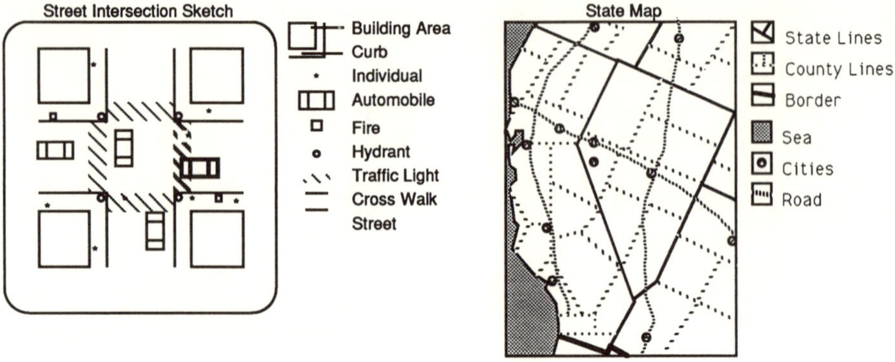

(a) (b)

FIGURE 1. Sample database pictures. (a) Traffic database; (b) geographic database.

Section 2 describes a data model providing basic picture interpretation semantics, i.e., associating pictorial primitives: pixels, lines, points, segments, ..., with corresponding real world concepts. Section 3 introduces rules for unambiguously interpreting query pictures. We show that even though certain requests can be described directly by pictures, in general, it is necessary for the user to provide additional nonpictorial information. Section 4 presents language extensions that enable the definition of queries involving aggregate and nonpictorial data. Section 5 describes an algorithm that translates query pictures into first-order predicate calculus expressions. Examples given throughout the text are drawn from two problem areas, "sketch" and "geographic," that have different processing requirements. "Sketch" refers to any simplified picture that models a more complex image, e.g., Fig 1a; "geographic" to map and satellite image data; see Fig. 1b.

2. Data Model

This section presents the data model that serves as background to our work. We have chosen to describe pictures as collections of entities, relationships, and attributes. *Entities* are abstractions of real world objects as seen by the user; *relationships* are associations between entities; and *attributes* are abstractions of the properties that describe an entity or relationship. Entities or relationships described by the same set of attributes are grouped into *entity or relationship sets*. The basic constructs in this model are patterned after Chen's Entity–Relationship (ER) model 5; however, we provide extensions needed to support pictorial data management.

Pictorial database entities are associated with specific segments of a picture. These segments may be as simple as a point or a line, or as complex as the picture itself. Properties that determine the appearance of an entity in the picture, e.g., position, dimensions, and color, are captured in *primitive attributes*. Entity properties like the population of a city, the year a road was built, or a city mayor's name, whose values are provided separately from the picture, are organized in *nonpictorial attributes*. In addition, properties that can be measured or computed by applying mathematical functions to primitive attributes, e.g., the area of a city, or the length of a road, are captured in *derived attributes*. Table 1 illustrates entities and attributes in the sample pictures shown in Fig. 1b. The two values following an attribute name refer to its class (nonpictorial, pictorial primitive, or pictorial derived)

TABLE 1
Geographic Database: Entities and Attributes

Entity set	Attributes	Type[a]
STATE	Population	(NP, integer)
	Governor	(NP, string)
	Incorporation-date	(NP, string)
	Boundaries	(PP, polygon)
	Area	(PD, float)
	Border-length	(PD, float)
LANDMARKS	Name	(PD, string)
	Year-built	(NP, integer)
	Location	(PP, point)
COUNTIES	Name	(NP, string)
	Population	(NP, integer)
	Boundaries	(PP, polygon)
	Area	(PD, float)
CITIES	Name	(NP, string)
	Population	(NP, integer)
	Mayor	(NP, string)
	Date-founded	(NP, date)
	State-capital	(NP, string)
	State-revenue	(NP, money)
	Annual-visitors	(NP, integer)
	Boundaries	(PP, polygon)
	Area	(PD, float)
ROADS	Number	(NP, string)
	Built	(NP, date)
	Type	(PP, constant)
	Path	(PP, line-set)
	Length	(PP, float)

[a] NP, nonpictorial; PP, primitive pictorial; PD, pictorial derived.

and domain. We assume that "point," "line set," and "polygon" are abstract data types described elsewhere in the data dictionary.

Relationships are associations described by a list of the entities involved. Optionally, attributes that further characterize the association may be included. We distinguish two types of relationships:

- *Nonpictorial relationships* are associations between entities that exist independently of their position in the picture. For example: "[city x] shares-police-departments-with [city y]," "[city x] (natives) hate [city y] (natives)," or "[city x] receives-money-from [county y]." They are established by the user, and recorded separately from the picture.
- *Pictorial or spatial relationships* are associations established between pictorial entities because of their relative position in the picture. For example, "left," "right," "above," "below," "close," "far," "distance," "intersection," "overlap."

We use $R(E_1, ..., E_N, A_1, ..., A_M)$ to denote a relationship R, involving occurrences in the entity sets $E_1,...,E_N$, characterized by attributes $A_1,...,A_m$; N is the degree of the relationship: relationship set *names* are upper case, *occurrences*, lower case. We use "spatial" synonymously with "pictorial." Henceforth, "relationship" refers to "spatial relationship." The list of spatial relationship in the geographic database is shown in Table 2.

Spatial relationships are position dependent or position independent. A relationship set $R(E_1, E_2)$, is *position independent* if and only if the relationship $R(e_1, e_2)$ exists for every $e_1 \in E_1$ and $e_2 \in E_2$. For instance, the relationship set

TABLE 2
Geographic Database: Relationships

Position dependent	Position independent[a]
ADJACENT (City, City)	RELATIVE_DISTANCE (City, City, rdistance)
ADJACENT (County, County)	RELATIVE_LOCATION (City, City, rlocation)
ADJACENT (State, State)	DISTANCE (Landmark, Landmark, distance)
BELONGS-TO (City, County)	DISTANCE (City, City, distance)
BELONGS-TO (City, State)	DISTANCE (Road, City, distance)
BELONGS-TO (County, State)	DISTANCE (Road, Landmark, distance)
BELONGS-TO (Landmark, City)	
CROSSES (Road, County)	
CROSSES (Road, City)	
CROSSES (Road, State)	
INTERSECTS (Road, Road, intersection_point)	

[a] rdistance, {"FAR", "NEAR"}, rlocation, {"N", "S", "E", "W", ...}.

DISTANCE(Landmark, Landmark) is position independent: it exists regardless of the location of the entities in the picture. If there exist $e_1 \in E_1$ and $e_2 \in E_2$ such that $R(e_1, e_2)$ is false then R is *position dependent*. *BELONGS-TO*(*City*, *State*) exists between "Los Angeles" and "California," but not between "Los Angeles" and "Nevada," therefore it is position dependent.

Position-independent spatial relationships are always characterized by attributes. For example, the distance between landmarks is described by a numeric value; the relative distance between two cities by the constants FAR or NEAR; and their relative location by a constant in the set {N, S, E, W, NW, SW, NE, NW, ...}. Position-dependent relationships, on the other hand, may be characterized by attributes, e.g., *NEAREST_CITY*($City_1$, $City_2$, *distance*), or simply by the entities it associates, e.g., *BELONGS-TO(City, state)*.

We note that spatial relationships are built into the picture. For instance, the fact that Anaheim "belongs-to" Orange County may be deduced from the picture; it does not have to be recorded separately. Because of this property, spatial relationships may be described by functions that identify their ocurrences in a picture. For example, the relationship *DISTANCE*($City_1$, $City_2$, D), could be evaluateıd with the Euclidean formula, using as reference points the centroid coordinates of the polygon that describes the city boundaries as reference points. Other metrics (e.g., city block) or parameters (e.g., the closest point) could also have been used. In general, the number of potential relationships in a picture is too large to be predefined. Functions provide an extensible mechanism that allows tailoring visual structure to the needs of particular users or application domains.

Having categorized entities and relationships, we now extend the model with the concepts of aggregation and generalization presented in Ref. 12. Aggregation is a mechanism used to construct complex objects from simple ones. For example, cities are composed of streets and blocks; counties of cities; and states of counties. Generalization is a form of data abstraction that enables clustering of entities that share some subset of their attributes. The result is an object at a higher level. For example, cities, metropolis, and mountain resorts are all population centers characterized by common properties such as name, location, and population. In addition, cities may be further described by the mayor's name or date they were founded. A metropolis might be characterized by the name of the local ballet company, or a list of its predominant ethnic groups. Likewise, a mountain resort could be labeled by the number of ski lifts it has, or a list of airlines that fly near it: a detailed description of data abstraction is provided in Ref. 8.

Aggregation is defined by "is-composed-of" (ICO) relationships: existing entities are grouped to form complex objects. They may be nonpictorial, when they are user defined and stored as separate information; or spatial, when they are derived from the picture. For example, "the great lakes" is an object composed of five large bodies of water on the U.S./Canada border. Since there are no generic spatial relationships among them, their association must be established by the user outside the picture. Conversely, composite objects of type metropolis, defined as clusters of cities sharing boundaries, may be located by means of the city-adjacency relationship.

Generalization is a fundamental component of virtually every semantic data model, manifested by the ability to represent "is-a" (ISA) relationship. These associations enable the organization of database objects into classes and subclasses. The immediate implication of this organization is that class attributes are automatically *inherited* by the subclasses. For example, if city "is-a" population center, and population centers are characterized by their location and population, then cities automatically inherit these attributes.

TABLE 3
Traffic Database: Entities and Attributes

Entity set	Attributes	Type[a]
Building area	Boundaries	(P)
	Area	(D)
Traffic Light	Location	(P)
	Signal	(N)
Truck	Position	(P)
	Speed	(N)
	Direction	(N)
	Weight	(N)
Curb	Boundaries	(P)
	Area	(D)
Fire hydrant	Location	(P)
Street	Boundaries	(P)
	Area	(D)
	Name	(N)
	Direction	(N)
Automobile	Position	(P)
	Speed	(N)
	Direction	(N)
People	Position	(P)
	Speed	(N)
	Direction	(N)
Crosswalks	Boundaries	(P)
	Area	(D)
Motorcycle	Position	(P)
	Speed	(N)
	Direction	(N)

[a] (P), primitive; (D), derived; (N), Nonpictorial.

The class hierarchy established by generalization relationships determines a priori the class and subclasses to which an object belongs. Several studies (6–8) distinguish a second form of ISA relationship called *specialization*. This new form of association establishes class membership based on attribute values, enabling entities to change from one class to another dynamically. For example, two classes of cities may be "large" and "small," depending on whether the population exceeds one million people; i.e., cities may change class over time.

We illustrate data abstraction in pictorial databases by entities and attributes for "traffic" (see Table 3), distinguishing three basic classes: "fixed," "stationary," and "transient," via associations shown in Fig. 2a. "Fixed" entities represent objects whose position never changes, e.g., streets, curbs, and crosswalks. "Stationary" entities refer to objects that have the potential for moving (perhaps not on their own), but that are still when the picture is created, e.g., fire hydrants and traffic lights. "Transient" entities represent objects that may be changing positions when the picture is created, e.g., pedestrian and automobiles. The actual generalizations resulting from this entity characterization appear in Fig. 2a. Notice that we introduced the class of "vehicles" to which automobiles, motorcycles, and trucks belong. We also list attributes describing each class next to its box.

Two forms of specialization in the traffic database are illustrated in Fig. 2b using the attributes "speed" for vehicles and "shape/color" for traffic lights. These attributes determine membership in classes such as "moving" and "stopped" vehicles, or "left-arrow," "right-arrow" or "red-light," traffic lights.

Finally, spatial relationships shown in Fig. 3a are associations defined on entities at the highest level of the class hierarchy. Multiple relationships can be derived from them for each subclass combination. There are 48 different spatial relationships that can be obtained by specializing the "on" association for vehicles and fixed objects: 4 fixed objects × 3 vehicle classes × 4 specialization for each vehicle class; these are indicated by an example in Fig. 3b.

3. *Picture Interpretation*

This section presents a method for generating first-order logic predicates from query pictures. The method is based on the following rules that resolve potential picture interpretation ambiguities:

Rule 1: Query pictures are occurrences of the database schema.

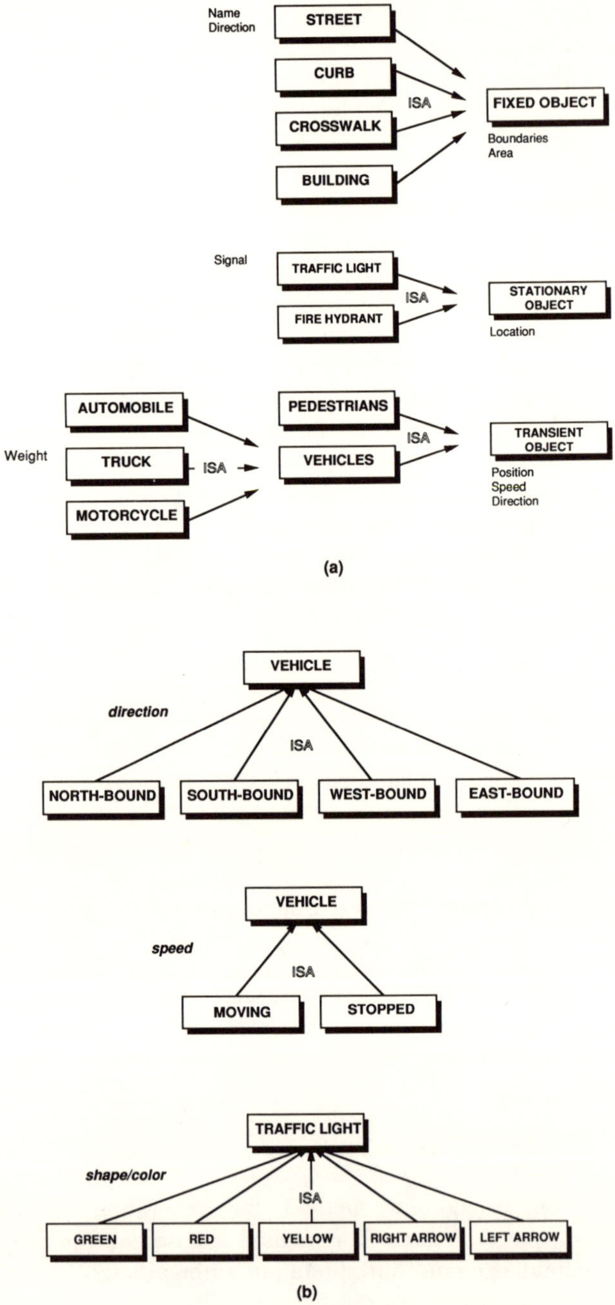

FIGURE 2. Data abstraction in the traffic database.

a

[Transient-Object]	"ON"	[Fixed-Object]
[Stationary-Object]	"CLOSE-TO"	[Stationary-Object]
[Stationary-Object]	"CLOSE-TO"	[Transient-Object]
[Transient-Object]	"CLOSE-TO"	[Transient-Object]
[Transient-Object]	"IN-FRONT-OF"	[Vehicle]
[Stationary-Object]	"IN-FRONT-OF"	[Vehicle]
[Fixed-Object]	"INTERSECTS"	[Fixed-Object]

b

[Transient-Object] "ON" [Fixed-Object]

- [Vehicle] "ON" [Crosswalk]
- [Vehicle] "ON" [Street]
  - [Truck] "ON" [Street]
  - [Automobile] "ON" [Street]
    - [South-Bound-Automobile] "ON" [Street]
    - [North-Bound-Automobile] "ON" [Street]
    - ...
  - [Motorcycle] "ON" [Street]
- [Vehicle] "ON" [Curb]
- [Vehicle] "ON" [Building Area]

[Stationary-Object] "CLOSE-TO" [Stationary-Object]

- [Traffic-Light] "CLOSE-TO" [Traffic-Light]
- [Traffic-Light] "CLOSE-TO" [Fire-Hydrant]

FIGURE 3. Traffic database relationships. (a) Basic relationships; (b) sample relationship specializations.

Rule 2: The query result contains all the pictures in the database satisfying the following predicate:

$$\exists(e_1 \in E_1, e_2 \in E_2, \ldots, e_n \in E_n) \mid (r_1(e_{i_1}, e_{j_1}) \wedge r_2(e_{i_2}, e_{j_2}) \wedge, \ldots, \wedge r_m(e_{i_m}, e_{j_m})) \quad (1)$$

where e_i, an occurrence of the entity set E_i, and $r_k(e_{i_k}, e_{j_k})$, an occurrence of the spatial relationship set $R_k(E_{i_k}, E_{j_k})$, are present in the picture.

Rule 1 establishes the interpretation context. It links the contents of query pictures to the underlying database schema. Rule 2 indicates that the query picture depicts examples of the entities and relationships to be found in the pictures being searched for. Formula (1) also provides the basic template for the first-order predicate calculus expression to be generated by the language. Extensions to this template, including the use of OR conditions, are presented in Section 4.

To illustrate the use of Rules 1 and 2, consider the condition

P1 automobiles and pedestrians simultaneously in a crosswalk

and the picture in Fig. 1a, where this condition is met. The application of Rule 1 lets us associate components of this picture with the database schema. In this case the picture contains an automobile (a); a crosswalk (x); a person (p); and the relationships $On(a, x)$ and $On(p, x)$. The following string results from substituting this into (1):

$$\exists(a \in Automobiles, x \in Crosswalks, p \in People) \mid (On(a, x) \wedge On(p, x))$$

Since this expression is logically equivalent to P1, we have shown that it is possible to associate query pictures with string descriptions of query predicates. At this point, however, we cannot say that this association is unique, because there are many other strings that can be derived from Fig. 1a while still satisfying Rules 1 and 2—for instance, "automobiles in a crosswalk"

$$\exists(a \in Automobiles, x \in Crosswalks) \mid (On(a, x))$$

or, "pedestrians in a crosswalk"

$$\exists(p \in People, x \in Crosswalks) \mid (On(p, x))$$

The remainder of the section concerns resolving these ambiguities: i.e., guaranteeing that query pictures describe one, and only one, predicate, is a necessary condition in the development of the language.

Ambiguities in the interpretation of Fig. 1a are due to lack of information about the set of entities and relationships involved in the query description. While the formula introduced with Rule 2 determines the way in which query picture components should be arranged to describe the desired result, neither this rule nor Rule 1 indicates exactly which components should be taken into account when building the string; thus we constructed multiple picture descriptions by arbitrarily selecting entities and relationships.

We now introduce two new rules that distinguish picture components relevant to the query from those that should be ignored. We first note that a minimum set of entities should be present in any picture that satisfies the query predicate. For example, a picture that satisfies P1 must contain at least an automobile, a person, and a crosswalk. We call these *participating entities*. Equivalently, there is a minimum set of *participating relationships* that must be present in a picture satisfying the query condition. For P1, this set includes *On*(*Automobile*, *Crosswalk*) and *On*(*Person*, *Crosswalk*). We may eliminate ambiguities, if during the interpretation of a query picture we agree to consider *all* the entities and relationships that conform to it—that is, if we take into account every piece of information available in the picture. Under this premise, we may uniquely describe a spatial query by drawing a picture containing *only* participating entities and relationships. For example, we may specify P1 with the sketch shown in Fig. 4. This sketch contains every entity needed to satisfy the condition, and no other. Furthermore, these entities are arranged in a way such that every participating relationship—in this case *On*(*Automobile*, *Crosswalk*), and *On*(*Person*, *Crosswalk*)—is true.

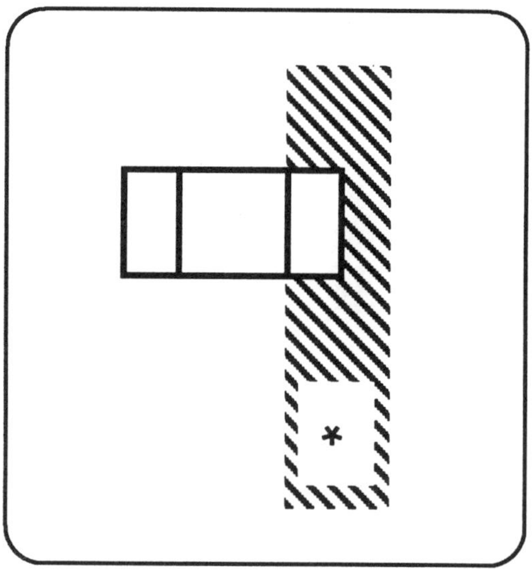

FIGURE 4. Unambiguous query description.

We have now established a unique association between P1 and the picture in Fig. 4. This demonstrates that it is possible to describe spatial queries through pictures. In the remainder of the section we will study the range of queries that can be expressed with this language. We begin by addressing in more detail the concepts of entity and relationship participation.

3.1. Entity Selection

Considering all entities and relationships in a query picture during its interpretation effectively eliminates ambiguities in the language. However, this approach forces the user to draw pictures that do no contain nonparticipating entities. Thus, the sketch in Fig. 1a, where a person and a car are in the crosswalk, therefore satisfying P1, could not have been used to describe the condition. Its interpretation, assuming complete entity and relationship participation, would result in a complex expression including many more entities and relationships than those required.

An alternative to drawing pictures that contain participating entities alone is to rely on visual effects to distinguish between nonparticipating and participating entities. Essentially, we propose using the picture editor to highlight the latter. For example, Fig. 5 shows a variation of Fig. 1a in which participating and nonparticipating entities are drawn with solid and dashed lines, respectively. Since the solid-line objects in this figure match exactly those in Fig. 4, the interpretation of both pictures is identical.

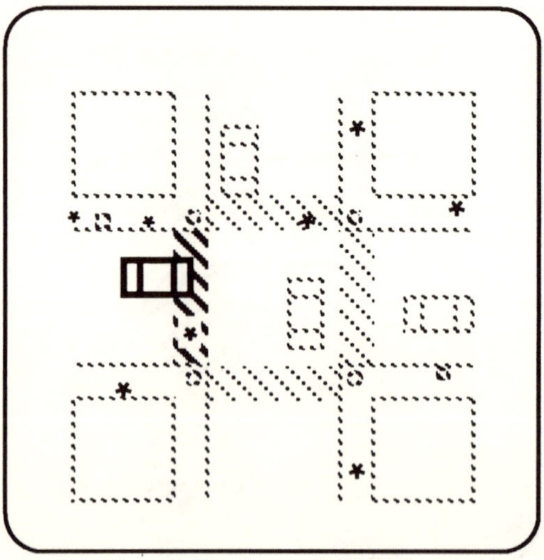

FIGURE 5. Query description using entity highlighting.

The following rule formalizes the entity participation concept:

Rule 3: An entity in a query picture should be used in its interpretation if and only if it is a participating entity.

3.2. Relationship Selection

Having established a rule for entity participation, we now proceed to discuss relationship participation. We begin by noting that spatial queries involve only spatial relationships; therefore we may eliminate nonspatial relationship from the list to be considered in the interpretation of query pictures. We also note that position-independent relationships cannot be discarded from the picture by changing the location of the entities involved. For example, if the query picture contains two automobiles, the relationship *DISTANCE*(*AUTOMOBILE*, *AUTOMOBILE*) will always be present, regardless of the entity positions.

This last observation is important, because it implies that we cannot assume, as we did before, that every relationship occurring in a query picture should be considered in its interpretation. Doing so would limit the expressive power of the language. For instance, in order to express the condition

P2 Two automobiles entering a street intersection simultaneously

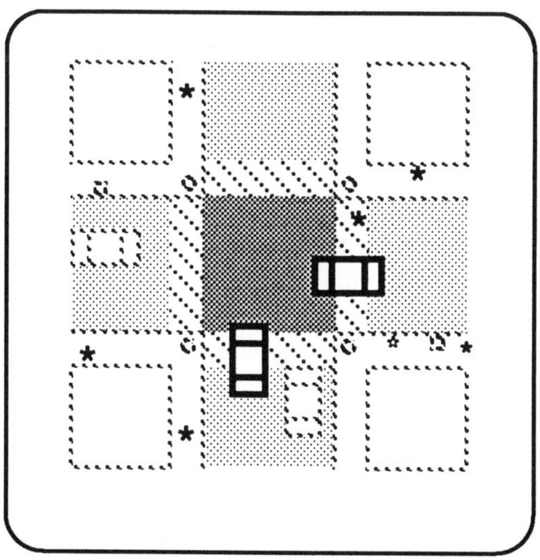

FIGURE 6. Position-independent relationships in query pictures.

we would need to draw a sketch showing two automobiles (a_1, a_2), and two streets (a_1, s_2). Furthermore, these entities would have to be positioned in such a way that the relationships *On*(a_1, s_1), *On*(a_1, s_2), *On*(a_2, s_1), *On*(a_2, s_2), and *Intersects*(s_1, s_2) were all true. While it is possible to draw a map meeting these requirements (see Fig. 6), such a map would also contain the relationship *Distance*(a_1, a_2), which could not be removed by changing the location or shape of a_1 or a_2.

The key to establishing relationship participation is based on the observation that conditions on position-independent relationships are always expressed in terms of their attribute values. For example, any reference to the distance relationship must be qualified with a value, or range of values, defining the separation between two objects, as in "the distance between two cars should be at least 5 yards." Hence in this design we have chosen to ignore position-independent relationships for which none of its attributes has been explicitly *qualified.* Essentially, an attribute is considered qualified when the user assigns it a value. A detailed discussion of relationship qualification is given in Section 4. The following rule formalizes the relationship participation concept:

Rule 4: Associations among participating entities are used in the interpretation of the picture if and only if they are spatial, and either position dependent or qualified position independent.

The following expression results from applying Rules 1–4 to the sketch shown in Fig. 6; it is logically equivalent to P2:

$$\exists(a_1, a_2 \in \mathit{Automobiles}, s_1, s_2 \in \mathit{Streets}) \mid$$
$$(\mathit{On}(a_1, s_1) \wedge \mathit{On}(a_1, s_2) \wedge \mathit{On}(a_2, s_1) \wedge \mathit{On}(a_2, s_2) \wedge \mathit{Intersect}(s_1, s_2))$$

We conclude this section with three examples of predicates that can be visually described in our language. They involve the following conditions:

P3 People on street intersections
P4 Automobiles parked on the curb
P5 People close to moving automobiles

Figure 7 show pictures that define these conditions; the corresponding strings are

P3: $\exists(p \in \mathit{People}, s_1, s_2 \in \mathit{Streets}) \mid (\mathit{On}(p, s_1) \wedge \mathit{On}(p, s_2) \wedge \mathit{Intersect}(s_1, s_2))$

P4: $\exists(a \in \mathit{Automobiles}, c \in \mathit{Curbs}) \mid (\mathit{On}(a, c))$

P5: $\exists(p \in \mathit{People}, a \in \mathit{Moving\_Automobiles}) \mid (\mathit{Close\text{-}to}(p, a))$

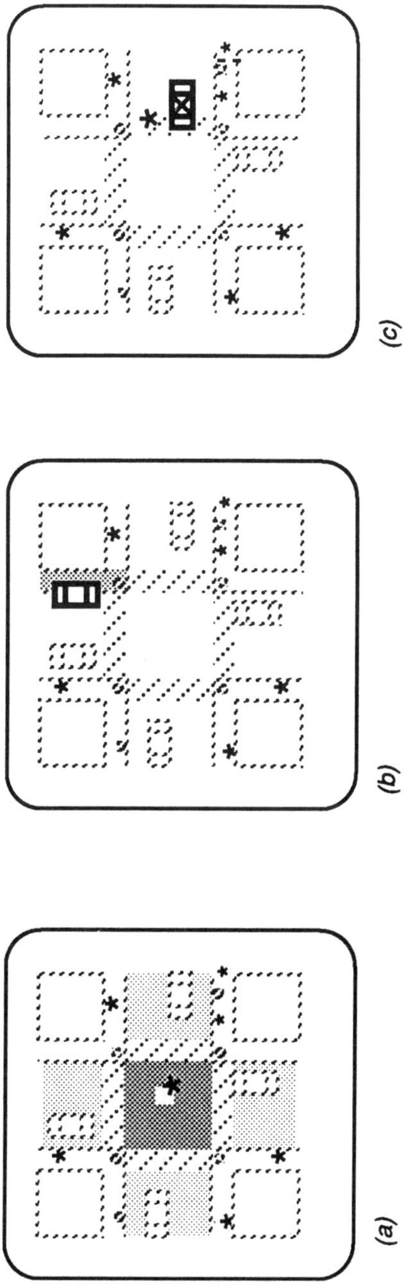

FIGURE 7. Examples of visually described spatial queries. (a) pedestrians on street intersections; (b) automobiles parked at a curb; (c) pedestrians close to moving automobiles.

Notice that automobiles in Figs. 7b and 7c are represented with different icons. This is to indicate that one is a generic "automobile," while the other, classified as a "moving automobile," is a subclass specialization. In general, the coupling of the graphics editor with specialization and abstraction concepts embedded in the database model will enhance the expressive power of the language. If differences between objects classes are visually representable, they can be used to describe a wider variety of predicates without having to rely on nonpictorial means. Since the picture editor in the previous example provides separate icons for each automobile class, we did not need nonpictorial means to distinguish moving automobiles.

4. Language Extensions

This section describes language extensions that enable the definition of queries involving aggregate and nonpictorial data. We also describe a mechanism for defining requests that involve comparison of attribute values among entities.

4.1. Aggregates

We have dealt with queries concerning individual entity-set occurrences. For example, the request "people close to moving automobiles," applies individually to each member of the "people" and "moving-automobiles" entity subsets. In this section we present a language facility that enables the definition of queries involving several entities of the same type simultaneously. For example, the evaluation of

P6 Curbs that have more than four fire-hydrants

requires that we consider all the fire hydrants in a curb at once.

Conditions such as the above cannot be expressed following the rules given in Section 3, because it is not possible to draw all the cases that satisfy the predicate, i.e., 5, 6, 7, ... curbs, in a single picture. In order to satisfy this limitation, we propose to use a new language construct called *aggregate entities*. These entities are drawn only once in the picture, but represent properties associated with an entity group, as opposed to individual members of an entity subset.

Aggregate entities are always associated with aggregate operators. Typical cases of such operators are COUNT, MAX, MIN, SUM, and AVERAGE. We note, however, that only the COUNT operation is defined across all domains. Other aggregates in this set apply to alphanumeric data,

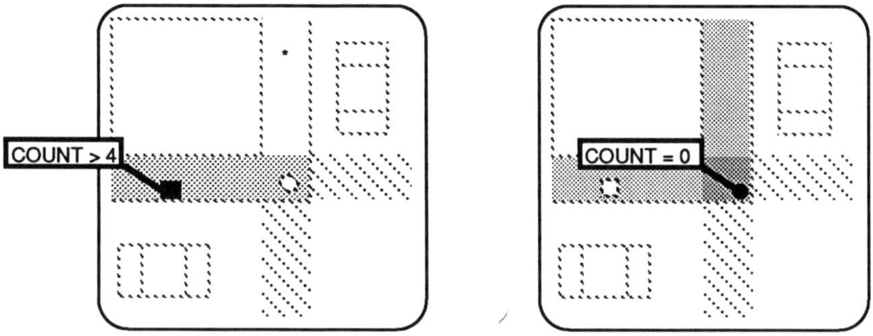

FIGURE 8. Query description using aggregate entities.

but their meaning is not well defined in pictorial domains. For instance, the concepts of MAX and MIN do not exist in the coordinate domain. On the other hand, we may think of the "largest" and "smallest" aggregates for this domain, defined in terms of the vectorial magnitude of a point. In general, pictorial domains will have their own specialized aggregate operators defined as part of the corresponding abstract data type.

Aggregate entity definitions must be accompanied by a nonpictorial description of the associated operator. This information is given to the system in a separate window linked to corresponding entity as shown in Fig. 8a. The nature of the language also requires the distinction of aggregate entities from other picture objects by visual means. Since normal video and highlighting have already been used for describing nonparticipating and participating entities, a third visual effect, say reverse video, must be used.

The formula introduced with Rule 2 must be modified to accommodate the interpretation of aggregate entities. In addition to listing entities and relationships, we must now include the semantics of aggregate operators. We observe that these operators are always defined within the context of a relationship. For instance, in P6, COUNT applies to the set of fire hydrants $\{f\}$, satisfying the relationship $On(f,c)$ for some curb c. Based on this observation, we propose the following interpretation of a request involving an aggregate entity e, satisfying the relationship $R(e,f)$ for some entity f, and associated with an aggregate operator O via the condition $O \times c$:

$$\exists(f \in F) \mid (O(e \in E \mid R(e,f) \wedge B) \times c) \qquad (2)$$

In this expression B represents conditions that further characterize e, e.g., attribute values, or relationships with other entities. Applying this formula to the picture in Fig. 8a, where an aggregate fire hydrant f is associated with a curb c via the relationship $On(f, c)$ and the aggregate condition COUNT > 4, yields the following string:

$$\exists(c \in Curbs) \mid (COUNT(f \in Fire\text{-}Hydrants \mid On(f, c)) > 4) \tag{3}$$

We conclude this section by noting the importance of entities associated with the predicate COUNT = 0. This predicate, interpreted as "the non-existence, or lack, of an entity," can be used to express queries of the form *not* $\exists f \mid P$). For example, the predicate

P7 Street corners without traffic lights

may be described as shown if Fig. 8b. The following string results from interpreting this picture:

$$\exists(c_1, c_2 \in Curbs) \mid (COUNT(t \in Traffic\text{-}Lights \mid On(t, c_1) \wedge On(t, c_2) \wedge Intersects(c_1, c_2)) = 0)$$

4.2. Subsets

In previous sections we implicitly assumed that query pictures entities were representatives of all occurrences in their entity sets. This level of abstraction was achieved by ignoring all the properties of an entity, with the exception of its existence. For instance, during the interpretation of Fig. 8b, the location and shape of the traffic light and curbs were considered irrelevant, while their actual locations were important only in regard to the existence of the *On*(*Traffic-Light*, *Curb*) and *Intersects*(*Curb*, *Curb*) relationships. In this section we discuss language extensions that enable the definition of queries affecting specific entity or relationship subsets. For example, the predicate

P8 State highways that cross state boundaries

affects only those roads whose type is "Interstate Highway."

4.2.1. Entities. Entity subsets are determined by the values of one or more of their attributes. Consider an entity set E, with pictorial attributes $A = \{a_1, ..., a_m\}$. Now assume that we want to express a condition that affects only the members of E in the subset

$$E^* = \{e \in E \mid a_i^* = v_i^* \wedge \cdots \wedge a_m^* = v_m^*\} \tag{4}$$

where a_i^* belongs to A, and v_i^* is a pictorial constant. The mechanism we propose to define E^* involves two steps: first an entity e is drawn in such a way that its pictorial attritbute values satisfy (4); then these values are *marked* to distinguish them from attribute values that are of no relevance to the condition. For example, the subset of "State Highways" in the *Road* entity set, could be defined by drawing a road with the pictorial properties of a state highway (e.g., a solid line), and then marking the attribute *type* as the distinction criterion. The value of the attribute *Road.path* is ignored in this case.

The user/system interaction needed to mark an attribute takes place in a *qualification window*. This window displays a nonpictorial representation of the attribute values, and provides a marking space. Figure 9 illustrates the use of this mechanism to describe P8. We note that qualification window expressions are reflected in the string generated during the picture interpretation. This occurs in the form of attribute predicates, as shown in the following expression describing P8:

$$\exists(\mathit{mark}\ r \in \mathit{Roads},\ s_1, s_2 \in \mathit{States}) \mid (\mathit{Road.type} = \mathit{Interstate} \wedge \mathit{Crosses}(r, s_1) \\ \wedge\ \mathit{Crosses}(r, s_2))\ \mathit{COUNT}(x{:}^{*} \mid (x \in \mathit{Cities} \wedge \mathit{Belongs\text{-}to}(x, c))))))$$

We note that the logical expression given in (4) always defines a subset in terms of (1) AND conditions, and (2) equality predicates involving attribute values and constants. The following extensions support the definition of more general entity subsets.

The first extension involves the inclusion of nonpictorial attributes. These attributes are not described as part of the query picture, and therefore they do not offer a value to be singled out. Entity subsets distinguished

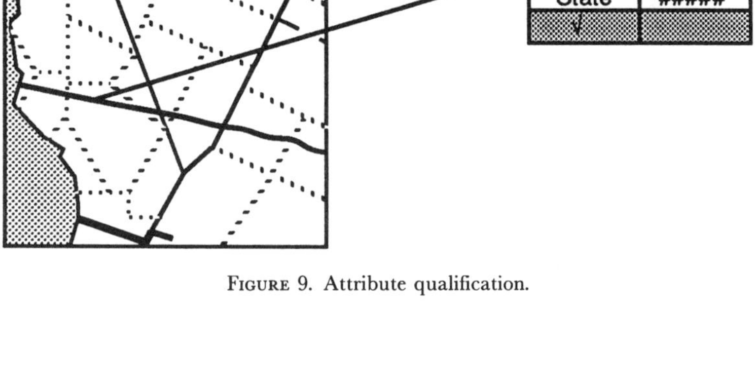

FIGURE 9. Attribute qualification.

Roads

Type	Path	Number	Built
'Interstate'	####	-	-
√			1965

FIGURE 10. Nonpictorial attributes in the qualification window.

through their nonpictorial attributes will be defined with values provided by the user separate from the picture. The qualification window mechanism may be extended to provide space for nonpictorial attribute values. This is illustrated in Fig. 10 with the definition of the subset "Interstate Highways built in 1965."

The second language extension involves the use of relational operators in the set R: $\{=, <, \leq, >, \geq, <>\}$. These operators may be used to modify marked attribute values, thus allowing the description of subsets of the form

$$E^* = \{e \in E \mid attr_{i_1} \times value_{i_1} \wedge \cdots \wedge attr_{i_1} \times value_{i_1}\} \tag{5}$$

where $x \in R$. Figure 11 illustrates the use of this facility.

We note that the operators in R are well defined only for values in the conventional alphanumeric domains. However, in pictorial domains these operators are not defined (e.g., $A < B$ if A and B are points in the plane); or their evaluation involves more than a straightforward comparison (e.g., to test the condition $A = B$ both coordinate values must be compared). In general, the set of operators that may be used to compare values is domain

Landmarks

Location	Name	Built
(x,y)	-	-
		<1950

(a)

Cities

Boundaries	Area	Name	Founde	...
####	100	-	d	...
	>			...

(b)

Roads

Type	Path	Number	Built
`State'	####	-	-
=			<1965

(c)

Counties

Bound.	Area	Name	Population
####	1000	-	-
	<		>1000000

(d)

FIGURE 11. Sample entity subsets. (a) Landmarks built before 1950; (b) city area above 100 miles; (c) state highways built before 1965; (d) counties smaller than 1000 miles and population above 1 million.

dependent. Thus a more accurate definition of the set R for a domain D, is given in the following formula:

$$R(D) = \{o \mid \textit{is a binary operator in D, and} \quad D(A, B) = TRUE \vee FALSE \; \forall A, B \in D\} \tag{6}$$

We assume that these operators are provided as part of the definition of D, as described in Section 2.

The third language extension involves the incorporation of the logical operator OR ($\vee$). This operator can be used to describe entity subsets of the form:

$$\{e \in E \mid (attr_{i_1} \times value_{i_1}) + \cdots + (attr_{i_1} \times value_{i_1})\} \tag{7}$$

where $x \in R$, and $+ \in \{\vee, \wedge\}$. The inclusion of the OR operator to the language requires changing the entity qualification window, since this mechanism has the AND operation built in. The modification we propose entails a two-dimensional syntax similar to that of QBE. This syntax is based on a matrix where values are filled in. OR conditions are specified in separate rows of this matrix, while AND operators join conditions involving values in the same row; see Fig. 12.

4.2.2. Relationships. The mechanism for defining entity subsets may also be used to define relationship subsets. This section includes two examples that illustrate the use of this language facility.

EXAMPLE 1: Consider the condition

P9 Cities near San Francisco that receive at least 1,000,000 visitors a year

In order to express this predicate we need two *City* entities: one representing "San Francisco," the other representing a city that receives more than one million visitors a year. We also need the position-independent relationship *Relative_Distance*, qualified by the value "near," and no position-dependent

Roads

Type	Path	Number	Built
`State'	####	-	-
=			1965
=			1966

(a)

Counties

Bound.	Area	Name	Population
####	1000	-	-
	<		
			>100000

(b)

FIGURE 12. Entity-subsets involving OR conditions. (a) State highways built in 1965 or 1966; (b) counties smaller than 1000 miles or population above 100,000.

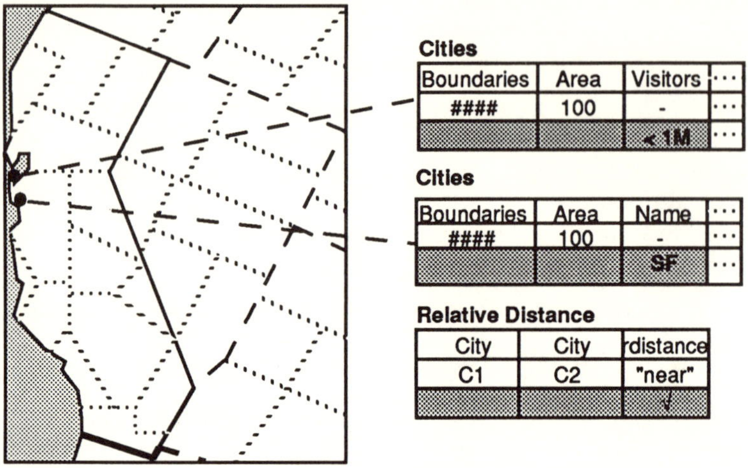

(a)

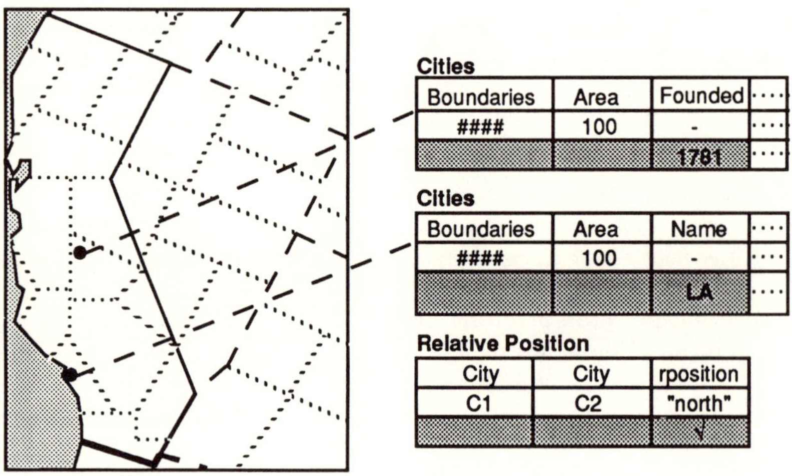

(b)

FIGURE 13. Position-independent relationship qualification. (a) Cities near San Francisco that receive at least 1 million visitors a year; (b) cities north of Los Angeles founded after 1971.

relationship. Figure 13a presents a query picture where these components are put together to express P9; the following expression is derived from this picture:

$$\exists(c_1 \in Cities,\ c_2 \in Cities)$$
$$\mid (c_1.name = San\ Francisco \wedge c_2.visitors < 1{,}000{,}000$$
$$\wedge\ Relative\_Distance(c_1, c_2, near))$$

EXAMPLE 2: Consider the predicate

P10 All cities north of Los Angeles founded after 1781

This condition is described in Fig. 13b. Its participating entities involve a city (c_1), qualified as "LA," and another city (c_2) qualified as built "after 1781." There are no position-dependent relationship among these entities; and the only qualified position-independent relationship is *Relative_Position*$(c_1, c_2, north)$. Notice that the relationship *Relative_Position*$(c_2, c_1, south)$ also exists in the picture because the set *Relative_Position* is symmetric. This relationship, however, is not qualified, and therefore it does not become part of the string generated from the picture; see below:

$$\exists(c_1 \in Cities,\ c_2 \in Cities)$$
$$\mid (c_1.name = Los\ Angeles \wedge c_2.founded < 1871 \wedge Relative\_Position(c_1, c_2, north))$$

5. *String Generation Algorithm*

This section presents an algorithm that takes query pictures as input, and creates their string descriptions. The procedure first builds an occurrence graph, then systematically traverses it generating portions of the string in each step. The following are the basic data structures:

```
struct Node {
     string   id,            /* Entity id */
              set,           /* Entity-set name */
              aggr_op,       /* Aggregate operator */
              aggr_attr,     /* Aggregate attribute */
              aggr_cond,     /* Aggregate condition */
              aggr_value;    /* Aggregate value */
     integer  class,         /* Entity class */
};
```

```
structArc {
    string  node_1,   /* First node name */
            node_1,   /* First node name */
            rel_set;  /* Relationship-set name */
};
```

For the sake of simplicity in this exposition, we chose to describe the algorithm in terms of binary relationships. If the database schema contains higher-order relationships, step 2 has to be modified to consider groups of three or more entities. Since the arcs that represent these relationships connect more than two nodes, the graph constructed would become a hypergraph. The graph generation portion of the algorithm works as follows:

/* BUILD NODES */

1.- For every participating entity create a node e. Label this node as follows:

$e.id = e_i$; $e.set = E_i$; $e.class = SIMPLE/AGGREGATE$;

In addition, if $class = AGGREGATE$, also fill in the name of the operator, the attribute it affects, and the condition it should meet.

/* BUILD ARCS */

2.1.- Using the date dictionary, determine the set of potential spatial relationships between e_i and e_j. Let $\{R_k\}$ be this set.

2.2.- For each member of $\{R_k\}$ evaluate $r_k(e_i, e_j)$. If the relationship exists, create an arc a; label this arc as follows:

$$a.node\_1 = e_i;\ a.node\_2 = e_j;\ a.rel\_set = r_k.$$

2.3.- Repeat steps i and ii for each pair of nodes (e_i, e_j).

Once the graph is built, we may proceed to generate the string. Let $\mathcal{N}$ be the set of nodes in the graph; let A be the set of arcs. The following steps are involved in this process:

/* IDENTIFY NONAGGREGATE ENTITIES */

3.- Let $X = \{e \in \mathcal{N} \mid e.class = SIMPLE\}$.

/* GENERATE EXISTENCE CLAUSE */

4.- If $X \neq \varnothing$ output $\exists($, otherwise go to 5.

/* GENERATE MEMBERSHIP CLAUSES */

5.- For each node $x \in X$ output $x.id \in x.set$

/* GENERATE "Such that" CLAUSE */

6.- Output ")|("

/* IDENTIFY RELATIONSHIPS AMONG SIMPLE ENTITIES */
7.- Let $A_x = \{a \in A \mid a.node\_1 \in X \vee a.node\_2 \in X\}$.
/* GENERATE RELATIONSHIP LIST */
8.- For each $a \in A_x$ output $a.rel\_set(a.node\_1 \rightarrow id, a.node\_2 \rightarrow id)$
/* IDENTIFY AGGREGATE OBJECTS */
9.- Let $Y = \{e \in N \mid e.class = AGGREGATE\}$
/* GENERATE AGGREGATE CONDITION */
10.- For each $y \in Y$,
Output:

$$y.aggr\_op(y.id \rightarrow y.aggr\_attr \mid (y.id \in y.set$$

Let $Z_y = \{a \in A \mid (a.node\ 1 = y.id \wedge a.node\ 2 \in X) \vee (a.node\ 1 \in X \wedge a.node\ 2 = y.id)\}$ (arcs representing relationships of y with nonaggregate entities).
For each $z \in Z_y$ output:

$$\wedge\ z.rel\_set(z.node\_1, z.node\_2)$$

Output:

$$)\, y.aggr\_op\ y.aggr\_value.$$

Output"

")"

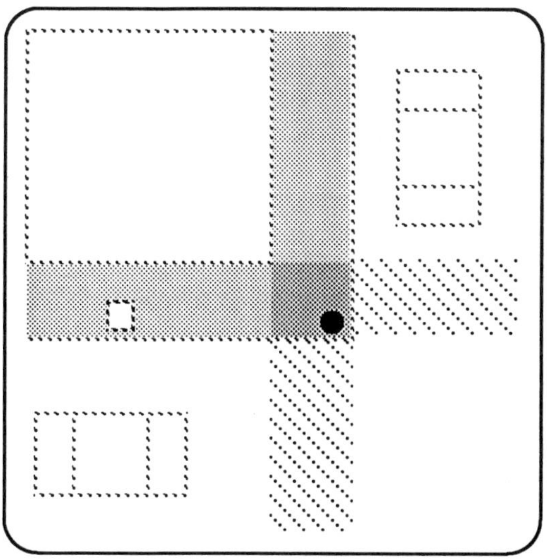

FIGURE 14. Sample query of curbs with at least one traffic light.

STEP 1: Ocurrence Graph

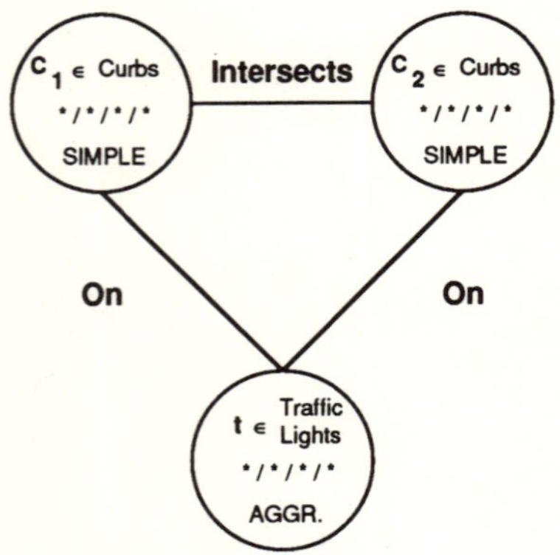

STEP 2: String Generation

STEP	SET	STRING
1	**X**={C1,C2,T}	
2		∃ (C1 ∈Curbs
3		∃ (C1 ∈Curbs, C2∈Curbs
4		∃ (C1 ∈Curbs, C2∈Curbs, T∈TrafficLights) |
5	**Ax**={Intersects, On, On}	
6		∃ (C1 ∈Curbs, C2∈Curbs, T∈TrafficLights) | (Intersects (C1, C2)
7		∃ (C1 ∈Curbs, C2∈Curbs, T∈TrafficLights) | (Intersects (C1, C2) ∧On(T,C1)
8		∃ (C1 ∈Curbs, C2∈Curbs, T∈TrafficLights) | (Intersects (C1, C2) ∧On(T,C1)∧On(T,C2))

FIGURE 15. String generation algorithm when applied to the query "curbs with at least one traffic light."

Figure 15 illustrates the result of applying the previous algorithm to the condition "curbs with at least one traffic light" (Fig. 14)

6. *Conclusions*

This work has investigated a novel approach to defining queries in pictorial databases. We have demonstrated that it is possible to use pictures as the vehicle for describing predicates involving spatial relationships among objects in the database. We defined a series of rules that enable unambiguous picture interpretations, and applied them in an algorithm that generates first-order predicate calculus expressions describing user requests. A simple visual language was developed with these ideas, and later extended to allow the definition of queries involving nonpictorial attribute values. We showed the generality of the method through examples in two areas, geographic and sketch, that have different modeling requirements.

References

1. D. Badal, Visual recall query language, Hewlett-Packard Laboratories, Technical report No. STL-87-R, 1987.
2. S.K. Chang, N. Donato, B.H. Mc Cormick, J. Reuss, and R. Rochetti, A relational database system for pictures, Proc. IEEE Workshop on Picture Data Description and Management, 1987, pp. 142–149.
3. N.S. Chang and K.S. Fu, Query-by-pictorial-example, *IEEE Trans. Software Eng.* **6**, 519–524, 1980.
4. S.K. Chang, Visual languages: A tutorial and survey, *IEEE Software* **4**, 29–39, 1987.
5. P. Chen, The entity–relationship model; towards a unified view of data, *ACM Trans. Database Syst.* **1**, 9–36, 1976.
6. E.F. Codd, A relational model of data for large shared data banks, *Commun. ACM* **13**, 377–387, 1970.
7. U. Dayal and H.Y. Hwang, View definition and generalization for database integration in a multibase system, *IEEE Trans. Software Eng.* **10**, 628–644, 1984.
8. R. Hull and R. King, Semantic database modeling: Survey, applications and research issues, USC Computer Science Department Technical Report No. TR-86-201, 1986.
9. T. Joseph and A. Cardenas, PICQUERY: A high level query language for pictorial database management, *IEEE Trans. Software Eng.* **14**, 639–650, 1988.
10. J.A. Orenstein and F.A. Manola, PROBE spatial data modeling and query processing in an image database application, *IEEE Trans. Software Eng.* **14**, 611–629, 1988.
11. N. Roussopoulos, *et al.*, An efficient pictorial database system for PSQL, *IEEE Trans. Software Eng.* **14**, 630–638, 1988.
12. J.M. Smith and D.C.P. Smith, Database abstractions: Aggregation and generalization, *ACM Trans. Database Syst.* **2**, 105–133, 1977.
13. D. Woelk and W. Kim, Multimedia information management in an object oriented database system, Proceedings of the 13th VLDB Conference, 1987.
14. A. Yamamoto and M. Takagi, Extraction of object features from image and its application to image retrieval, Proceedings of the 9th ICPR, IEEE, 1988, pp. 988–991.
15. G.Y. Tang, A Logical Data Organization for Integrated Databases of Pictures and Alphanumeric Data, *Proc. IEEE Picture Data Description and Management Workshop*, 1980, pp. 158–166.

MNEMONIC ASSISTANCE AND USER PRODUCTIVITY

ALLEN KLINGER

1. Introduction

This paper concerns designing computer software to obtain improved human performance in executing information-management tasks. We survey aspects of the existing software and consequences of wide use of computers in society in order to examine whether there is a basic problem with memory overload in users caused by current interface designs. The main contributions of this work are in

1. Displaying problems associated with increased dependence on computer technology; and
2. Discussing potential solutions.

We begin with a series of questions that we seek to answer in the material below:

a. What are the mnemonic characteristics[1] of general applications programs including word processors and text editors?
b. How effective are interface designs that use graphically displayed windows and icons? Specifically how do they perform at
   i. Reminding users about their environment?
   ii. Teaching interaction shortcuts?
   iii. Facilating development of a useful workspace model?
c. What kinds of software might advance user productivity?
d. How might "error-correction algorithms" of information retrieval systems be extended to general software?

We use "mnemonic assistance" as an abbreviation for "human–

ALLEN KLINGER • Computer Science Department, University of California, Los Angeles, California 90024-1596.

computer interfaces that accelerate data retrieval or commands." Computerization is associated with efforts to increase user productivity through automation of burdensome information-handling tasks. We seek to assess how well that has become an actuality, and where we should go from here. The "art of memory" described in Ref. 1 was a basic approach to knowledge for prepaper societies. Today mnemonic organization is only a curiousity because of numerous tools for managing paper records, books, and files. Examples range from "table of contents" and "indices" to "abstracts," "key words," and "Dewey decimal system" techniques.

To date computerization goes along with great expansion in the amount of needed compressed-information knowledge. (For example, compare printed railroad time tables and computerized airline schedules in reservation systems like SABRE.) Computer-based software tools generally violate principles regarding the number of alternatives a user can easily manage.[(2)] Instead of the "magical seven" suggested by Miller, there are so many mnemonic options that many software packages (Microsoft Word, APL, and many others) "help users learn" by distributing adhesive film labels attachable to keys. This creates the impression of user power by providing single stroke actions that move through documents. However, the reality is that the wide number of alternatives degrades capabilities: Many users make little use of the available power and *all have difficulty in acquiring skill, something that is true in all computer application areas.*

It would be desirable to have quantitative evaluations of interfaces with regard to three basic factors:

1. Acquisition of new capability;
2. Needed user interaction; and
3. Requirements on human memory.

However, the technology is relatively new and that kind of information is now unavailable. We begin instead with a broad discussion that reviews current interfaces. We use this review to develop evidence that there is indeed a problem with user memory overload.

2. *Anecdotal Review of Interfaces*

2.1. *Keyboard and Text Entry Interfaces*

Text consisting of alphanumeric symbols is the principal method for human interaction with software. Interfaces to information retrieval systems, word processors, text editors, and such applications programs as computer graphics, spreadsheets, and databases involve large numbers of options and voluminous documentation. Most users avoid studying or even scanning documentation. Instead they begin immediately, referring to error messages, on-line help, or abbreviated guides when difficulties arise. Error messages often leave the user frustrated because they are obscure or fail to correspond to the individual's assumptions. On-line help may be preset or user selected: e.g., library information retrieval systems' display of text screens specifying alternatives (preset); personal computer word processor pull-down menus (user selected). Abbreviated guides may be concise or lengthy. The former include "reference cards," e.g., two-sided manuscript pages; the latter, "how to" books. Condensed documentation is difficult to use: for typical text editors such as "vi" or "emacs," or word/graphics processors (such as Word, MacWrite, Wordperfect, MacDraw, and MacPaint) a reference card is often three columns per page, 20–30 command abbreviations per column, for a total of about 150 entries. The difficulty is twofold:

a. There are far too many alternatives[2];
b. Such summaries give inadequate descriptions of
   1. Effects from executing a command; and
   2. The alternatives.

On the other hand, on-line help consumes user time and requires active involvement.[3].

The word "mnemonic" signifies "a construct for representing an item to be recalled," and originated as a tool for training human memory in rhetoric.[1] A glance at many tools for assisting a computer user, such as the reference cards for word processors or the abbreviations for spreadsheets, database management software, graphic programs, or many programming languages, confirms the ubiquitous nature of mnemonic devices in computing. For one example, see the list of terms in use in a program for electronic mail handling, shown in Fig. 1. We can put the current situation in the form of a statement:

The use of computers involves, and demands much of, human memory.

The remainder of this paper seeks to answer the following questions:

How do current computer interface mnemonics work in practice?
Do they aid human productivity?
Can the interface technology be improved?

The persistence of labeled keys as mnemonic tools in computing indicates the conceptual clutter involved with current computer interfaces' numerous alternatives and terse commands. This tool is well established in

The mh (message handler) electronic mail program abbreviations are:

mh - Message Handler **ali** - list mail aliases **anno** - annotate messages **comp** - compose a message **dist** - redistribute a message to additional addresses **forw** - forward messages **inc** - incorporate new mail **mark** - mark messages **msgchk** - check for messages **next** - show the next message **packf** - compress a folder into a single file **pick** - select messages by content **prev** - show the previous message **refile** - file messages in other folders **repl** - reply to a message **rmf** - remove folder **rmm** - remove messages **scan** - produce a one line per message scan listing **send** - send a message **show** - show (list) messages **sortm** - sort messages **whom** - report to whom a message would go

The abbreviations fit into these mnemonic categories:

8 word items (show, next)
2 words extended by a single letter (packf, sortm)
3 word strings with deleted letters (msgchk, rmm, rmf)
1 the initial letters of two words (mh)
1 a segment of a keyword in a phrase (dist)
7 word truncations (repl, prev, inc,)

An experienced computer user found:

only 6 of the 21 functions regularly employed after one year.

FIGURE 1. Mnemonic devices in an electronic mail handler.

computing. The APL programming language special command character set was an early system using labeled keys. Recently, visual computer interfaces have been designed to overcome limitations in human ability to deal with information in the form of text strings, abbreviated or not. These interfaces function by means of two processes:

1. Utilizing visual icons' mnemonic qualities; and
2. Restricting storage and retrieval actions.

We now consider the visual role.

2.2. Mouse and Visual Interaction

Visual interfaces upgrade humans' ability to deal with computer data. Their main form is as graphic displays of windows and icons to remind individuals of office items. In computer terms, a window/iconic environment's entities are blocks of computer code enabling an "object-oriented" approach to information. This advance is based on data entry via the mouse device supplementing the keyboard, for data access, entry, and command actions.[4, 11] This point/select-enabling device accomplishes two things: it enables the user to draw upon spatial abilities in positioning and locating data, and it replaces text or symbol strings with graphic images. Two-dimensional graphic displays overcome the abbreviated-text-orientation of most software. By enabling actions on iconic items resembling office entities, such as pages, folders, file-drawers, trash cans, etc., it uses visualization as a mnemonic tool.

In essence, user interaction involves communicating with items connected to experience and memory. Unfortunately this "advance" involves an additional burden.

The simplicity of graphic interfaces makes them easy to learn. They are widely seen as "user friendly." Commercial visual interfaces (Apple's Macintosh and Microsoft's X-Windows) are very popular. However, the gain in facility or learning they offer occurs at significant increases in user interaction time. Each time an option is offered, the user must point and click or select a menu or a menu item. This is probably beneficial overall, since it allows more rapid control actions: it is the source of graphical interfaces' popularity. Nevertheless, the strong user-to-computer coupling required introduces difficulties. Should a highly trained and very busy specialist such as a radiologist be spending time in this manner? Is it efficient? Does the ease associated with learning to use a visual interface warrant the expenditures of time needed to change from keyboard to mouse actions? Do menus and their system overhead serve a useful purpose for skilled users who employ keyboard equivalents to point and select commands, or are they a burden?

2.3. *Obtaining Skill*

Most software in the hands of a skilled user advances productivity. Upgrading human computer skills has been part of the information-processing field since design of the first higher-level languages and library retrieval systems. But it is well known that such upgrading is difficult: it is a bottleneck for new users of applications software, including information retrieval systems.[3]

We believe that the thoughtful and conservative use of mnemonics in the design of a computer interface or the control aspects of an interactive application can reduce the number of user interaction steps needed, hence improving productivity, and foster learning, that is, movement to higher skill levels. One way to attain such improvements is to reapportion tasks currently assigned to human memory between new software modules and individuals.

The variety of computer mnemonic aides—reference cards, on-line help, and complex control sequences interpreted by software—either supplant, augment, or prod human memory. Can anything else be done? What are the experiences for moving from one system to another? What amount of effort is needed to make this possible? Can the general workplace support system changes, or produce effective use of computer software, in view of the general low-level skills found there? It is clear from considering the time effort and volume of documentation that must be absorbed even to utilize an essentially similar computer-based system, that some basic software design flaws exist.

One user notes that manufacturer-prepared documentation is frequently written "in a form that only makes sense once you already know how to use it"—in other words, is in reference manual format. Manufacturers often prepare workbooks of equal size to their reference manuals; users find the combined documentation sufficiently difficult to use that there is a commercially successful publishing industry preparing instructional manuals. (Such commercial documentation frequently is large, having numbers of pages equivalent to manufacturer's reference manual and instructional materials combined).

Reducing documentation needed is clearly desirable. However, the importance is increased when we consider the societal impact of computerization. General users are simply unable to devote the significant amount of time needed to gain skill in new application software, under current interface arrangements and documentation. Unless something is changed, necessary training costs will overcome productivity increases, owing to a combination of the complexity of computer application software, the modest skills of general users, and the normal rate of personnel turnover. What does the historic record indicate regarding adoption and acceptance of computerization? Figure 2 compares the ORION on-line library information retrieval system at UCLA to card catalogs. Here user–computer interaction steps are abbreviated. The notation indicates screens displayed by the computer by a single descriptive word. The list of computerized actions is longer than the card catalog set. However, few people would prefer the latter when they understand the advantages of the computer—namely, accessibility to large databases, and greater power. Full use of that power requires easily comprehended and applied computer utilities.

How is the computer software industry acting to provide needed utilities? What current products and directions deal with the changes required as systems move from tools used by computer-knowledgeable

Noncomputerized	Computerized
Find Library Card Catalog	Find Unused Computer Terminal
Find Subject Heading	Find Browse Subject Mnemonic (B SU)
Find Cards on Spain	Enter B SU Spain
Scan Cards and Make Notes	Error, Help, Enter B SU Spain Travel
Choose Call Number	Choose Subject Key
Look For Book	Choose Book Entry Key
	Note Call Number (One? Repeat?)
	Look For Book

FIGURE 2. User interaction: Computerized library information retrieval.

people into the general community? Three predominate: icons in interfaces (discussed in part above, and elsewhere in this volume), Unix operating systems, and on-line CD-ROM documentation combined with search and menu displays.

The notoriously unfriendly Unix system offers one potential basis for change, because it supports numerous utilities that allow skilled users both flexibility and power. Movement of Unix into the general user community is emerging: commercial Unix-oriented hardware (personal computers, workstations) products are now in the marketplace. Many individuals, even skilled computer people, find Unix sufficiently difficult to learn and use, to prefer personal computer general-user application software. However, others favor investing significant amounts of time in Unix training, and recommend creating new shared-computing workplace cultures. For such environments to succeed, organizations need to develop working modes compatible with varied staff skill levels, secretarial to professional, and a system—such as via bulletin boards, computer clubs, training, and support networks—to facilitate sharing useful computing routines.

But even in the best of possible situations available today—icons, menus, Unix, high-quality graphics and sound, and CD-ROM storage of documentation—the current technology has mnemonic bottlenecks. Anyone other than a computer professional with time available would find rapid display of on-line documentation in a graphics window of a Next computer a poor substitute for what could be developed in the way of automatic routines, or that which is available now via skilled-user assistance. Staff turnover and diverse working styles cause continuous difficulties in real office and manufacturing plant situations. New employees take months to become modestly skilled. Long-term employees prefer to use familiar software. Hence advances via different utilities—mail-handlers, editors, etc.—are difficult to disseminate. Diversity of the available range of application software, and the personal choices made by skilled users in learning only a small subset of the tools, causes "experts" to be sometimes unavailable or documentation (on-line help, hardcopy manuals) to be outdated. Few users are able to take advantage of options that save time or lower costs.

3. Current and Future Technology

3.2. Computer Utilities

New computer utilities, software modules that implement widely accepted functions, are frequently developed, but the effort involved has not sufficiently addressed either obtaining skill by primary users or the general

CARDIFF
UWCC LIBRARY

issue, the ultimate transfer of utilities to large and less computer-skilled user populations. Computer utilities, word processors, spreadsheets, electronic messaging mail-handlers, and editors have in common that they provide new modes of data handling. Clearly ease of use is the key to the wide acceptance in all cases, whatever the data domain. Whether in dealing with numbers, text, or mail, computer software improves user productivity—once assimilated. That is why the rate of adoption of computer utilities is so rapid. Spelling correction, grammatical and stylistic improvements to writing, outlining, and presentation visuals are now commonly used in the workplace and in the home; Ref. 6 from a 1980 research journal describes spelling programs now widely available in personal computers. File-location utilities now available provide simple forms of mnemonic assistance. Their very existence testifies to the difficulty of managing one's own mnemonically named computer files, a situation where information retrieval is simply recall of a file written *by the same person.*

File creation can have been moments, hours, days, weeks, or months earlier. But "mnemonic" user-defined structures are so idiosyncratic that retrieval is often obstructed by forgetting what one has previously seen as a natural abbreviation or association. "Memory extender"[(5)] was designed to overcome this automatically. It is an instance of a general class of programs that overcome small errors. In this case the capability is focused on file name recall. Some utilities allow extension of a few characters to provide the actual file or directory name. In Unix, any directory name existing in the present file structure can be completed by typing an "escape" after a few distinguishing initial letters. Personal computer file-finding programs function by string search among the list of established terms. Reference 5 describes research on automatic extending. Similar pattern matching routines are used in spelling correction[(6)]; likewise, there are information retrieval aides that extend inputs to similar terms, or succeed by changing letter case (upper/lower).

Software systems with utilities included for extending, finding, and allowing small command varations raise basic research issues involving analysis and tracking of user-input information: the Orwellian 1984 "big brother." There is a substantial history of designs that deal with user input error. The "Do What I Mean—DWIM" features of Interlisp[(8)] and recent similar suggestions for personal computers[(9)] are "self-correcting" in the description of six possible methods, colorfully described in Ref. 10 as *Gag, Warn, Do Nothing, Self Correct, Teach Me,* and *Let's Talk About It.* The self correct technique is more desirable for widely disseminated software. It is also significantly more applicable to certain kinds of user imputs.

We distinguish two types of user-input data: Type A is a narrowly specified set of alternatives. For example, capitalized or lower-case letters anywhere in a known file name; concatenations of two or three descriptive words or abbreviations; adjacent, space, or separated-by-period organization; dates in month/day/year or day/month/year order; and hyphen in place of slash. Type B is nested sets that are linked by information content—for example, the sets of alternate abbreviations used by an individual. These data could be recorded for one user in general or on a particular day, for different users, or even for classes of users. Such data would then be grouped in hierarchies and used to govern search. Heuristics to provide ranking/ordering data so that most likely alternatives can be evaluated first could be based on statistics of how a given user actually functions in a computer interface environment.

Unix and Lisp environments provide utilities to sense similarity (string-matching), small differences, yield character/word/line counts, and obtain file keys like date (e.g., grep, fgrep, diff, wc, and find in Unix). These enable automatic methods that eliminate or greatly reduce the needed human effort to function in many practical data-handling situations. Such methods are in existence in some cases, e.g., software routines that automatically extend user- specified strings that are actually partially specified files, directories, or even complete directory paths.[11] However, the general approach needs to be more widely applied to reduce the burdens placed on the user in current interfaces.

In information retrieval, software to overcome user's small text-string differences implements "error-correction algorithms."[12,13] One routine automatically inverts first and second author names, triggered by "no matches" on first search. Another error-correction program searches first word of entered author name followed by first letter of the second word; this retrieves additional entries when author alone produces "no match." Other techniques decrease error rates by searching unmatched terms in an index, then displaying those close alphabetically; passing unmatched terms other than author names through a spelling check routine; and forcing users into a help routine after some number of identifiable errors.[13]

Software building on these systems that employs mnemonic aiding techniques could use a record of individual user's employed mnemonics, or characteristic patterns of employing interface resources.

3.3. *Networks and Learning*

Current computer utilities accomplish much more than was envisioned earlier when information retrieval was seen as dominating computing.[14,15] Improvements in interfaces and general aspects of computer technology enable exceeding the capabilities of many existing installed information retrieval systems with low-cost systems. For example, the UCLA library uses the ORION information retrieval system for book catalog information. This is a mainframe based system with strongly defined computer screens that constrain a user's search actions. However, a PC-based analog has recently been installed for magazines, and a skillful graduate student has accessed a more powerful abstracting system running at a remote university via Internet to gain information about specialized journals. The combination of PCs and workstations with computer communication networks is a technology change that markedly improves user productivity. Mnemonic assistance would allow information resources to become more widely used.

Individuals working via computer networks, e.g., Internet, or employing powerful operating systems such as Unix, have significant interface and computational advantages over existing installed information handling systems. Writing shell commands, redirecting output to files, and performing remote log-in all result in qualitive differences in the results obtained. These changes impact productivity in several ways including of course the facilitation of information sharing.[12]

The ways that computers have changed data handling to date result from three fundamentally different, and predominantly hardware, interface technologies:

a. Graphic interaction modes[10, 11];
b. Remote site dialog; and
c. Personal computers and workstations.

The primary input tool enabling graphic interaction is the mouse hardware device.[4, 13] The availability of a pointing capability along with improvements in display quality—both those in the marketplace today and those envisioned as a result of development of high-definition television—introduces a different quality to computing environments.

The growth of computer-communication networks and the transmission of computer-originated files within offices over LANs (local area nets), and between distant locales, also changes the nature of the workplace. Besides introducing "telecommuting"—work based at home via computer—this enables greater interaction of ideas among individuals with no ability to synchronize their activity (as needed for in-person meetings or phone calls).

Likewise, the low cost and wide availability of computers and workstations has societal implications. Design decisions in both hardware and software have impact over a very large population of individuals. Changes from replacement of writing, paper records, and physical files, by keyboards, digital memories, display screens, and computer systems based on alphanumeric symbols, can be viewed as having both positive and negative consequences.

These changes in information handling are destined to continue to depend on human learning of computer abbreviations. This is partially a consequence of Unix becoming more widespread owing to its portability, and also the result of the way that "simplification"-by-abbreviation pervades computing. Figure 1 shows a set of mnemonic abbreviations used in a mail-handling computer utility that runs under Unix. This indicates the significant memorization burden needed to use just one utility. Figure 2 contrasts interaction with library card catalogs with a computerized information retrieval system. Diverse abbreviation modes render software utilities difficult to assimilate. Many mnemonic modes are used to construct this electronic mail-handler's commands: truncation ("comp" for "compose"; "prev" for "previous"), commonly used English words or their extensions ("send" for "send a message"; and "sortm" for "sort messages"), and letter omission ("rmm" for "remove message"; "msgchk" for "check for messages"). Reference 18 gives a theoretical framework for how individuals learn synthetic language. Research emphasizing file name creation in computing appears in Refs. 19–21. All contribute to describing the complex linguistic structure created for computers. Analysis of human recall of mnemonic abbreviations in real computing situations is needed to develop software systems designed for rapid learning skill acquisition, and overall net gains in user productivity.

Finally, the abbreviations used by successive software designers has resulted in one obvious candidate for mnemonic assistance: stopping a running routine, a problem related to sequencing screen and menu changes. Currently, difficulties are caused by systems' rigid, and differing, requirements. Users sometimes waste time deciding whether to enter "quit," "q," "end," "halt," "stop," "off," "logout," "logoff," and other commands utilizing control or command keys or key-combinations. Currently, difficulties are caused by systems' rigid, and differing, requirements. A mnemonic assistance system could interpret potential quit-commands. Its goal would be replacing human trial and error, and sometime success, by a context-sensitive interpretor to save user learning and operating time.

4. *Conclusions*

Considering the rapidity with which personal computers have become a common tool in the home, relatively little has been done to make them "fool-proof" for basic tasks they are normally used for, high on the list being word processing. We see large amounts of documentation and fairly crude interfaces even when visual interfaces employing icons are available, and on-line help resides in pull-down menus. This contrasts with what could be done with systems based on well-known software concepts ranging from Lisp-based "Do What I Mean Routines" to Unix filters and utilities. Little is currently available in user application areas to foster learning by individuals with lesser computer skills. Systems are installed with "hacker," not "general-user," orientations: far too many options and alternatives are available. Far too little automatic interpretation of user errors occurs. The result is an excessive human burden requiring significant amounts of time to gain modest familiarity with software, and substantial requirements for memorization.

The basic situation is that memorization is just not feasible for the large number of commands available in general computer utilities. Information retrieval experience suggests that utilities are needed to help users more successfully perform their actual tasks. The goal of harnessing available computer power to practical tasks requires software to be designed that incorporates the desirable goal of providing "self correction" or "mnemonic assistance." This could be done with current systems, if tracking software were provided to allow the system to model the user's actual behavior: commands avoided, stylistic quirks in names of files and directories, being two examples. This could increase utilization of mnemonic abbreviations in software—i.e., yield effective shortcuts. The problem that needs to be addressed is to foster learning such shortcuts. Currently, substantial time is needed to learn them.

Is information access less free due to the complexity inherent in dealing with computer interfaces? The answer seems to be a qualified yes. There is a lot of work to be done to make it become a strong no.

ACKNOWLEDGMENTS

This paper benefited greatly from discussions with Christine Borgman, UCLA, and from her comments on earlier drafts.

References

1. F.A. Yates, *The Art of Memory*, Rutledge and Kegan Paul, London, 1966
2. G.A. Miller, The magical number seven plus or minus two: Some limits on our capacity for processing information, *The Psychol. Rev.* **63**, 81–97, 1956.
3. C.L. Borgman, Why are online catalogs hard to use? Lessons learned from information-retrieval studies, *J. Am. Soc. Inf. Sc.* **37** (6), 387–400, 1986.
4. J. Conklin, Hypertext: An introduction and survey, *IEEE Comp.* **20** (9), 17–41, 1987.
5. W.P. Jones, On the applied use of human memory models: The memory extender personal filing system, *Int. J. Man-Machine Stud.* **25**, 191–228, 1986.
6. J.L. Peterson, Computer programs for detecting and correcting spelling errors, *Commun. ACM* **23**, 676–687, 1980.
7. W. Teitelman and L. Masinter, The Interlisp programming environment, *Computer* **14** (4), 25–33, 1981.

8. J.W. Ratcliff and D.E. Metzener, Pattern matching: The gestalt approach, *Dr. Dobb's J.* **July**, 46–51, 1988.
9. C. Lewis and D.A. Norman, Designing for error, in D.A. Norman and S.W. Draper, Eds., *User Centered System Design: New Perspectives on Human–Computer Interaction*, Lawrence Erlbaum, Hillsdale, New Jersey, 1986, pp. 411–432.
10. T. Cheatham, Message decryption and spelling checkers, *Cryptologia* **April**, 115–118, 1987.
11. UNIX Users's Supplementary Documents (USD) 4.3 Berkeley Software Distribution, Virtual VAS-11 Version, April 1986, Computer Systems Research Group, Computer Science Devision, Department of Electrical Engineering and Computer Science, University of California, Berkeley, California 94720; or W. LeFebre, *The Sun Observer*: P.O. Box 399, Cedar Park, Texas 78613: October 1989, pp. 56–57.
12. T.W. Malone, K.R. Grant, F.A. Turbak, S.A. Brobst, and M.D. Cohen, Intelligent information-sharing systems, *Commun. ACM* **May**, 390–402, 1987.
13. C.L. Borgman, Human–computer interaction with information retrieval systems: Understanding complex communication behavior, M. Voigt and B. Dervin, Eds., *Progress in Communication Sciences*, Ablex, Norwood, New Jersey, 1986, pp. 91–122.
14. V. Bush, As we may think, *Atlantic Mon.* **July**, 101–108, 1945.
15. T.H. Nelson, Getting it out of our system, *Informtion Retrieval: A Critical Review*, G. Schechter, Ed., Thompson Books, Washington, D.C., 1967.
16. D.C. Englebart, A conceptual framework for the augmentation of man's intellect, in *Vistas in Information Handling, 1,* Spartan Books, London, 1963.
17. D.C. Englebart and W.K. English, A research center for augmenting human intellect, *AFIPS Conf. Proc. 33*, Part 1, Thompson, Washington, D.C., 1968.
18. G. Zipf, *The Psycho-biology of Language*, Houghton Mifflin, Boston, Massachusetts, 1935.
19. L.A. Streeter, J.M. Ackroff and G.A. Taylor, On abbreviating command names, Bell Laboratories, TM 80-3161-1, 1980.
20. M.H. Hodge and F.M. Pennington, Some studies of word abbreviation behavior, *J. Exp. Psychol.* **98**, 350–361, 1973.
21. A.S. Reber, Transfer of syntactic structure in synthetic languages, *J. Exp. Psychol.* **81**, 115–119, 1969, and A.S. Reber, Implicit learning of synthetic language: The role of instructional set, *J. Exp. Psychol.; Hum. Learn.* Mem. **2**, 88–94, 1976.

TECHNOLOGY, INFORMATION, AND THE INDIVIDUAL

ROBERT R. KORFHAGE AND KEITH S. JOSEPH

1. Introduction

The advancement of a species, or of an individual within a species, is closely tied to the ability of that species or individual to control its environment, rather than to be controlled by its environment. The control depends fundamentally on two things: a "tool kit" of means to effect that control, and information relevant to deciding when and how to use which tool from the kit. Control of the environment inevitably means increasing the complexity of the environment—adding physical objects that were not previously available, discovering new energy sources, creating microclimates that are not naturally present. At the same time the control information itself becomes more complex, as it must describe the increasing complexity of the environment. In an escalating process, control of the increased complexity demands more effective information access. Thus as better tools are invented, included among these are the tools necessary for handling information and presenting it to the individuals who need it.

The tools necessary for handling information developed very slowly until recently. Thousands of years elapsed between the invention of writing and the invention of printing; hundreds of years between the invention of printing and the development of electrical and photographic information processes; decades then passed before the advent of the computer and electronic communication technology. However, as this technology developed, beginning around 1950, we truly moved into an age of information. Information technology is developing at a sufficiently rapid pace that any new

ROBERT R. KORFHAGE • Department of Information Science, University of Pittsburgh, Pittsburgh, Pennsylvania 15260. KEITH S. JOSEPH • Communications Applications Research, Pacific Bell, San Ramon, California 94583.

product is well on the road to obsolescence shortly after it appears on the market.

Paralleling this escalating development of information technology has been a broadening of the class of people to whom good information is readily available. This class has potentially become all of humanity through development of the printing press, the telephone, radio and television, and particularly digital computers and electronic communications systems. The thrust of these developments has been to make any individual not only a user of information, but also a generator of information, as it becomes increasingly easy to interact with an information system, and to add to the information contained within the system.

Information handling falls into six categories; *creation*, *storage*, *transport*, *retrieval*, *display*, and *consumption*. Over the millennia there have been many desires, goals, needs, and trends in information handling. The primary goal has been to overcome time. People have striven to record and store information for later generations—and, on the other end, to use the information developed and preserved by our ancestors effectively. Second, information handlers have had the goal of overcoming place, that is, of recording information in order to move it over great distances, making it more readily available to the user at his or her site. The third great goal has been to make information readily available to increasingly larger groups of people. Although this movement to make information more widely available has often been controversial, it has also been a significant trend, especially since the onset of the Industrial Revolution. Finally, another important goal has been to store and display information with as much realism as possible. Any information that is stored is an abstraction from the real, physical world. That abstraction has always reflected reality, initially in very concrete and naturalistic ways, and later in increasingly abstract and metaphorical ways. To meet these goals many technologies, ancient and modern, have been developed.

We discuss five eras of information technology. In each era we shall find that the technologies of the age limited the methods of handling information, but at the same time information-handling technologies developed to meet the needs of the age. In each of these generations the economics of information—both the cost of recording and transmitting information, and the power provided by information access—has led to the appearance of a "priesthood" of information adepts, and the division of society into the "information rich" and the "information poor."

We begin with the technology before the invention of writing. In this period the only way to record information was pictorially. The second era corresponds to the invention of writing, which permitted scribes and scholars both to record the names of people and things, and also to develop and record abstract concepts. The mechanization of information handling, beginning with the printing press and carrying on through the Industrial Revolution, greatly augmented humanity's ability to store information in both text and nontext formats, increased information mobility, and saw the development of mass media. The fourth era is characterized by the digitization of information made possible by the stored program digital computer. The fifth era is developing now with the networking of computers and the integration of computers and eclectronic communication systems to provide a single powerful tool for the generation, transport, storage, retrieval, display, and consumption of information.

2. *Before Writing*

Throughout much of history humanity relied heavily on oral transmission of information from one generation to another. Even today the oral tradition plays an important part in virtually every culture. Yet the oral transmission of information has two obvious problems: the witting or unwitting alteration of information as it passes from one story teller to another, and the possibility that a story teller may die before he or she can pass along all of the accumulated knowledge and wisdom of the cultural group. Thus there was also a very early development of means for recording information. Once mankind developed the ability to shape and use tools, the carving of marks into wood and bone as a means of preserving information was not far behind. By 30,000 years ago, people were painting and engraving animal and human subjects on cave walls in Europe, Africa, Australia, and South America. This early technology preserved the information, but suffered from two major problems. First, the "vocabulary" was very limited, covering only indigenous animals, humans, and the interactions between them. Second, while the recording was permanent, cave paintings were also fixed in place, and hence accessible to a relatively small number of people. These factors limited the value of the technology.

3. *Written Language*

The development of writing and paper provided a major breakthrough in making information more widely available, in both time and place. Before written language, permanent storage of information was highly limited, and the main mode of information transmission remained the oral tradition. The only way to get information was to obtain it in person from another human.

While this personal contact is still important, it remains a very limited information channel.

The invention of writing was revolutionary for it enabled people (at least those who could read and write) to do things that previously were not only impossible, but even unthinkable. Writing is a recording method, and for many generations was the only recording method other than painting and sculpture. Writing made possible the retention of business records in ancient Babylon and Egypt. Writing also made possible the development of precise abstract concepts, as in the mathematics of ancient Greece. Written language provides a highly compact representation of information, making it possible to transport information over long distances and to save information for future generations.

Thus "recorded history" was born. (However, the first continuous historical dating system was not developed until 312 B.C.) The movement of information became something other than verbally transmitted folklore. As with many new technologies, early implementations were clumsy and expensive. The earliest pictographic writing developed in Sumer around 3500 B.C. For many centuries this remained the only style of writing. Reading and writing were difficult skills to master. For many centuries, access to written information was limited to an elite class of scribes, scholars, and priests.

More abstract ideographic scripts first appeared in China around 1500 B.C., at about the same time that "Linear B" script appeared in Crete and cuneiform was developed in Anatolia. These developments paved the way for the invention of the first alphabet script by the Phoenicians around 1100 B.C. These developments resulted in a script composed of fewer distinct symbols, hence easier to learn. Yet reading and writing skills still remained largely the province of the "priesthood." Today literacy is an accomplished fact for a major portion of the world population. We have responsed by developing a deeper level of information representation—ASCII and other binary codes—that is once again directly understandable only by the initiated.

The systematic accumulation of knowledge took a major step forward around 300 B.C. with the development of the historical dating system previously mentioned, and the founding of the Alexandrian library in 290 B.C. The idea of accumulating knowledge in a single location where scholars could go to use and study it became, and remains, a key factor in information handling. The technology developed within the last quarter century, however, has raised the possibility of storing and accessing information in a more distributed fashion. There is a growing sense that it is cheaper and more effective to bring distributed information together to bear on the problems of one individual wherever that individual may be, than it is to bring distributed individuals together to handle their disparate information needs from a single information source, wherever that source may be.

Early development of information handling was held back by the lack of cheap and simple writing implements and a good medium on which to record the writing. The brush and the stylus served as adequate implements until the development of paper in China around AD 105, and its introduction into the Muslim world, and thence into Europe around AD 750. Reed and quill pens became the prefered method of writing due to the ready availability of reeds and quills and the ease of making the pens. Since that time the implements of writing have gone through extensive development. Pencils, mechanical pencils, fountain pens, ball point pens, and typewriter are examples of important inventions to support the technology of writing. While the technology of making paper has also changed, the differences between modern and ancient papers are not as dramatic as those for writing implements.

As long as information was replicated by tedious and expensive hand copying, access to information was very strictly limited. Since each replication of a major piece of writing took years of work, providing copies of writings to more than a very few carefully selected individuals or libraries was out of the question. The invention of printing in China around AD 730 and the development of movable type some 300 years later provided a major step forward in mankind's ability to replicate information for distribution. Yet another 400 years passed before Johannes Gutenberg brought printing to Europe in 1445. Now the time required to produce a book could be measured in weeks or months, rather than in years, and it became reasonable to make information resources more widely available, at least in all of the major cities. Yet the process remained labor intensive, requiring a staff to set the type, man the press, produce the covers, and bind the books.

With books more readily available, reading and writing became skills of considerable more importance, and acquisition of these skills spread throughout the "upper class," merchants, and professional people. The first weekly newspapers appeared in Amsterdam in 1620, thus expanding the horizons of the world of information at the same time that the geographic horizons of the world were being expanded through colonization of the New World.

These developments did much to take information more readily available to masses of people. Yet the effect of the technological changes was largely one-sided. It remained difficult for the individual to generate and disseminate information. The printing and distribution facilities remained largely in the hands of governments or influential people, who could thereby control and limit the information available to most people.

4. Information Technologies of the Industrial Revolution

4.1. Mass Media

By the year 1800, hand-produced and machine-printed text had been available for years and even centuries in some parts of the world. However, reading was still limited to a small segment of society, copies of books were expensive and rarely accessible to the general public, and even though newspapers existed, much of the "daily news" was available only through posted notices and (the oral tradition again) the town crier. An elite group remained firmly in control of the most important sources of information. When information was made available to the public, it was often too late for the public to react effectively to the information.

The Industrial Revolution brought with it radical new means of recording and distributing information. During the nineteenth century information became available to a mass market through printed text in newspapers, books, and magazines. The cylinder printing press was developed in 1812; photography was demonstrated by Joseph Nicéphore Niepce in 1822, and made practical by Louis Jacques Mandé Daguerre in 1839. The telegraph, developed in Britain in 1838, and made feasible by Samuel F.B. Morse in 1844, became an important factor in the opening of the American West. The typewriter (Chritopher Sholes) appeared in 1868, the telephone (Alexander Graham Bell) in 1874, and the Linotype machine (Ottmar Mergenthaler) in 1888. Wireless telegraphy, magnetic recording, the phonograph, and motion pictures developed in a rush around the turn of the century.

By 1900 the technological groundwork had been laid for allowing people more direct and individual access to information handling. Much development remained before the technology (the typewriter, phonograph, and camera, for example) became inexpensive and easy to use. But the movement toward enabling any individual to generate information as well as use it easily had begun.

4.2. Multimedia

The major information development in the nineteenth century lay in the more ready availability and distribution of printed materials, in contrast to prior technology. However, some inventions of that period laid the groundwork for two trends in early twentieth century information processing—wider use of nontextual information, and facilitated information movement. Photography made the recording of images a mechanical process, far less dependent on the skill of an artist, with the result that by the Civil War photojournalism was well established. The phonograph and early experiments of magnetic sound recording led to the development of a whole new industry. The development of optical sound recording, when coupled with the development of motion pictures, led to the "talkies," in 1927.

Until the nineteenth century, text and drawings on paper were the only ways besides the human brain to store information. Both involved much hand labor. Still photography (1839), magnetic recording (1898), and moving pictures (1905) reduced the amount of hand labor, making it economically feasible to store images, sound, and moving pictures on permanent and widely distributable media. Photographs and sound recordings could replace descriptions of distant scenes and cultures and provide a much more complete and accurate representation of the reality.

4.3. Movement of Information

Until the 1800s, the only way to transport information was to move copies of printed text or pictures from one location to another. The telegraph and telephone enabled long distance transmission of textual or verbal information to individual destinations. Radio (1920) and sound movies (1927) became key sources of public information during the second quarter of the twentieth century, most particularly during World War II. Television was developed in 1936, but did not become a widely available information source until the 1950s. Much of the basic technology needed to support modern television broadcasting was developed as part of the war effort. This technology provided rapid, reliable, and inexpensive information movement across great distances. As a result, we have seen television broadcasting become a dominant force in social and political life throughout the world, to the detriment of some of the earlier media. This dominance has raised new concerns about the extent to which the information media influence world events, in contrast to just reporting on them.

5. The Digitization of Information

Development of the stored program digital computer began in Germany and the United States in the late 1930s. The first, relatively crude, electromechanical computers were used exclusively for numerical calculations in support of the war effort in the 1940s. Public statements by many involved with these developments, including such geniuses as John von Neumann, grossly underestimated the eventual impact of these machines. The first commercial electronic computer appeared in 1952, and within a few years it became evident that the power of these machines lay in their ability to represent all kinds of information, not just numbers, in digital form. Within less than a decade text and pictures were being represented and processed digitally within computers.

The history of digital computers has been characterized by increasing miniaturization and decreasing cost. The transistor (1948) and the continued development of integrated circuits starting in the late 1950s led to inexpensive, widely available personal computers in business and homes. At the same time the general development of electronic technology has led to other significant changes in information handling. Sound recording has moved from plastic phonograph records, to analog magnetic tapes, to digital magnetic tape and compact optical disk technology. The home movie camera has largely been replaced by the home video camera, and the first still camera based on video technology has appeared on the market. The technology for document reproduction envolved from mimeographs to the photocopy machine. Current refinements to this technology include facsimile transmission (FAX) machines and electronic scanners.

With each new technical development the cost of the hardware to process information has decreased and the application of that hardware has expanded to include information that would not previously have been handled. The major impact of these developments has been to increase the immediacy of information. Costly and time-consuming processing methods have been replaced by methods that permit distribution and use of information within seconds or minutes of its generation. Bulky media that are difficult and expensive to transport are being replaced by media that are more compact and less subject to damage during shipment. In many situations the physical shipment of information media has been eliminated, as the information is transmitted electronically, with a physical copy generated at the receiving end if necessary. The generation and use cycle for information has been shortened, with the result that more people can now use information in an interactive mode, generating new information while they are using already available information.

6. *Networking of Computers*

Communication technology advanced in parallel with computing technology, with the first space satellite (1957), microwave transmission (196?), and the development of high capacity fiber optic transmission systems in the 1980s. This has led to the integration of computing and communication systems until today within the United States, and increasingly throughout the world, communication systems are computer-based, and computer systems communicate directly with each other over high-speed digital data transmission networks. the technologies of photocopying and electronic communications have been combined in FAX machines to provide an exceedingly simple, yet rapid and efficient, means of transmitting documents great distances.

The current Information Age is characterized by the capability to store, manipulate, and transmit massive amounts of information rapidly, reliably, and efficiently. We can transmit the information contained in a book across the country electronically in less than a second. Computer speeds now measured in "mips" (millions of instructions per second) and "megaflops" (millions of floating point operations per second), and storage capacities measured in "terabytes" (trillions of characters) enable us to search and process massive amounts of data in all forms—numeric, textual, pictorial, and sonic. One result of this increased capability has been to focus attention on the problems of using all of the information that can now be generated and transmitted. People are finding themselves inundated with more information than they can reasonably use, and few effective ways to select the best information from that available at any time.

7. *Usability Issues*

In the past the economics of information—both the cost of recording and transmitting information, and the power provided by information access—has led to the appearance of a "priesthood" of information adepts. With the development of writing came the scribes, who were charged with properly recording that which was dictated to them. Widely available printed materials led to the development of public library systems, and to the need for librarians with the skill to organize and manage the library collections, and to aid users in searches for information. Each new technology—the telegraph and telephone, film, radio, and television—brought forth a group of professionals who could use the technology and make it accessible to those who were not so trained. Similarly, computer programmers and system developers have technical knowledge not held by the common user. They can develop the programs and user interfaces that enable ready use of the computer as an information tool. Now, telecommunications professionals are entering the same "priesthood" role, with the skills and knowledge necessary to create and manage the communications networks of today. Each new professional group develops a two-part language—an internal jargon useful to their craft, and an external language for communication with the user.

Each new information system presents a challenge to the user. The system and its information are useless until the user either learns to handle the system or finds someone who can do so. The problem of making an information sytem easily usable is difficult because of the wide range of background knowledge and capabilities of the system's potential users. The system developer must anticipate the characteristics of the user population, and design a system that is responsive to these characteristics.

Although getting information from printed text needs no special hardware, it requires a high skill level and significant user effor. In terms of usability, print technology is the most sophisticated of the information media. More ways to access the information are available than is the case for any other medium. Books, newspapers, and magazines provide page numbers, tables of contents, chapter headings, headlines, and summaries. The user can browse before reading and anticipate in advance what he or she is going to read. Thus, before reading printed material in detail, the user can get good information about usefulness, appropriateness, and reading time. Afterwards, the user can reread selected material. Indeed, many users feel that a major mode of information access is the ability to browse through the stacks of a library or a bookstore, observing the arrangement of the books and spotting items of interest.

The technology that makes print such a versatile medium also makes it difficult to utilize fully, because of the training that the user needs. Even many people who can read well are not comfortable with use of indices, card catalogs, and other aids to accessing printed materials. As a result, alternatives have been developed that lack some of the sophisticated access methods, but that require substantially less user training and involvement. Written literature has been reproduced in comic book versions, movies, and audio and video tapes, forms that are accessible to a wider audience, including visually impaired people as well as those with limited reading skills. These forms are not as fully accessible as a printed text with table of contents, indices, and other aids; but they do make the underlying information available to people who do not want, or cannot use, the more flexible access methods. At the same time, these alternative versions make sophisticated use of the information very difficult. A biblical concordance, for example, will enable the user to locate all references to Solomon in the Bible. How would one locate such references using only an audio tape of the Bible? The key concept behind the versatility of the print medium is that of nonlinear access, provided by indices, card catalogs, concordances, and other devices. This nonlinear access in largely missing if one is restricted to using audio or video tapes.

The same nonlinear access capabilities that exist for printed text are being built into computer-based information systems, indeed, developed into more sophisticated forms. But the ease of use that exists with print-based systems is not yet fully realized in the computer. The technology exists to make printed text as readily available by computer as it is on the shelf, but the economics of the situation do not permit this yet. One major factor is the cost of converting print text to a computer-usable form. Most modern printing systems are computer based, including the rapidly developing "desktop publishing" industry. Publishers are generally reluctant to make documents available to the public in computer-readable form. Even when the documents can be made available, they may include printing and control characters that must be ignored or removed by any system for retrieval and use of the information. Electronic scanners are now available for the conversion of information from print form to computer-readable format, but the process is tedious and expensive.

Another factor is that electronic access to information is still restricted to a relatively small (but rapidly growing) segment of the population. This access is probably most readily available in countries like the United States and Japan; it is largely missing in the developing countries of Africa, Asia, and South America. Even among people who have this electronic access, a relatively small fraction have high-quality access. The computer systems currently available to most users do not have the ability to display high-quality text and graphics, of the quality, say, that one finds in *National Geographic Magazine*, or in a scientific journal. The systems available to most users do not have the ability to store and randomly access large quantities of information. In addition, many information users who have access to computer-based systems, and who could make good use of the information available through them, do not have the training that enables them to feel comfortable with the system: they have the piano and know that striking a key produces a sound, but cannot play music.(6)

Increasingly sophisticated technology enables an increased emphasis on multisensory and multimedia systems. Even before printing was invented, text and drawings were integrated into single documents. Motion was included with the development of movies, and sound was added when "talking pictures" were developed. Now, all of these sensory components are being transferred to computer-based systems. At the same time the systems are being adapted to handle multiple information media—printed materials, photographic and video film, audio tape, and most recently compact video disks. simple systems are really federations of sybsystems to handle each of these media. Only with the more complex computer-controlled systems do we have the ability to control and use the interplay among the various media, to coordinate sight, sound, and motion from various sources.

Various information-recording media are currently in use: magnetic tape, photographic film, magnetic disk, and optical disk. Tape and film are sequential. The browsing and reviewing that is so natural with printed text becomes clumsy with these media. Access is effectively limited to "linear scrolling" rather than "flipping pages." However, the disk is a random access device. Hence browsing and reviewing are theoretically possible. For years the disk was used only to store and play back music and other sound. It was, in effect, a sequential storage medium, and its random access capabilities were not fully exploited. Even with magnetic disks random access is inhibited owing to the need to move the read/write heads mechanically across the disk surface (except for very expensive head-per-track disks). Currently optical disks and laser technology eliminate the need for mechanical movement, provide highly compact information storage, and can be coupled with a computer-based indexing and search system to provide full random access to the information.

Computing technology enables storage of text, still pictures, animation, and moving pictures as well as sound disks. Storing information on disk and accessing it with the computer enables *multimedia reading*, browsing and reviewing in several media information—text, pictures, and sounds—in a manner similar to reading a book or newspaper.

7.1. Access to Remote Information

Before the networking of personal computers, radio and television were the only remote information media. Radio and television supply "real time" information about an event while it is happening, wherever it occurs. Compared to text, the user has very little control over this access. While watching a television program or listening to a radio program, the user consumes information in a sequential fashion. The user can, of course, change stations; but the information on each station is only available sequentially. Since there are no page numbers, chapter headings, tables of contents, or index, it is impossible to browse, skip around, or return to a particularly entertaining or informative portion. In addition, books are very quiet and unintrusive, leaving the user in control, whereas radio and television intrude more heavily into the user's life.

Networked personal computers provide an information access medium with the immediacy of radio and television, and much of the storage capability and browsability of books. But the real improvement in information processing afforded by computer networks comes from combining these features with two-way access: the user is once again an actor in the information seeking process, not merely a passive receiver of information.

7.2. The Public Switched Network

The public switched network, viewed as a single entity, is the largest and most complex machine ever built. Its original purpose was to permit voice communication at a distance (by telephone) between two people. Today the network is used for many other purposes, e.g., facsimile transmission of text and pictures. Workers can communicate with each other in a live video teleconferences. Electronic and voice mail are new forms of communication. There are synchronous and asynchronous computer-based conferences to exchange ideas. Users can access large numbers of on-line databases. We use the network for banking, airline reservations, and to access library card catalogs.

Computerization and digital representation and storage of information combine with the public switched network to begin a fundamental change in handling information. The public switched network is envolving away from a telephone network to a system which in the future could supply most entertainment, education, information, and communication needs.

A major problem is that the technology is outpacing the perceived need. Some segments of the academic community are rapidly adapting to each increase in the electronic availability of information. "Scholar workstations" are being developed that provide both sophisticated local information handling and access to computer networks. Rapid advances are being made in producing user–computer interfaces that can be readily tailored to the needs of a particular scholar or group of scholars. Unfortunately, not as much effort has been put into making the use of computer-based networks as easy and uniform as use of the telephone network is. First electronic contact with a colleague across the country, even when her e-mail address is supposedly known, is still somewhat a matter of luck.

This same technology that scholars are increasingly using is available for providing high-capacity information access to every home and office (although the economics is not). The need for such access is not presently there, nor is it readily apparent when such a need will develop. Color television was available for a number of years before it became the home standard. Personal computers have now been widely available for a decade, but few people use them effectively in the home. A similar length of time may need to elapse before the need for high-capacity information access becomes apparent.

7.3. New Forms of Information

Each new information technology spawns new art forms. The technology of printing led to short stories, novels, newspaper reporting, the cartoon strip, and sports writing. Radio and television gave us radio drama, variety shows, TV sitcoms, and live sports broadcasting. Audio recording led to recorded music. Networked computer systems are no exception.

Computer forms of expression permit the user to participate actively. The ideas behind this date back to the pioneering work of Vannevar Bush,[2] Ted Nelson,[18] and others. Today, computer "bulletin boards" and newsletters are actively used by thousands of people throughout the country.

Some of this use, such as access to stock market information, has an immediate and serious application. Other access is to information that may be of interest in daily life, such as recipes and entertainment information, humor, and games. The bulletin boards allow each user to interact fully, reading whatever information is available, and adding his or her own comments to that information. We already see the beginnings of collaborative creative work via the computer network, as individuals participate in the development of games and works of fiction. This type of creative activity will produce new works of great variety that will evolve from today's computer art,[26] computer animation, games, simulations, interactive fiction, hypertext,[19] expert systems,[5] electronic mail, collaborative work and play,[1] and bulletin board systems.

Even without computer and communication technology, information is organized into "chunks" of closely related information. These chunks, in turn, are richly linked with each other as the relationships among the chunks develop. We might, for example, organize a chunk of information about the population of each individual city. These then might be linked to each other through geographic, political, or size linkages, as well as to other chunks of information dealing with pollution, education, transportation, and so forth. Each information creator, such as a scientist, writer, composer, or movie maker, is influenced by those who have gone before and by contemporaries and in turn has an influence on his or her audience. Sometimes these links are as explicit as a reference to a book, but frequently the links are not given explicitly. Computer and video technology, and the further development of the information network can provide access to more information than could reasonably be stored in any one library building. Furthermore, this multimedia information is available in a much more flexible time scale, enabling the user to follow both explicit and nonexplicit connections between chunks of information easily. A creative work, even if primarily the work of one person, can now be influenced not only by the ideas that preceded its conception, but also by the reaction to the work as it develops.

Consider, for example, a book on the history of jazz. There are pointers in the book to earlier writings on jazz, to musical recordings, to concerts, and perhaps even to movies and television shows. The reader can follow thes pointers only with some effort and very few readers other than the serious scholar will do so. More importantly, in traditional media these pointers are always backward in time, pointing only to earlier works.

As time passes other works may be written that reference this book on jazz history. Since the author was obviously unaware of these when the book was written, it is quite difficult for the conventional reader to know about them. Even the serious student has difficulty learning about these. Bibliographic references, by their nature, point only backwards in time.

However, within many science and technology areas, *Science Citation Index* now provides the forward pointing references—pointers to papers and books which reference a given article, thus enabling a reader to see what other information has been generated by the existence of a particular information item. This capability has not yet been extended broadly outside of the science and technology literature (and even there it is not available for all references) as the creation of the forward links is still expensive and highly labor intensive. Thus our jazz scholar still has difficulty in finding reviews, commentaries, and other works generated from a given piece of information.

Now suppose this book were to exist in a database on the computer network. We have the capability of including in this database much of the material that the author has referenced, as well as newer works referencing the book. As time passes there will be additions to the database with pointers to this book. Linking all of these pieces of information in a hypertext framework moves the user immediately from the jazz history book to the full panoply of secondary materials relevant to it, to materials relevant to these secondary materials, and on into a vast, rich informtion resource. In a sense, the book on the computer network may never be finished, as various users access it, react to it, and add their reactions to the store of information. Thus the book becomes "infinite."[27] As other writers and readers reference the book, these references are added as computer-supported links. The author's words are never changed, but additions to the book are made forever. The reader will begin with the book, navigate to past items that the author did not know about, and also to items added since the book was written.

7.4. *Individualized Access*—A Divided Road

Information systems are being faced with problems from three sides: (1) the quantity of information being made available to such systems is increasing rapidly, (2) more books and journals are being published every day, and (3) the number of information users is increasing. Automated financial, medical, production, and satellite monitoring systems add their information to the systems continually. The time available to react to this added information is becoming ever shorter, partly because of the increased volume, and partly because various "real time" situations demand a response to information more rapidly than humans can provide one. The larger volume of users strain the capacities of existing information systems. This triple threat—more information, shortened response time, and increased number of users—has created a situation that cannot be handled by conventional information sytems. A more powerful interface between the information and the user is needed.

The response to this has been twofold. On the one hand there has been a movement toward placing barriers between the user and the information, making it more difficult for the user to access any information. Unfortunately, denying the demand does not eliminate, or even reduce it. On the other hand, there have been efforts to make the user access easier, largely through enabling the user to specify more precisely the information that he or she needs. Proponents of this approach believe that systems can and should do a more effective job of filtering the information before it reaches users.

The barriers being placed depend primarily on whether a real time response is needed or not. If quick response is required, as in medical situations or space shuttle reentry, the solution has been to place more of the response burden on computer programs, to let the computer respond to the situation within the time parameters, and only provide the human user with information that is pertinent and usable, including information about the automatic response. Since immediate response times are below the human threshold, potential users are faced with the choice of relying on the system to correctly interpret and respond to the information, or giving up such demanding systems and reverting to simpler ones. The latter choice is frequently so unpalatable (would we really accept a manually operated telephone system, or eliminate modern medical monitoring systems?) that there is really no choice.

In situations where there is no critical time pressure, and the pressure comes largely from the volume of information to be handled, the response has been to isolate the user by interposing an "unfriendly" mechanism or system between the user and the information. For example, when faced with the large volume of information to be handled, some libraries are resorting to compact shelving, making it more difficult for the user to reach the books, or to robotic book handling systems, effectively closing the stacks to the users and preventing browsing. Other libraries limit access to information, particularly expensive journal information, through consortia, so that a desired journal may not be immediately available, but must be requested from another library.

Such barriers may ease the workload in the libraries; it is not clear that they have a positive effect on the accessibility of information. One might argue that through such devices more information is made available, albeit more slowly, than would otherwise be possible. This may be true, but most library directors would feel that other solutions to the problem are preferable. In particular, the barriers now used by libraries are based on the assumption that paper copies of the informtion sources must be available. More effective use of available storage space, and more efficient handling of the documents seems to be required. While having paper documents available certainly matches the habits and preferences of many users, it perhaps limits the range of responses to the problem. The use of electronic technology to organize, search, display, transmit, and copy documents may provide better solutions to some information access problems.

If we try to avoid placing barriers between the users and the information, we must address the sheer volume of information on another way. Approaches to handling this volume at the user end of the system fall into two categories: techniques that enable the user to have more direct control over the presentation of information, and techniques that are used behind the scenes to organize the information in a way that is more in line with the user's apparent preferences.

Giving the user more direct control over the presentation of information means redesigning the interface between the user and the computer. The first major step in this redesign occurred in the 1970s, when the then standard punched card and printout interface was replaced by the keyboard and video display terminal. As these devices became accepted the punched card disappeared from the computing scene, and the printout was relegated to a secondary role. However, the deficiencies of early terminal systems were readily apparent. They were slow, and the user had little or no control over the format of the display. These early difficulties are now well under control. Users have the choice between monochrome and color screens of varying size and resolution. Speed of the terminal systems has increased. And most importantly, the user can often tailor the display to his or her own preferences, including usage of color and sound. In addition "windows" are now available on all but the least expensive systems, giving the user the option of looking at more than one piece of information at once. This greatly reduces or eliminates the requirement that the user remember one piece of information while looking at a totally different one: both can be displayed simultaneously.

More recent work on the user interface has focused on extending the range of user options. Many efforts seek to provide the user with a fully flexible interface system. The user should have a choice of text, icons, images, or any combination of these, with or without sound. The user should have the option of constructing a system interface that conforms to his or her own view of the underlying data and operations, using a variety of inpunt and output devices. This option should be available dynamically, so that the user can adapt the interface to the needs of the moment.

Other research focuses on the behavioral aspects of the interface system. Not only is it important to provide many options for the user, but also it is important to provide the user with some guidance or assistance in interface design, rather than leave the selection from a large array of choices completely open. Behavioral scientists and graphic designers have much practical experience and some research data that characterize good and bad features of interface design. Increasingly these people are being brought into the design of visual interfaces that combine the best aspects of text, images, icons, color, and motion. Concepts developed in human engineering are being applied to interface design. There remains a large amount of inertia to overcome before some of these ideas bear fruit. (How many Qwerty keyboards have been replaced by Dvorak keyboards, despite the improved efficiency of the latter?) The main point is that behavioral and graph design factors are now actively considered in human–computer interface design, resulting in improved accessibility of the computer systems.

Improvements in interface design do not directly alter the amount of information being presented to the user. But by changing the form and organization of the information through windows, improved graphics, and other techniques, it becomes possible for the user to tune the selection of information more finely, with the result that much information that would have been displayed in older interfaces now is not normally displayed, although it remains accessible on demand.

Behind-the-scenes control of the information displayed to the user is a new approach to the problem of information overload. The idea is to more finely tune the information to the needs and modes of operation of the user through knowledge of the user and a careful study of his or her information habits. The concept is simple: better knowledge of the user leads to an improved ability to fit the presentation of information to the user's needs. Experiments have demonstrated that knowledge of the user can be an effective factor in selecting the information to be displayed, with the result that the user sees more useful information and less information that has little or no relevance to current needs. This same user knowledge enables the system to organize the information display to meet the user's needs better, through on-screen spatial arrangement, use of graphics, use of color, and lexical devices for emphasizing parts of the information likely to be of greatest use.

It has been said that the ultimate in "friendly" interface design is to have a system that will anticipate most, if not all, of the user's needs and respond to them even before the needs are fully articulated. This utopian vision is far from realization, even in information systems that rely on the services of a highly trained human intermediary. Indeed, one area of active research is on the merits and problems associated with an interface system that has such a detailed, intimate knowledge of the individual user. Such an interface involves difficult ethical and privacy questions.

8. Conclusion

Music, voice, still images, moving pictures, animation, and text can be represented as sequences of binary digits. The binary digits can be stored in a magnetic or optical storage device. The stored symbols can be recovered, displayed in a variety of ways, transported over a digitally switched optical fiber network, copied, modified, and then stored again. This process is reliable and inexpensive for the individual user. During the past two decades we have seen the introduction and rapid popularity of the handheld calculator, VCR, personal computer, portable computer, cellular telephone, and facsimile machine. These new information technologies are gaining public acceptance at an increasingly rapid rate, as they are becoming less expensive and easier to use.

The "priesthood" of information adepts still exists; but it is being hidden from the users under layers of interface technology. At one time a user needed to go to the story teller, the reader, or the scribe to make use of the information technology; now most users of a computer-based card catalog in a library, for example, are at best vaguely aware that they are accessing the technology through a very high level interface. The "priests" who have the skill to develop the computer and communication technology and the software to support its use are invisible to the users. As this information technology develops toward maturity, the users have increasingly easy, rapid, and inexpensive access to information of all kinds, wherever it is generated and stored.

The ease of access creates many opportunities for the users, as they are able to control the design of their individual interfaces to information systems, and respond more rapidly and directly to information that they find in the system. Dynamic interfaces between human and computer lead, in effect, to a single information system with both human and machine components, resulting in more and better access to information than was ever available under any preceding technology. But along with the opportunities these systems present come increased responsibilities and difficulties, as the systems become more complex and sophisticated. The users gain control over an increasing portion of the information system; but if the system simultaneously grows at a rapid rate, do the users really gain increased control of the relevant part of the world modeled by these data?

References

1. S. Ambron and K. Hooper, Eds., *Interactive Multimedia: Visions of Multimedia for Developers, Educators, and Information Providers*, Microsoft Press, Redmond, Washinton 1988.
2. V. Bush, As we may think. *Atlantic Mon.* **7**, 101–108, 1945.
3. S.K. Chang, Ed., *Visual Languages*, Plenum Press, New York, 1986.
4. K.E. Dowlin, *The Electronic Library*, Neal-Schuman, New York, 1984.
5. E. Feigenbaum, P. Mc Cordick, and H.P. Nii, *The Rise of the Expert Company*, Times Books, New York, 1988.
6. W. Gardner, The electronic archive: Scientific publishing for the 90's. *Psychol. Sci.* (unpublished).

7. R.R. Korfhage and C. Hemphill, Natural language interface, Proceedings IEEE 1984 Workshop on Languages for Automation, New Orleans, November1984, pp. 173–177.
8. R.R. Korfhage and M.A. Korfhage, The nature of visual languages, Proceedings IEEE/CS 1984 Workshop on Visual Languages, Hiroshima, Japan, December 1984, pp. 177–183.
9. R.R. Korfhage, Two-dimensional languages for geographic information systems, Proceedings of First Scandinavian Research Conference on Geographic Information Systems, June 1985, pp. 134–142.
10. R.R. Korfhage, Constructs for iconic languages, 1985 IEEE Workshop on Languages for Automation, June 1985, pp. 164–169.
11. R.R. Korfhage, Intelligent information retrieval: Issues in user modelling, Proceedings of the Symposium on Expert Systems in Government, October 1985, pp. 474–482.
12. R.R. Korfhage, Browser: A concept for visual navigation of a database, 1986 IEEE Workshop on Visual Languages, June 1986, pp. 143–148.
13. R.R. Korfhage and M.A. Korfhage, Criteria for iconic languages, in *Visual Languages*, S.K. Chang, Ed., Plenum Press, New York, 1986, pp. 207–231.
14. R.R. Korfhage, T. Ichikawa and E. Jungert, Eds., *Visual Languages and Applications*, Plenum Press, New York, 1990.
15. T. Kuhn, *The Structure of Scientific Revolutions*, University of Chicago Press, Chicago, 1970.
16. S.H. Myaeng and R.R. Korfhage, Towards an intelligent and personalized information retrieval system, Proceedings of the International Symposium on Methodologies for Intelligent Systems, October 1986, Knoxville, Tennessee.
17. S.H. Myaeng and R.R. Korfhage, Integration of user profiles: Models and experiments in information retrieval. *Inf. Process. Manage.* **26**(6), 719–738, 1990.
18. T.H. Nelson, *Computer Lib/Dream Machines*, published by the author, 1974.
19. T.H. Nelson, *Literary Machines*, published by the author, 1987.
20. D.A. Norman, *The Psychology of Everyday Things*, Basic Books, New York, 1988.
21. B. Nussbaum, *The World After Oil: The Shifting Axis of Power and Wealth*, Simon and Schuster, New York, 1984.
22. I. de S. Pool, *Technologies of Freedom: On Free Speech in an Electronic Age*, Belknap Press of Harvard University Press, Cambridge, Massachusetts, 1983.
23. H. Rheingold, *Tools for Thought—The History and Future of Mind-Extending Technology*, Prentice-Hall, Englewood Cliffs, New Jersey, 1986.
24. B.R. Schatz, *Telesophy: A System for Manipulating the Knowledge of a Community*. Bell Communications Research monograph, 1987.
25. D. Schmandt-Besserat, The earliest precursor of writing, in *Language, Writing, and the Computer. Readings from Scientific American*, W.H. Freeman, New York, 1986.
26. L.F. Schwartz and L.R. Schwartz, *The Computer Art Book*, Norton Press, New York, 1990.
27. H. Wheeler, *The Virtual Book* (unpublished).
28. T. Winograd and F. Flores, *Understanding Computers and Cognition: A New Foundation for Design*, Addison-Wesley, Reading, Massachusetts, 1987.
29. B. Winston, *Misunderstanding Media*, Harvard University Press, Cambridge, Massachusetts, 1986.
30. R. Darnton and D. Roche, Eds., *Revolution in Print: The Press in France 1775–1800*, University of California Press, Berkelee, 1989.
31. *The Times Atlas of World History*, Hammond, Inc., Maplewood, New Jersey, 1989.

COMMUNICATION TOOLS FOR HUMAN–COMPUTER KNOWLEDGE TRANSFER

JODY PAUL

Intelligent information systems, such as those built using knowledge-based system (KBS) technology, stress the human–computer interface in new ways. Such systems use knowledge as well as lower-level information. The application of KBS technology depends on two aspects of the communication of knowledge (*knowledge transfer*). First, it depends on the ability of humans to explicate their knowledge in ways that can be incorporated into the computer system. Second, it depends on the ability of the computer to divulge its knowledge to humans in a palatable and understandable form.

The use of knowledge imposes additional burdens on the human–computer interface. Knowledge constitutes a higher level of information than the data typically used by computer systems (see the information spectrum depicted in Fig. 1). It is more complex and concerns conceptual rather than concrete information. In practice, knowledge is often comprised of heuristics or rules of thumb rather than simple values or data structures. An example of knowledge is the nature of the relationship between the value of a product liability claim and the claimant's personal characteristics (age, sex, etc.). In contrast, a bank account record that includes personal data (name, address, etc.), current balance, and transaction history is an example of lower-level information. Standard communication techniques (displaying a number, printing a report, etc.) are not adequate to cover the more complex conceptual level of communication (describing a rule of thumb), explaining a chain of reasoning, etc.). Existing tools, while sufficient for applications like the banking example, are ineffective for describing the complex interaction of factors such as those involved in evaluating liability claims.

JODY PAUL • Social Policy Department, The Rand Corporation, Santa Monica, California 90406.

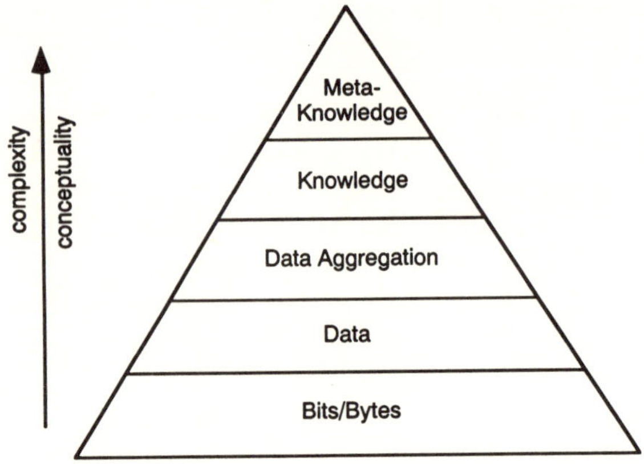

FIGURE 1. Classes of information

The nature of the tasks to which KBS technology are applied compounds the problem. KBSs provide a means for applying computer technologies to solve problems for which algorithmic and formal solution approaches do not currently exist. The tasks addressed involve complex, often subjective, reasoning. It is more difficult to express the domain knowledge and expertise than to express numeric data and algorithms. The complexity also increases the potential for minsunderstanding; yet the users of such systems need greater understanding of the system internals since they must assess the accuracy, applicability, and validity of recommendations.

This chapter concerns the human–computer knowledge transfer process and interface. It describes the nature of the problem and a method for mitigating the difficulties. Examples from the System for Asbestos Litigation (SAL) expert system are used to illustrate key points. SAL was developed at the RAND Corporation to investigate models of legal reasoning and the benefits of enhanced explanation. It demonstrates application of the method for enhancing knowledge transfer in a real-world sociolegal problem domain: asbestos product liability litigation. SAL is described in the Appendix.

1. *Background*

The special needs of humans who must interact with intelligent systems, as they are built and used, have not been adequately addressed. Although the importance of knowledge transfer (as explanation and knowledge acquisition) has been pointed out often in the literature[8, 14, 19, 32, 62, 67, 70] and significant local advances have been made,[6, 10, 11, 15, 31, 40, 49, 71, 85] little in the way of broad progress has been achieved. An overview and comparison of important contributions to expert system explanation is found in Ref. 52. Each of the research efforts addresses a particular aspect of the problem. The techniques are closely tied to specific building tools and exhibit limited applicability (e.g., to a specific representation and inference paradigm[(75, 84)]). Greater value and utility is realized only when these are integrated as tools within an overall schema. Such a framework is presented in this chapter.

A method that facilitates human–computer knowledge transfer can mitigate many interface problems. This chapter introduces a new method for improving the transfer of knowledge between humans and computers. The improvement results from enabling the computer system to disclose its internal composition, behaviors, and expectations. Such software divulges how it arrived as its conclusions, what knowledge it possesses, the soundness and justification of its reasoning, and aspects of its user interface and protocol. We call such systems *self-revealing*. The method provides a concrete approach to generating such systems.

2. *Knowledge-Based Systems*

Knowledge-based systems comprise a subdiscipline of artificial intelligence that focuses on a special class of heuristic problem-solving techniques. These systems employ elaborate reasoning about abstract concepts in complex domains.[(66, 73)] The development techniques involve encoding large amounts of knowledge about a particular domain or problem area and form what has come to be known as the *knowledge-based approach*. This approach is applicable to problems for which algorithmic and formal solution approaches do not exist. Typical characteristics of tasks that require the use of KBS technology are shown in the first column of Table 1. Contrasting characteristics of conventional software are shown in the second column.

TABLE 1
Comparison of KBS Tasks and Conventional Software

Common characteristics of KBS tasks	Contrasting characteristics of conventional software
Many possible solutions	Single solution
Conceptual	Formulaic and numerical
Incomplete, uncertain, subjective, inconsistent and time-varying information	Data must be complete and unambiguous
Disagreement among experts	Accepted solution approach
Changing and evolving problem	Stable task

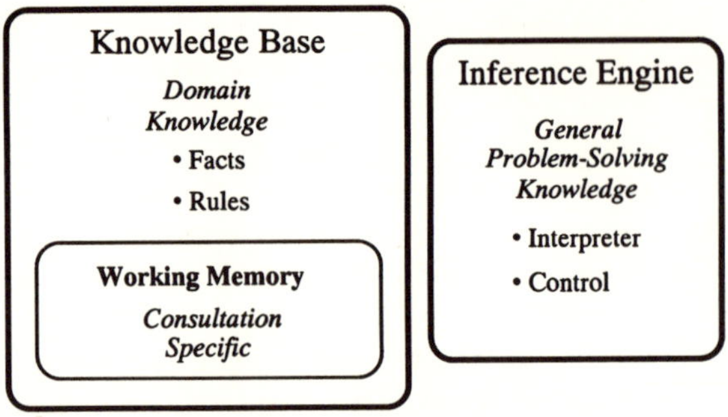

FIGURE 2. Knowledge-based system structure

The overall structure of a typical KBS is shown in Fig. 2. The key aspects that distinguish it from conventional software are (1) reliance on a large base of explicit knowledge, and (2) separation of that knowledge from the inference engine that interprets it. The problem-solving power of such a program comes from the knowledge it possess, as well as the formalisms and inference schemes it employs.[(77)] *Explicit knowledge* is that to which the programmer and the system have direct, manipulatory access.[(2)] Because explicitly represented information is in a form that allows global interpretation, the same data can be used for multiple purposes. In MYCIN, for example, the rules that form the system's knowledge about how to diagnose possible causes of infectious diseases are used not only by the diagnosis module, but also by the routines that explain the diagnosis module's reasoning and those that acquire new rules.[(16)]

The process of building a KBS is called *knowledge engineering*. This typically involves the interaction between a system builder, called the knowledge engineer, and a domain expert. The knowledge engineer interviews the domain expert, organizes the knowledge, and determines how it will be represented in the computer. The formalized knowledge is turned into a working computer program, which is tested, put into use, and continually modified throughout its lifetime.

Sometimes the tool used to build a knowledge-based system is confused with the system itself. As shown in Fig. 3, the tool includes both the language for representing and accessing the system's knowledge and the support environment. Confusion often arises because the support environment is also part of the completed knowledge-based system.

An *explanation facility* is a component of the support environment that provides explanations of how and why particular actions were taken or conclusions were reached. It may also provide explanations of why those actions or conclusions are justified. Such a facility allows the system to explicate its knowledge and activity and to justify its reasoning and conclusions.

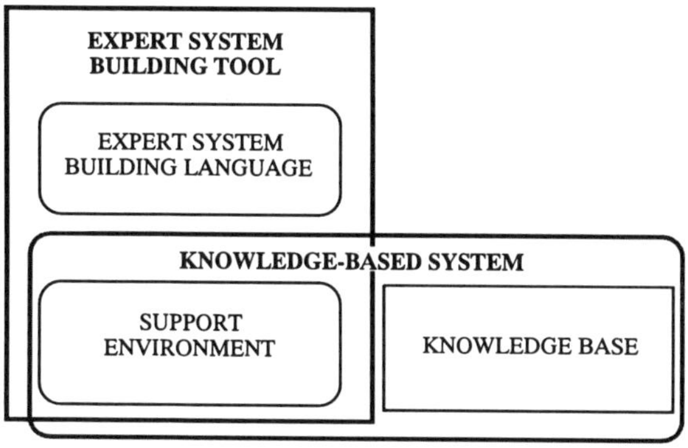

FIGURE 3. Expert system building tools and knowledge-based systems

Explanation is critical to the construction and utilization of a KBS. Explanation facilities significantly affect user acceptance and the speed of system development, although the facilities in most current expert systems are primitive.[53] Explanation helps make a system acceptable to users by inspiring confidence in its performance and reasoning. Explanation facilitates development by assisting in debugging, testing, and refinement of the system.

3. *Understandability*

The ability to understand the behavior of a KBS is more difficult to achieve than for conventional software due to the increased *complexity* of the problems being addressed and the software tools used. For example, a user can readily verify the correctness of an algorithm used to calculate mortgage payments for a given annual percentage rate. Determining the merit of an estate tax plan proposed by an expert system is not so straightforward. The tasks addressed by knowledge-based systems require heuristic knowledge and expertise because simple or algorithmic approaches are not available. Each case processed may result in a complicated and unique chain of reasoning. Since there is no single algorithm embodied in the software, the reasoning for each situation needs to be explicated and justified. Because a KBS addresses problems by employing heuristics rather than algorithms alone, a user needs more convincing that the system's behavior is proper. This is best accomplished by explaining how it arrived at its conclusions and why its reasoning was justifiable. The interactive behavior of a KBS is similar to a consultation among humans, where the utility of a conclusion

TABLE 2
Benefits of Understandable Software

Utilization	Development	Domain
Appropriate usage	Testing	Expertise enhancement
Acceptance	Debugging	
Persuasion	Refinement	
Education	Validation	
	Verification	

or recommendation depends on convincing the client of the soundness of the reasoning.[8, 25, 62, 65, 71]

Increasing people's understanding of a software system involves revealing to the designers, developers, maintainers, and users (1) the program's behavior, (2) how that behavior is effected, (3) the appropriate protocol for human–computer interaction, and (4) why the program's actions are reasonable. Improving the understandability of a software system accrues benefits in each of the areas shown in Table 2.

A user who understands the operation of and logic behind a system is more likely to use the system for its intended purpose. Explanations of results that validate the soundness of the reasoning increase user acceptance of and confidence in the system. Apprehending the operation and reasonableness of a complex system may persuade users that unexpected results are appropriate. Novices in a domain may become educated where their knowledge is weak and acquire expertise as well. Revealing the system knowledge and operation facilitates development and maintenance by assisting in testing, debugging, and refinement. The system's logic and beliefs need to be examined and their correctness verified if errors or uncertainty arise. An expert in the domain of a system who comprehends what an existing systems knows and does can also grasp its limitations. This provides a basis for designing natural improvements to the system. Such comprehension is prerequisite for formalizing and codifying additional domain knowledge, processes that further enhance an expert's understanding of the domain.

4. Human–Computer Knowledge Transfer

The transfer of knowledge between humans and computers is an extension of the data input/output activity of conventional computer usage. The existing problems with achieving effective human–computer interfaces are compounded by the inherent difficulty of communicating complex high-

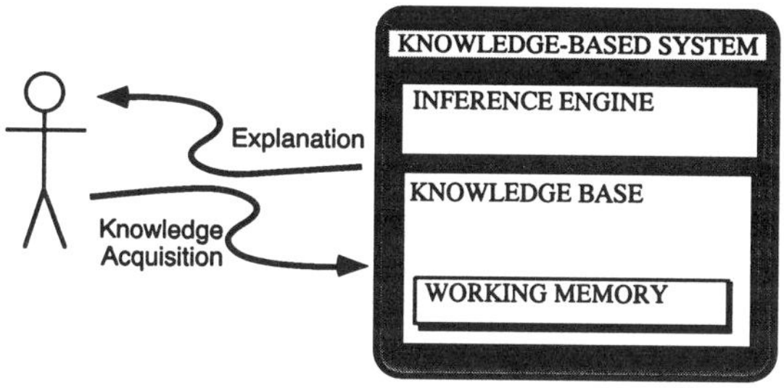

FIGURE 4. Human–computer knowledge transfer.

level concepts. Effective communication of conceptual knowledge between humans and computers is predicated on the human's comprehension of the system. The lables *explanation* and *knowledge acquisition* correspond to the two directions of knowledge flow as shown in Fig. 4.

Explanation refers to the transfer of information from a KBS to the user of the system. Explanations concern how and why particular actions were taken or conclusions were reached and why those actions or conclusions are justified. They allow the system to explicate its knowledge and activity and to justify its reasoning and conclusions. This ability is critical to the construction and utilization of a KBS.

Knowledge acquisition refers to the transfer of information from humans to the knowledge base of a KBS. It is the acquisition of the conceptual knowledge for *constructing* KBSs and *not* the case-specific data that such systems process during execution. Knowledge acquisition is generally acknowledged as the primary bottleneck in the construction of a KBS.[7, 32, 33] The current state of affairs has been summarized as follows: "The curve of development effort has leveled off around five person-years for all but the simplest consultation systems. ... Knowledge acquisition remains a bottleneck on the way to passing the five-year barrier."[18]

There is a reflexive and synergistic relationship between the knowledge-transfer processes of explanation and knowledge acquisition: knowledge acquisition depends on explanation, yet explanation requires knowledge acquisition. A primary requirement for effective knowledge acquisition is keeping the provider of new knowledge aware of what has already been built—the very function of explanation. Yet the information used in constructing explanations must first reside in the KBS. That is, it must have been acquired before it can be presented. Some of that information is not necessary to perform the task of the target system, but is required only to support justification and explanation. The explanation function itself thus imposes some additional knowledge requirements. A flexible and robust explanation facility design can mitigate the added knowledge-engineering burden. For example, allowing surface-level explanations obviates the need for cumbersome, nonintuitive encoding of concepts from the domain.[50, 52, 72]

Those who add information need to know a great deal about the current system. System builders need to understand not just what information is contained in the knowledge base, but also how it fits together, how it is used, how it is intended to be used in given situations, how knowledge is represented, why certain things are represented as they are, etc. A key aspect of knowledge acquisition is entering new information into the knowledge base. Knowledge acquisition support is needed to (1) help system builders to avoid inconsistent encodings of knowledge, (2) facilitate the entry process, and (3) enhance coverage completeness. Each of these concerns can be addressed by exploiting the explanatory capability of a system.

Information must be uniquely encoded in the knowledge base and similar representations used for similar concepts. This requires the information provider to be fully aware of the information currently resident in the knowledge base. The system must be able to reveal its internal knowledge and behavior clearly to the user.

The acquisition process can be facilitated by direct access to and combination of existing concepts. An enhanced explanation facility could provide mechanisms for accessing, displaying, and combining individual and aggregate concepts from the knowledge base. This technique of using *conceptual building blocks* also mitigates the problems of inconsistent representation by ensuring that the same encodings of concepts are used.

The problem of achieving comprehensive coverage of the domain similarly involves informing the knowledge provider of the state of the knowledge base—here, what the system *needs to know*. Tests are needed to determine whether all relevant domain concepts have been incorporated into the knowledge base. Fundamental research is required to develop a means for programs to determine what is missing from their knowledge bases—a paradigm for *knowing what you do not know*. Once such opportunities have been identified, the system can elicit the missing information using the explanation-interface protocols.

5. *Self-Revealing Software*

Software systems that disclose their reasoning and the underlying justification for that reasoning by exploiting an awareness of their own structure and activity may be said to be *self-revealing*. Such systems require the ability to examine their internals and generate understandable explanations. This section contains an overview of methodological considerations for developing self-revealing software and a description of a particular method.

The objective of the methodology is the production of more understandable knowledge-based systems. The means involves designing in the ability to generate explanations of their own structures and activities, i.e., making them *self-revealing*. Improved understandability results from enabling the system to disclose its internal composition, behaviors, and expectations. Such software divulges how it arrived at its conclusions, what knowledge it possesses, the soundness and justification of its reasoning, and aspects of its user interface and protocols. The particular method presented, called REVEAL, provides a concrete approach to generating such systems.

The fundamental principle of REVEAL is applying knowledge-based technology for self-explanation, a paradigm we call *inferenced generation*. This is a significant departure from previous expert system explanation facilities.(52) A system constructed following the REVEAL method applies its reasoning mechanisms to self-knowledge to consider its reasoning and its interaction with users. Communication of conceptual knowledge between computers and humans is not a well-understood process. However, a knowledge-based approach can take advantage of the extant empirical knowledge and rules of thumb (e.g., "Don't overload the audience with excessive quantity or detail.").

Inferenced generation enables the system itself to generate explanations by the same techniques it uses to effect task-performance reasoning. This increases the power of the explanation facility to that of the full KBS building tool itself. The explanation function is implemented in the same fashion as other reasoning. This allows dynamic creation of explanatory text and flexible user models. The explanation process, implemented in the same formalism, integrates with the system as a whole. This contrasts with the typical approach of building a separate "explainer" that exists outside the taks-performance system and attempts to interpret its behavior.

Manipulation and presentation of explanatory information is accomplished using primarily the same reasoning tools and structures employed to address the domain problem. New structures support the additional information necessary to effect competent explanation, such as "compiled-out" knowledge,(51) by adding to the system knowledge base or enhancing the user–system interface.

There is no established theory or methodology for software system explanation. The paucity of explanation in current expert systems is attributable primarily to this lack. As discussed previously, the need for improved understandability is significantly greater for knowledge-based systems than for conventional software. The pressing need was recognized early in the development of expert system technology, but only recently has the field matured to the point where the development of such theory is feasible.

TABLE 3
REVEAL Method Design Steps

1. Establish coverage goals.	Decide which needs the system is to address.
2. Incorporate explanatory information.	Encode and integrate data needed to accomplish the selected goals.
3. Integrate explanation behavior with task performance.	Provide support mechanisms for the system to a. Recognize and identify a need for explanation; b. Select information to satisfy that need; c. Reformulate the information with respect to active context; and d. Display the resulting reformulation.

The primary aspects of a method for producing "self-revealing" systems include (1) users' needs, i.e., what the explanations are to accompish; (2) knowledge requirements, i.e., the information to be imparted; and (3) the actual information transfer process. These are realized in the REVEAL method by the three design steps shown in Table 3.

The design of a self-revealing software system depends on the types of explanations that it must be capable of producing. The key involves assessing the needs of users—ascertaining what assistance is required. A characterization of the types of explanations that a software system may have to provide can help the system recognize and identify the needs. It may also be used as a measure of explanation coverage and competence. Actual systems each address a subset of these needs according to their unique goals.

The information necessary to effect the desired explanations must next be determined and incorporated. For example, rationale information must be available in order to justify the application of a particular heuristic to a generic class of problems. The information may come from a model of the underlying mechanism, a corpus of knowledge in the field, or directly from a domain expert's experiences. Whatever its source, it must be encoded and integrated with the task-performance knowledge. Constructs for representing the information and incorporating it into the knowledge base should support (1) retrieval based on content, and (2) association with task-performance information.

The actual process of explanation is comprised of need recognition, information selection, reformulation, and presentation as shown in Fig. 5. These behaviors should be integrated with the task-performance aspects of the system. The characterization of users' needs from Step 1 can be used to help the system recognize those needs. Information selection involves mapping identified needs to the information resources developed in Step 2.

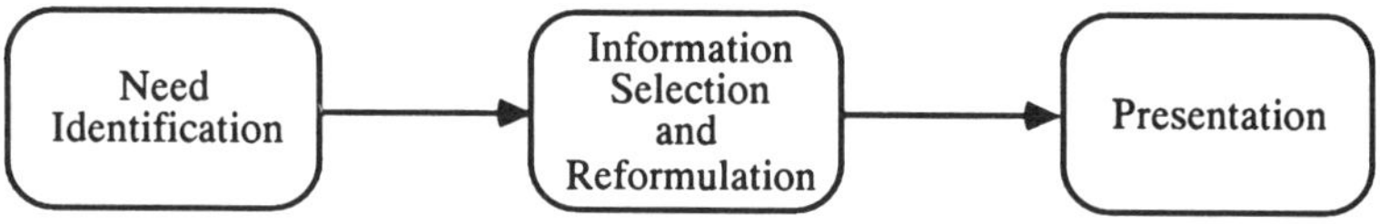

FIGURE 5. Explanation Tasks

Reformulation covers filtering the selected information on the basis of active context: what is known about the user, the consultation, and the current state of the dialogue. Presentation concerns display and interaction with the user. This activity conveys the information to achieve the desired goals.

We next explore each of the three aspects of the methodology in detail. Examples drawn from the SAL expert system illustrate the application of the principles described. Most of the concepts and features of REVEAL are realized in SAL. Notably, SAL utilizes inferenced generation. It employs its own reasoning procedures, methods, and mechanisms for constructing the explanations it presents.

We will spotlight one example in particular as we discuss the various aspects of the method: the production of the general explanation for why SAL concluded that "AGE is important" during a consultation shown in Fig. 6.

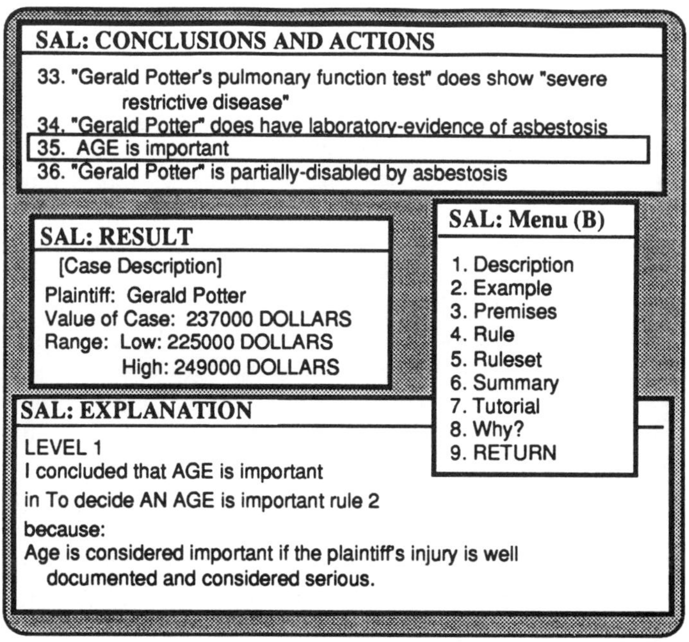

FIGURE 6. Explanation of importance of age (SAL).

We will examine the knowledge and reasoning used to generate it. In the illustrations that follow, text shown in a plain sans serif font (like this) indicates direct computer output. Text shown in the same font in italics (*like this*) is input from the user, either typed or selected from a pop-up menu. A box around a menu item indicates that that item has been selected by the user. Text shown in italicized serif font (*like this*) is annotation to the figure.

5.1. Coverage

First and foremost, the primary goal of explanation is to address unsers' needs for information. The set of needs to be addressed defines the explanation *coverage* of a system. The desired coverage determines what information the explainer must have available to achieve the necessary degree of competence. The needs and information resources, along with knowledge of the process of communication, govern the generation and presentation of explanations, thereby achieving the fundamental goal.

The needs of users can be characterized by looking at (1) the different types of users, (2) explanation request types, and (3) the functions of explanation. Each of these categories spans the space of user needs. They are not orthogonal but very much interrelated. Responses to each kind of request may accomplish several functions. Different types of users have diverse functional requirements and stereotypical requests. This multiple decomposition is useful because each characterization supplies different implications for the information-resource and explanation-behavior characteristics. For example, a request-type analysis emphasizes the underlying reasoning that is required to fulfill the request.

5.1.1. User Types. An explanation facility must support different users with different needs; thus a simple "one size fits all" solution is not adequate. Most explanation facilities cannot adjust quantity or level of information for presentation. They present everything rather than selecting only what is relevant. A theory of explanation must allow for different explanations corresponding to the needs of its various users. A characterization of users and their explanation needs must be at the heart of any such theory. Typical users of knowledge-based systems and their relationships are shown in Fig. 7. The explanations shown in Fig. 8 demonstrate how differences in user type can be reflected in the explanations presented in response to the same request. (The explanations were generated by SAL for "partial disability" and depend on the user type determined earlier in the consultation session.) Note that the explanation for end users provides a general overview; for domain experts, an evaluation of supporting evidence; and for system builders, the procedural steps followed.

Needs of particular users are not static, but change over the course of

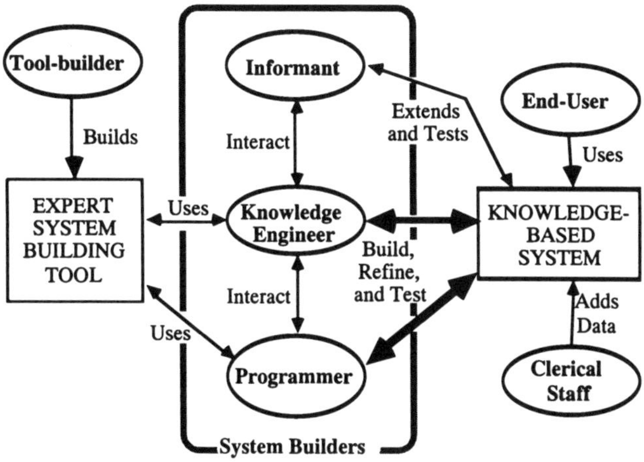

FIGURE 7. Roles and relationships of knowledge-based system users.

For End-Users:

John Smith is partially-disabled by asbestosis
Explain
I did conclude that John Smith is partially-disabled by asbestosis because
Since John Smith is not totally disabled but does have some disability due to asbestosis, we assert that the plaintiff is partially disabled.

For Domain Experts:

John Smith is partially-disabled by asbestosis
Explain
I did conclude that John Smith is partially-disabled by asbestosis because
John Smith doesn't exhibit sufficient evidence for total disability, but does have enough for a claim of partially disabled.

For System Builders:

John Smith is partially-disabled by asbestosis
Explain
I did conclude that John Smith is partially-disabled by asbestosis because
We first rule out the cases of no disability and total disability due to asbestosis. If neither of these apply, complex conditions of activity and medical evidence are checked (see RULE source).

FIGURE 8. User-dependent explanation in SAL.

a consultation. An explanation method must support a flexible approach to need recognition and assessment. For example, a user that requires extensive support in early stages of a consultation may become comfortable with the interface and no longer need such assistance or may become encumbered by it. Furthermore, needs are individualized—fixed explanation strategies for each class of user are inappropriate. Each user and consultation has unique requirements which must be dealt with individually. Thus, a static explanation schema is not effective.

The specific types of users and their requirements may further depend on the domain, target problem, and fielding environment. Thus, tools should accommodate system builders' customization of explanations and not artificially restrict the user classes and set of addressable needs. A methodology for producing understandable software must provide for tailoring explanations to various users and their particular needs. It should provide a general, customizable facility. Explanation dynamism, based on inferenced generation, permits a system to deal with a user's needs that change within a consultation session.

SAL maintains information about users types in the same manner as other conceptual knowledge. Basic knowledge about user types is shown in Fig. 9 and realized by the included ROSIE code.

SAL's first interaction with the user, shown in Fig. 10, requests a characterization of the user's familiarity with the domain and the SAL

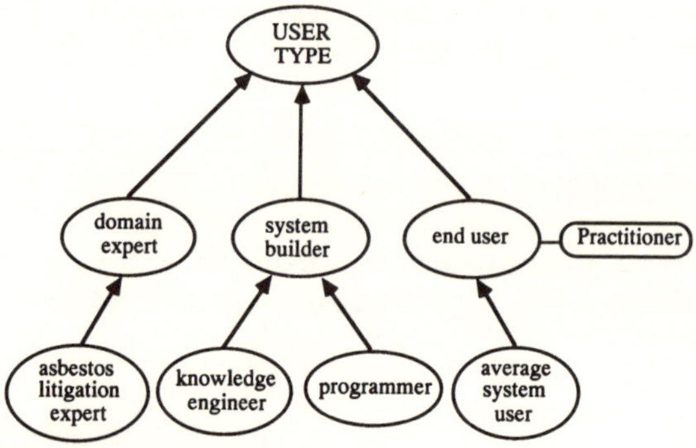

Assert any domain expert is a user type.
Assert any system builder is a user type.
Assert any end user is a user type.
Assert "Asbestos Litigation Expert" is a domain expert.
Assert "Knowledge Engineer" is a system builder.
Assert any programmer is a system builder.
Assert "average system user" is an end user.
Assert "Practitioner" is a synonym for end user.

FIGURE 9. Conceptual knowledge: user type.

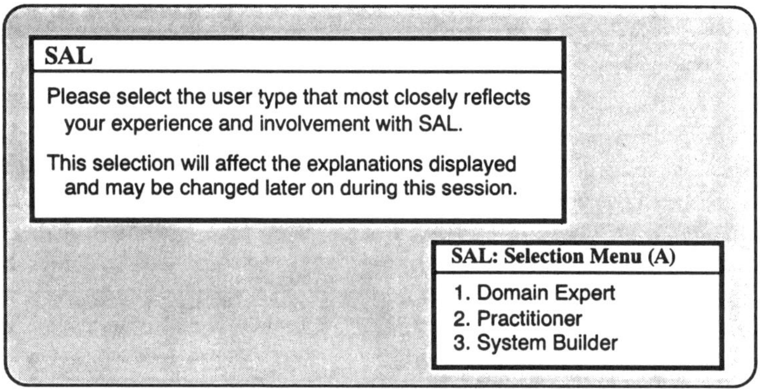

FIGURE 10. Consultation initiation dialogue.

system itself. This is the initial piece of information on which the system bases assumptions about the user's knowledge and expectations. Each of the menu items has additional elaboration information associated with it that will be presented to the user upon request. This is derived from the concept hierarchy shown in Fig. 9.

When the user makes a selection from this menu, an assertion is made to working memory that indicates the user level. Thus, if the user selects "Practitioner" the knowledge base will be augmented by the proposition "END USER is a user-level" and any previously assumed user level will be deleted. By recording this information in the knowledge base, it is available at any time for use in explanation. It is augmented by additional information about the user discerned during processing.

5.1.2. Request Types. *Request types* provide a natural and intuitive characterization of explanation needs. This characterization aids the need recognition and identification task. It is also particularly useful as a basis for assessing the coverage of an explainer.

A set of explanation request types is shown in Table 4. Each is accompanied by a generic formulation as a question. The use of the question formalization is for illustrative purposes only; actual requests may not be phrased as questions and often are implicit. In contrast to general categorizations of questions such as those proposed by Lehnert[41] or Schank,[60] this set exhibits a finer granularity, which aids in need recognition and explanation capability assessment. An expert system building tool should provide the system builder with sufficient power to enable the system to generate explanations in response to any of the indicated types of requests, although not every fielded system needs to support all request types.

SAL's explanation support mechanisms are capable of addressing each of the request types shown with the exception of counter-factual requests.

TABLE 4
Explanation Request Types[a]

Motivation	Why is the system asking for this particular information?
Amplification	How should this question be answered?
Enumeration	What are elements of the set of possible responses to this request?
Definition	What is the meaning of this term?
Analytic	How was this result obtained?
Representation	What knowledge is encoded by this structure?
Description	What is the general function or intent of this structure?
Example	What is an example of the function of this structure in the context of the current consultation?
Planning	Why did the system consider applying this structure?
Applicability	Why was this structure applicable to the current consultation?
Occurrence	Why was this conclusion made? Why was this action taken?
Alternative	What alternative structures were involved in making inferences about this result?
Intention	What is the function and intent of the set of structures used to make inferences about this result
Rationale	What is the rationale for the set of structures used to make inferences about this result?
Counterfactual	Why didn't the system make this conclustion? Why didn't the system take this action?
Usage	How does the system use this structure?
Hypothetical	What if the value of this datum was changed?

[a] Recall that the term "structure" appearing in the questions can be replaced by "rule," "object," "procedure," "frame," or any system building block as appropriate to a given representation paradigm.

This lack is due to a corresponding lack of theory of how such questions should be answered (independent of the context of knowledge-based systems).

5.1.3. Explanation Functions. Characterizing needs by *functions of explanation* emphasizes the types of knowledge required to effect the various functions. In addition, this formulation is very similar to a formal statement of the goals of explanation. Reasons for an expert system to explain its reasoning are shown in Table 5. Users need to understand the knowledge in the system and the line of reasoning in order to maintain it and use it effectively. Explanations provide a means for system builders to examine the

TABLE 5
Functions of System Explanation

Application	Development and maintenance	Domain
Appropriate usage	Testing	Expertise enhancement
Acceptance	Debugging	
Persuasion	Refinement	
Education	Validation	
	Verification	

The pulmonary-function-test of John Smith does show "moderate restrictive disease".

Explain *(Requests general explanation of conclusion)*

I did conclude that the pulmonary-function-test of John Smith does show "moderate restrictive disease" because vital capacity or lung volume readings in the 60s indicate moderate restrictive disease.

Summary *(Requests generic information about a PFT result)*

The Pulmonary Function Test (PFT) results indicate a degree to which the plaintiff suffers from a restrictive lung disease.

Tutorial *(Requests additional tutorial information)*

Vital capacity or lung volume below 80 percent of normal indicates restrictive disease. Readings in the 70s indicate slight restrictive disease; in the 60s moderate disease; in the 50s, serious disease; below 50 percent, severe restrictive disease.

FIGURE 11. Summary and tutorial explanations in SAL.

program's reasoning if errors or uncertainty arise. A self-revealing system can expose its structure and activity to external scrutiny for validation and verification. Explanations also inspire confidence in the performance and reasoning processes of the system. They may persuade users that unexpected advice is appropriate and educate users in areas where the users's knowledge may be weak. In many areas, the formalization of knowledge, a process inherent in developing an expert system, has provided a clearer picture and new insights into the domain.

The diverse roles that explanations play complicate the operation of an explainer. The system must allow users to dynamically examine the reasoning history and underlying knowledge at various levels of detail. The multiple functions imply exposition of different types of knowledge according to what is being explained and for what purpose. When asked about *Pulmonary Function Tests*, SAL provides a user who is seeking tutorial information with a description of the meaning of various test results as shown in Fig. 11. A user whose interest is to determine what aspects of the active context caused the conclusion to be made is shown those premises from the particular rule that were responsible for that rule's firing, as seen in Fig. 12.

The pulmonary-function-test of John Smith does show "moderate restrictive disease".

Explain *(Requests general explanation of conclusion)*

I did conclude that the pulmonary-function-test of John Smith does show "moderate restrictive disease" because vital capacity or lung volume readings in the 60s indicate moderate restrictive disease.

Premises *(Requests information about the context active at the time the conclusion was made)*

The most-important pft-finding is equal to FIFTY-TO-SIXTY-PERCENT-LUNG-VOLUME.

FIGURE 12. Context explication in SAL.

The actual space of explanation needs is viewed in its entirety along each of the dimensions just explored. Although each spans the complete space, the viewpoints provide different foci and implications. For particular systems there are specific points or regions in this space that are of special interest; for example, a chemistry tutor does not share the same mix of needs as a job scheduler or a weather forecaster. Since this is domain and problem specific, explanation mechanisms should address the full spectrum of needs, providing system builders with the tools they need to tailor the system. The supporting mechanisms implemented for SAL are therefore not oriented toward individual needs, but are more general in their applicability. These characterizations provide the basis for assessment techniques to measure the competence of an explainer. In addition to their assessment value, the characterizations of users' needs provide guidance in building supports for accomplishing the task of recognizing explanation requests.

5.2. Incorporating Information Resources

The selection of information for use in an explanation requires the system to determine what knowledge the user lacks that, if communicated, would achieve the desired understanding. Each type of information potentially required by an explanation facility must be available and organized effectively. A methodology for explanation must recognize the importance of explanatory information and define constructs for recording, retrieving, and manipulating such data.

A methodological approach suggests explicit recognition and identification of the types and levels of information that must be available and specifying constructs that support the acquisition, representation, organization, and maintenance of the information. The support constructs must handle: knowledge about the domain and the specific problem; problem-solving methods; user interactions; and how the system performs its functions. They must also record the dynamic reasoning process and the chain of reasoning followed by the inference engine as it works on a given problem. The method should encourage awareness of the importance of explanatory information, especially that which is not necessary for task performance.

Explanatory information is often not required by the system for producing its conclusions. Most expert systems neglect non-task-performance information completely. Such *compiled out* underlying information remains only implicitly in the performance program.[51] A primary motivation for Swartout's automatic-programming aproach to explanation[72] is capturing this type of information explicitly. Clancey[9] describes an explanation strategy that requires other types of nonperformance information, such as metaknowledge that describes high-level strategies used by the system or deep knowledge that represents fundamental principles or causal relationships among objects in the system. Techniques employing such knowledge are effective in areas within medicine, electronics, engineering, and other application domains that have well-understood underlying models.

The methodology must account for complex control structures, at least as much as is found in contemporary high-level programming languages. Traditionally, expert system building tools provide limited control structures: simple rules of production and a forward- or backward-chaining rule interpreter. Newer languages, such as ROSIE 3.0,[38] allow nested recursive procedures, iterative rulesets, and other sophisticated control contructs. Explanation facilities for typical rule-based languages, such as EMYCIN[74] or M.1[12] are not sufficient to explain the behavior of tools with more complex control processes. A methodology for generating understandable software must apply to systems built using sophisticated language constructs.

Explanations depend on the *context* in which they occur. For a KBS, this context includes: a model of the user, including assumptions of the user's knowledge and beliefs; the active state of the reasoning; and the active state of the explanation dialogue. The user model and active state of explanation dialogue support the system's recognition and assessment of the user's needs and the selection and reformulation of information to best address those needs. The active state of the reasoning is a primary source of explanatory information.

Context is not static; thus the construction of explanations must also be dynamic. The recognition of the need for *dynamic explanation construction* is a significant departure from previous work in explanation and the type of explanation used for conventional software systems. In particular, the pervasive "canned text" approach is not effective precisely because it is not dynamic and does not permit the flexibility necessary to produce uniquely tailored explanations appropriate to an immediate context.

Appropriate constructs and techniques for the encoding, recording, and attachment of explanatory information must be available to system builders. The ability of a system to answer a variety of questions or to provide diverse kinds of answers depends on whether or not it has access to a broad range of types of knowledge. The presence or absence of different knowledge types relates directly to the competence of an explainer.

We next detail the kinds of information needed to support the coverage goals, present means for encoding such information, and describe provisions for associating explanatory and task-performance knowledge.

5.2.1. The Knowledge Resource. At the moment an event occurs, information about that event and the context surrounding that event must be recorded. (In our example the event is SAL making the conclusion that age is an important factor in the particular case.) Such information adds to the

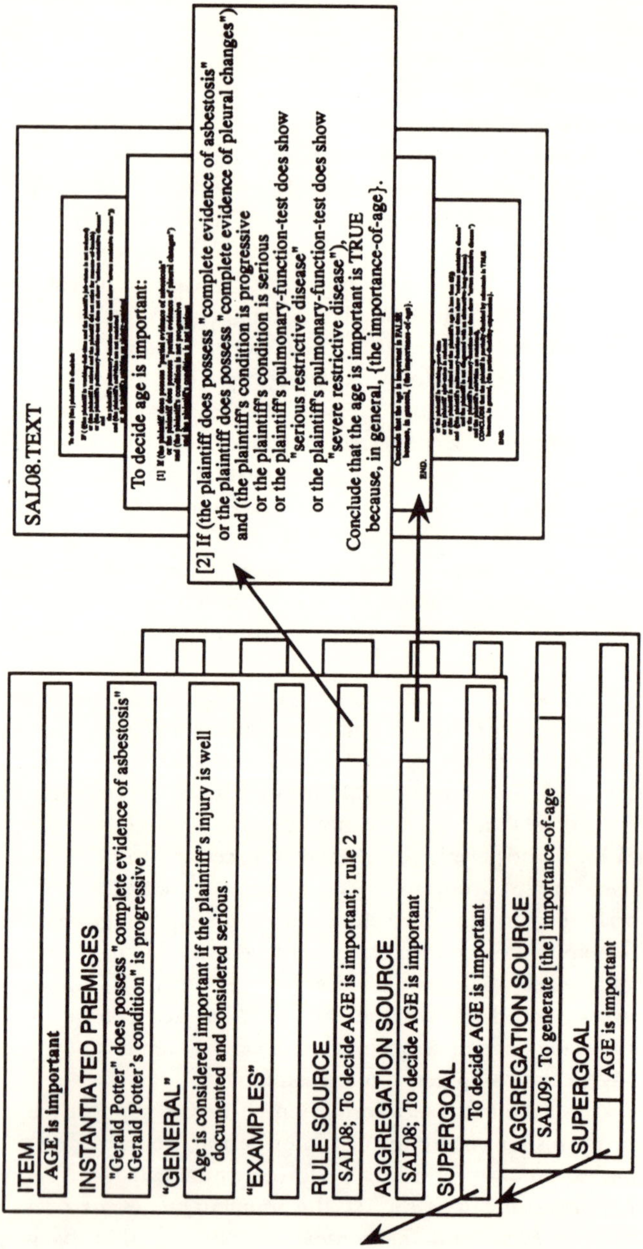

FIGURE 13. History explanation structure (SAL).

existing resource of explanatory knowledge that is drawn upon in constructing explanations. In SAL, much of this information is recorded using a history explanation structure, like that shown in Fig. 13.

First, the system decides whether or not to record any explanatory information for the event, i.e., it looks to see if the inference has been distinguished for explanation. In our example, the conclusion was the result of a successful invocation of rule 2 from the rule set "To decide AGE is important" (the text of which is shown on the right-hand side of Fig. 13).

The information stored in the history explanation structure includes the conclusion itself (ITEM), the knowledge structure that made the conclusion (RULE), the local context that enabled the structure to make the conclusion (INSTANTIATED PREMISES), the reasons why the structure was activated (SUPERGOAL and its link), the aggregation of which the structure is a part (AGGREGATION), and how to locate the source code for the structure and its aggregation (RULE SOURCE and AGGREGATION SOURCE).

Figure 13 shows the particular information recorded for the conclusion that "AGE is important." The immediate supergoal link points back to the history list element for the event that activated this rule. Here it is the ruleset "To decide AGE is important." The full reasoning context is realized through the chain of supergoals. For example, by following the link once more we would find that rule 2 from the rule set "To generate the age-factor of the plaintiff" (shown in Fig. 14) is responsible for establishing the higher goal. Note that this rule is also responsible for activating the rule set "To generate the age-range of an age."

At this time, general and specific explanations may also be generated, thereby taking advantage of the full knowledge of the current state of processing. Alternatively, this processing may be deferred until a request for the explanation is made, substituting a delayed goal for an active one.

In our example the explanation structure attached to the conclusion clause causes the system to post an explanation goal of determining the value of "the importance-of-age." This goal results in the invocation of the rule set "To generate the importance-of-age." The particular explanation form

[2] If age is important, select the age-range of (the plaintiff's age):

<LESS-THAN-50-YEARS-OF-AGE>	Produce 1.33 as the age-factor of the plaintiff;
<50-TO-54-YEARS-OF-AGE>	Produce 1.20 as the age-factor of the plaintiff;
<55-TO-59-YEARS-OF-AGE>	Produce 1.00 as the age-factor of the plaintiff;
<60-TO-64-YEARS-OF-AGE>	Produce 0.80 as the age-factor of the plaintiff;
<65-TO-69-YEARS-OF-AGE>	Produce 0.67 as the age-factor of the plaintiff;
<OVER-69-YEARS-OF-AGE>	Produce 0.50 as the age-factor of the plaintiff;
Default:	Produce 1.00 as the age-factor of the plaintiff.

FIGURE 14. Rule 2 from To Generate the Age-Factor of the Plaintiff.

[1] Select the user-level:	
<END USER>	Produce the brief-explanation of age as the importance-of-age;
<SYSTEM BUILDER>	Produce the detailed-explanation of age as the importance-of-age;
<DOMAIN EXPERT>	Produce the expert-level-explanation of age as the importance-of-age;
Default:	Produce the brief-explanation of age as the importance-of-age.

FIGURE 15. Rule 2 from To Generate the Importance-of-Age.

used indicates that the resulting explanation should be associated with the GENERAL facet of the history structure. The posted explanation goal is treated like any other goal—the reasoning engine applies its power as though this were furthering a task-performance objective. The explanation is generated and linked into the knowledge base by association with the history explanation structure. The system could also choose to modify working memory or to display an explanation.

The explanatory text returned by the generator rule set depends upon the active context. Rule 1 from "To generate the importance-of-age" shown in Fig. 15 employs a selection criterion to determine the level of detail desired. The selected action posts an additional goal that causes the derivation of the appropriate explanation. The explanation further depends on whether or not age is important. If age is important, the generator ruleset "To generate the brief-explanation of age" returns:

Age is considered important if the plaintiff's injury is well documented and considered serious.

If age is not important, the generator returns either

Age is considered unimportant because the plaintiff's injury is not well documented.

or

Age is considered unimportant because the plaintiff's injury is not considered serious.

depending on the instantiated premises of the original rule.

5.2.2. Types of Knowledge. Information used to reveal the knowledge and behavior of a software system may be viewed along the following dimensions: System Characteristics, Level, Specificity, Conceptual/Performance, and Structure. A system must have access to a variety of types of knowledge in order to answer diverse questions or to provide different kinds of answers.

System characteristics. A convenient system for categorizing explanatory information is by the related system characteristics they concern. These include: Interface, Knowledge, Behavior, Effector, and Justification.

Interface characteristics concern how people and the system interact. A key element is informing the user of expectations that the system has about the dialogue. *Knowledge* characteristics pertain to the structure and content of the information used by the system, including working memory, domain-specific knowledge, and general problem-solving knowledge embodied in the inference engine. *Behavior* characteristics relate to what the system did, is doing, or is capable of doing. *Effector* characteristics refer to how the behavior is brought about, i.e., the means for its accomplishment. *Justification* characteristics include the system's motivation and rationale for its activities and conclusions—why they are reasonable and credible.

Explanations usually combine several of these aspects of knowledge. An explainer must have access to such information, at whatever levels of detail necessary to achieve the desired goals. The following discussion describes these categories of explanatory knowledge and demonstrates an application of each.

Interface characteristics are aspects of how to interact with the system, especially making plain the expectations that the system has about the dialogue. Examples of explanations involving interface characteristics include: informing the user of acceptable responses to system-issued requests, advising the user as to the interpretation of possible input, and explaining the meaning of a given term. As shown in Fig. 16, SAL can display a menu of possible responses for the user to select from that SAL is prepared to consider. By elaborating upon possible user responses, as shown

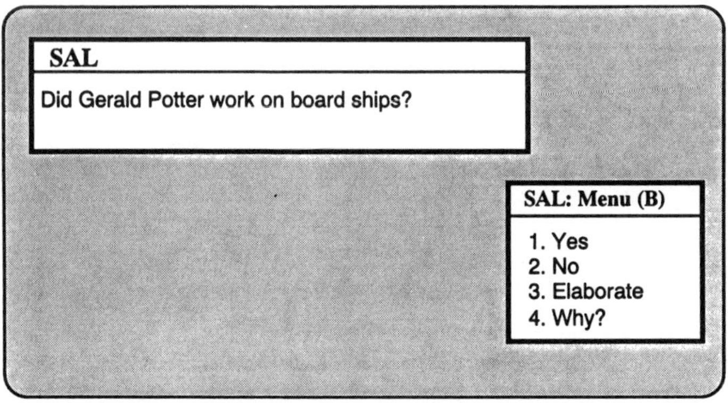

Displaying a pop-up selection menu to show acceptable responses to a question.

FIGURE 16. Use of selection menu in SAL.

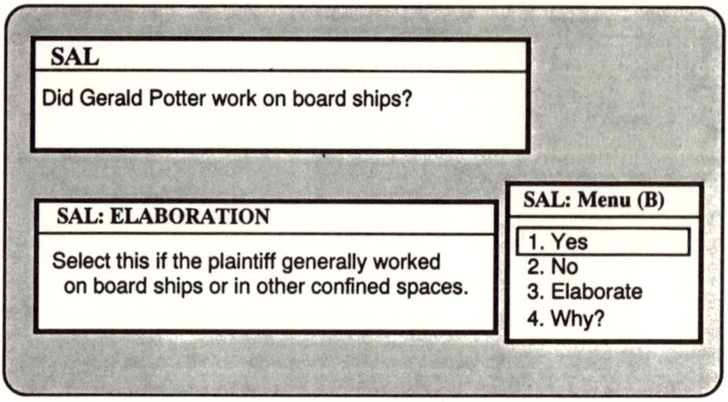

Elaborating on the interpretation of a "Yes" response.

FIGURE 17. Selection elaboration in SAL.

in Fig. 17, SAL advises the user as to the particular meaning that it will attribute to a given input. SAL explains the meaning of a term or phrase to affirm its interpretation, such as shown in Fig. 18.

The system must be able to explicate the *knowledge* it possesses, including consultation-specific, domain, and general information. This involves presenting users with select contents of working memory and the knowledge base as well as characteristics of the inference engine and support environment. Figure 19 shows explanations given by SAL about its static knowledge (above the line) and working memory (below the line).

The *behavior* characteristics tell what the system did, is doing, or is capable of doing. An explainer must have access to a historical record of

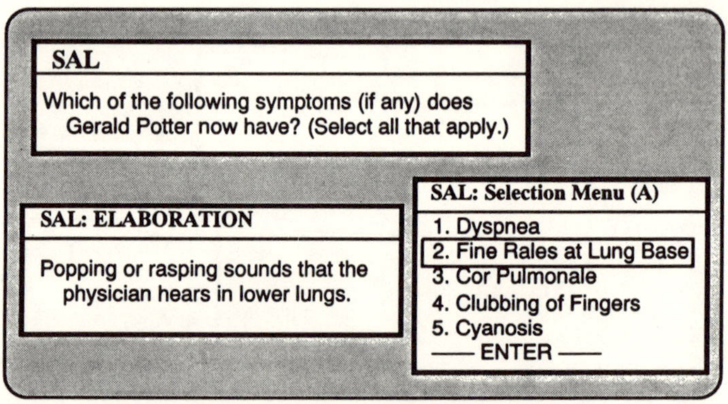

Describing the meaning of "Fine Rales at Lung Base."

FIGURE 18. Phrase amplification in SAL.

SAL: EXPLANATION

The value of the case is the sum of the amounts awarded for injury, disability, and aggravation, as adjusted by considerations of the personal characteristics of the plaintiff.

Here we find the value of the case to be 237300.0 DOLLARS by combining the injury value (100000 DOLLARS), the disability compensation (97750.0 DOLLARS) and the aggravation value (0 DOLLARS).

This total (197750.0 DOLLARS) is adjusted (multiplied) by the personal characteristic adjustment (1.2) yielding the final value.

FIGURE 19. Explanations from SAL's knowledge base and working memory.

actions and conclusions as well as the current state of a consultation. *Effector* characteristics encompass how the behavior is brought about—the means for its accomplishment. For this, an explainer must have access to control information and a trace of the operation of the system during consultation. Some motivational aspects are included, such as a specific behavior explained as a subgoal for achieving another. Figure 20 shows an explanation generated by SAL that uses both behavioral and effector knowledge to satisfy a user's desire to know why a given line of reasoning was pursued.

Justification is comprised of the motivation and rationale of the system's activities and conclusions, showing why the actions and decisions are credible. The knowledge may arise from a formal model of the underlying mechanism or directly from surface-level explanations provided by the informant. Figure 21 shows explanations generated by SAL to advise the user of the rationale for a given conclusion.

SAL: EXPLANATION

I needed the conclusion that "Gerald Potter" does have "a history of exposure to asbestos" in order to conclude that "Gerald Potter" does possess "complete evidence of asbestosis"

We consider evidence to be complete if the plaintiff has a history of exposure to asbestos, symptoms of asbestosis and both laboratory evidence and radiographic evidence of parenchymal asbestosis.

FIGURE 20. Behavioral/effector explanation in SAL.

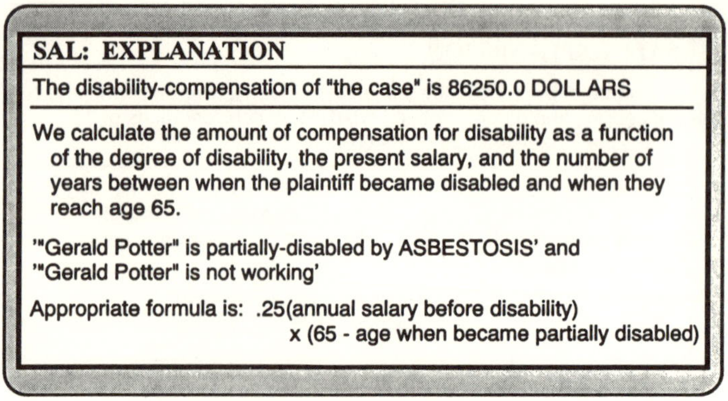

FIGURE 21. Justificatory explanation in SAL.

Level. The level of knowledge refers to the degree of granular detail and accounts for epistemological differences such as between "reasoning from first principles," "empirical knowledge," and "reasoning about reasoning." Figure 22 shows a set of levels of knowledge used in explanation. The SURFACE level corresponds to the basic level of knowledge used by a reasoner, and typically encodes associations between observations and possible significant mechanism states. A SURFACE-level rule from SAL is shown in Fig. 23. Knowledge at the MECHANISM level concerns the target entity being reasoned about. This includes fundamental domain

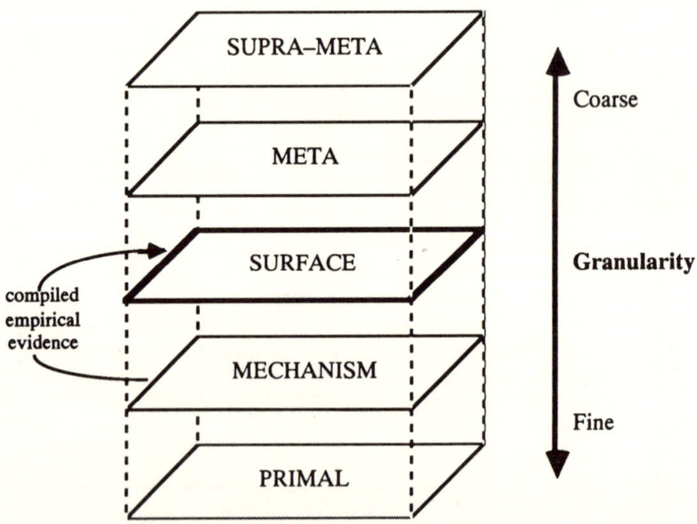

FIGURE 22. Levels of knowledge.

If the plaintiff does have (a history of exposure to asbestos) and the plaintiff does manifest radiographic-evidence for pleural-changes and the plaintiff does have dyspnea as a symptom and (the plaintiff does have laboratory-evidence of asbestosis or the plaintiff does demonstrate symptoms of asbestosis),

Conclude that the plaintiff does possess "complete evidence of pleural-changes", Also return.

FIGURE 23. SURFACE level: Evidence rule (SAL).

knowledge representing the mechanism under discussion and is the level of deep models. Figure 24 shows a portion of the "exposure history" semantic network that constitutes part of the deep model within the SAL system. META-level knowledge is comprised of knowledge about how to use SURFACE-level knowledge, i.e., how to control reasoning. This includes strategic and planning information,[13] and is used by SAL to manipulate explanatory information as well as domain knowledge. The PRIMAL level corresponds to basic first principles of a general nature. At the other extreme is the SUPRA-META level, which accounts for knowledge about the organization an application of metaknowledge.

Most expert systems represent knowledge at the SURFACE level because (1) typical expert system domains rely heavily on empirical knowledge and experiential expertise, and (2) experts generally express their reasoning at this level. For domains where a formal model of the underlying mechanism exists, formal representation of such a model is a more robust source of explanatory information. These models can be used to answer a broad range of questions whose foci are slightly different from one another. Deep models are invaluable where they are available.[22, 39, 72]

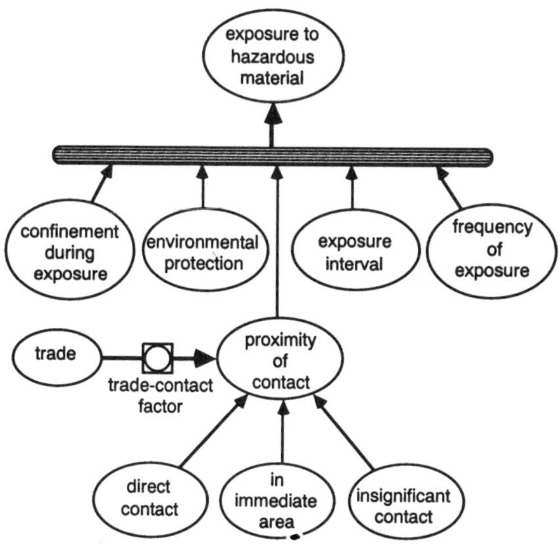

FIGURE 24. MECHANISM level: Exposure history network (SAL).

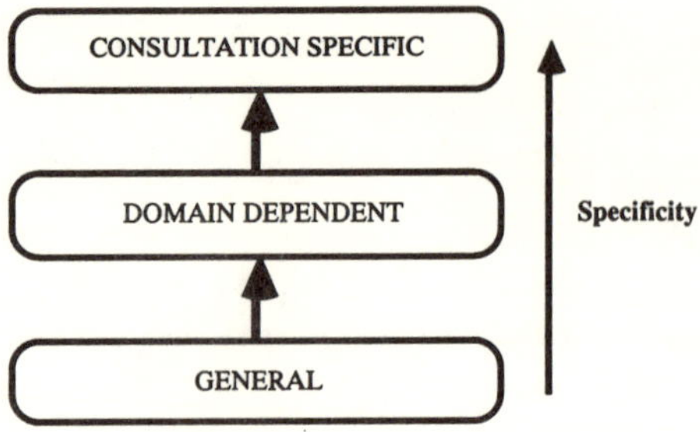

FIGURE 25. Specificity of knowledge.

Many expert system task domains, like law, lack such models. It is often precisely this absence that suggests the application of knowledge-based-system technology in the first place. While the development of an explicit domain model is a laudable goal, it is not necessary nor generally feasible for the construction of an expert system. Instead, the surface-level explanations provided directly by informants can be adopted, i.e., the rationale that domain experts give when queried directly. Thus, one may achieve the desired level of explanatory performance without incurring an excessive knowledge-acquisition effort required to explicate and codify the domain. In contrast, the Explainable Expert Systems approach taken by Swartout and co-workers,[(50)] an extension of his XPLAIN system, is predicated on the explication of such a model.

Specificity. Specificity relates to the general applicability of knowledge and addresses the differences between general information, information about a domain, and data specific to a consultation (see Fig. 25). In most knowledge-based systems, some general information is embedded in the inference engine. The permanent knowledge base is comprised of general and domain-dependent knowledge. Consultation-specific information makes up the bulk of working memory and includes the case data and set of assertions that a consultant comes to believe about a particular case either through observation or reasoning. Figure 26 shows data of several different degrees of specificity from the SAL system.

Conceptual/Performance. The distinction between *conceptual* and *performance* knowledge contrasts the *usage* of information. Conceptual knowledge relates to information about the case, the domain, and problem-solving techniques. On the other hand, performance knowledge is concerned with the operation and functioning of the system—what is often referred to as control. Figure 27 shows some conceptual knowledge from the SAL system. The diagram depicts relationships among concepts that is realized by the ROSIE code below it. Figure 28 shows performance knowledge, also from

Case Specific:

```
"John Smith's blood gas test" does show "reduced CO diffusion capacity".
```

Domain Dependent:

```
If the plaintiff's activities are not restricted,
        Conclude that the plaintiff is not disabled-for-work.
```

General:

```
To generate an answer to a question:
        Output {the question}.
        Read {anything (bind the response), cr}.
        Produce the response.
End.
```

FIGURE 26. Various specificities of knowledge (SAL).

the SAL system. The diagram indicates the flow of control implemented by the ROSIE code following it.

Structure. Information required for explanation may be described by its relation to knowledge structures, as shown in Table 6. The following discussion concerns implications for the incorporation and utilization of these types of explanatory knowledge.

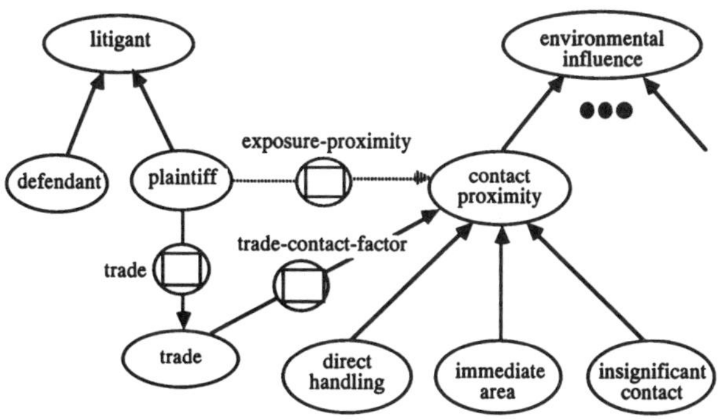


```
Assert any defendant is a litigant.
Assert any plaintiff is a litigant.
Assert any contact proximity is an environmental influence.
Assert any direct-handling is a contact proximity.
Assert any immediate-area is a contact proximity.
Assert any insignificant-contact is a contact proximity.
Assert any exposure-proximity of a plaintiff is a contact proximity.
Assert any trade-contact-factor of any trade is a contact proximity.

To generate an exposure-proximity of a plaintiff:
  Let the exposure-proximity of the plaintiff be the trade-contact-factor of (the plaintiff's trade).
End.
```

FIGURE 27. Conceptual knowledge (SAL).

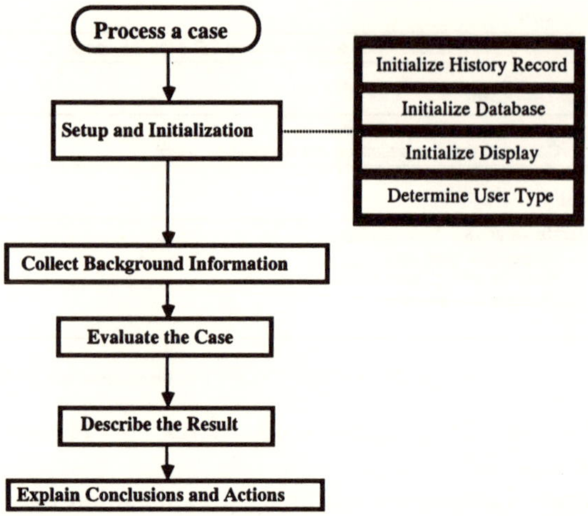


```
To process-a-case:
        :
  Cancel each of XPL.PROFILE and XPL.RSFORMS.
  Clear database.
  Clear every window.
  Close every window and switch to SAL window.
        :
  DO Collect background...
  DO Evaluate the case...
  DO Describe the case...
End.
```

FIGURE 28. Performance knowledge (SAL).

Source code is information possessed by the system expressed in the original format in which it was presented by the system builders. The most accurate description of a structure's operation in isolation is the actual source code, not massaged or translated. The code also provides key infor-

TABLE 6
Explanatory Knowledge Types

Source code	Knowledge in the format in which it is presented to the system
Deep model	Model of the mechanism addressed by the system
Activity	Historical record of events that occurred and their relationships to one another; an extended trace including all levels of knowledge
Rationale	The reason why a given structure exists; its justification
Intent (purpose)	What a given structure is supposed to accomplish
Function	What a given structure does; how it accomplishes its purpose
Operation	How a given structure accomplishes its function

```
[(anticipate*o (toxicity*f (drug*r !pv)))
 [input↑ #e [(dose*r (drug*r !pv)) #m variable]]
 [output↑ #e  [((dose*r (drug*r !pv))*f
                  (adjusted*for (get-all-matches*c aspect1,)))
                #m variable
                #t (dose*r (drug*r !pv))]]]
```

FIGURE 29. Source code (digitalis advisor).

mation about the representation of knowledge when viewed in combination with the function and intent of the structure.

The value and utility of source code for explanation depends on the code being understandable to the user. The principle of *understandable code* requires the program source to be easily read and comprehended by users who are not deeply steeped in computer programming nor have substantial background or experience with the knowledge engineering language. Program source code is not generally understandable to any people but those who have spent significant effort in learning the language and who have background knowledge in computer science and programming. For example, examine the source code from the Digitalis Advisor[71] shown in Fig. 29. This is the only level of code understandability provided by most expert system building tools.

An attempt at bridging the gap between computer programs and programming-language-naïve users is to provide machine translation from the source code into a more palatable form. Two problems with this approach are (1) the difficulty of defining and performing the translation, and (2) the lack of verifiability of the validity of the translation.

ROSIE, selected for SAL's implementation, is English-like and readily understood. A primary goal of the development of ROSIE was to allow system builders to write understandable code.[78] Sowizral and Kipps[68] claim that "a person unfamiliar with programming languages can read and undersand a ROSIE program almost as if it were in English."

However, ROSIE programs are not necessarily any more readable than those written in any other programming language. In fact, because of its natural-language-like flavor, it is possible to write programs that are more likely to be *misunderstood.* The common interpretation of text is not necessarily consistent with the interpretation that the ROSIE interpreter has. The problem with using ROSIE is the difficulty in writing readable code. Unless the code is carefully written, ROSIE-inexperienced readers may find (1) that the program code is opaque to them, and (2) that their intuitive interpretation of the code does not coincide with the actual interpretation made by the ROSIE interpreter.

SAL's source code has substantial explanation power as a result of writing and reviewing code from the perspective of making the system's

```
If the plaintiff does exhibit pleural-findings in test
  and the plaintiff's pleural-findings are not explained by alternatives
  and the plaintiff's pleural-findings are not contested,
Conclude that the plaintiff does manifest radiographic-evidence for pleural-changes.
```

FIGURE 30. Source code (SAL).

interpretation coincide with that of a naïve user. By using a restricted subset of the ROSIE language and carefully crafting the rules and rule sets, readability of the resultant code is enhanced. For example, compare the Digitalis Advisor source (Fig. 29) with that from SAL (Fig. 30). In addition, since knowledge-base assertions, conclusions, and actions are basic ROSIE clauses they too are readable and understandable. Thus, through careful use of

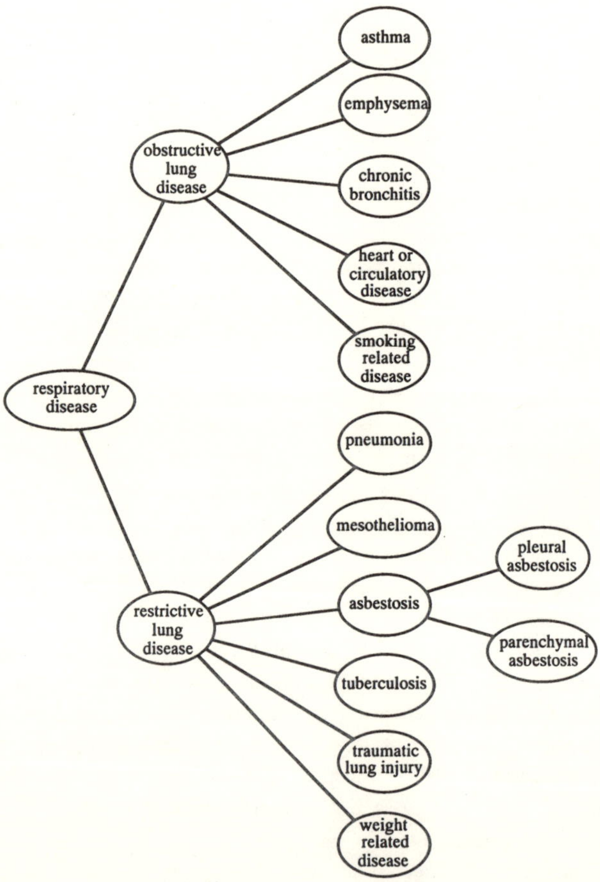

FIGURE 31. Deep model: Respiratory disease taxonomy (SAL).

ROSIE, SAL is able to present the actual structures stored in the history list to the user and use the rules themselves as explanations of their function. SAL's source code is by and large understandable by all of its intended users.

Deep models correspond to the MECHANISM-level knowledge already discussed. Recall that typical expert system problems and domains, such as those addressed by SAL, do not have comprehensive deep models available. A sample of basic MECHANISM-level knowledge incorporated into SAL is shown in Fig. 31. This knowledge demonstrates the ability to handle deep knowledge for explanation and was not needed to drive the task-performance aspects of the program.

Activity refers to a record or trace of the execution itself, a kind of performance knowledge that integrates conceptual information. Trace information comprises the recorded history of the system's reasoning process for a particular case. Two aspects of such a trace are (1) the internal architecture of the individual data structures which controls the kinds of information stored, and (2) the organization of the structures into a coherent history of the chain of reasoning. SAL uses the history mechanism to automatically record and organize the information about an action or conclusion and store it on a history list. The history list is accessible by the system itself, a prerequisite for a self-revealing system.

The remaining four types of information relate a knowledge structure to its *usage* in the system: *rationale* is the justification; *intent*, the purpose; *function*, the way the purpose is achieved; and *operation*, how the function is accomplished. Consider the structure shown in Fig. 32, a rule from the rule set used by SAL to decide whether or not a plaintiff is "totally disabled" by asbestosis. The *intent* of this structure is to determine if the plaintiff is totally disabled by asbestosis. The *rationale* is complex and includes items such as the medical implications of "cor pulmonale" and more obvious relationships such as the requirement that a totally disabled person's activities must be restricted. The *function* involves looking at the plaintiff's characteristics, such as the plaintiff's retirement status and the pulmonary function test results, evaluating the items collectively according to the given conjunctions. The *operation* of this rule involves examining first working memory then the knowledge base to determine the truth values of the individual clauses in the order in which they are specified. If any clause of a disjunct is true,

```
If (the plaintiff is not working or (the plaintiff is retired and the plaintiff's age is less than 65))
  and (the plaintiff's pulmonary-function-test does show "severe restrictive disease" or COR-
          PULMONALE is a symptom of the plaintiff)
  and the plaintiff is disabled-for-work
  and the plaintiff's activities are restricted,
Conclude that the plaintiff is totally-disabled by asbestosis.
```

FIGURE 32. A rule for determining disability (SAL).

clauses appearing after that clause are not tested and the entire disjunct is considered TRUE. If the entire conjunct evaluates to TRUE, the indicated conclusion is posted to working memory and the ruleset returns the value TRUE to whatever rule invoked it.

Each type of information presented in the preceding discussion is necessary for achieving the full set of goals developed in Step 1. Such information must be collected, encoded in the knowledge base, and accessible by the explainer.

5.2.3. Integration—Explanation Structures. Once the necessary information has been identified and accessible knowledge sources located, for this knowledge to be exploited it must be (1) encoded in the representation paradigm, (2) linked to the related task-performance knowledge, and (3) made accessible to the explainer under that organization. That is, the knowledge representation tool or knowledge engineering language must allow for representing explanatory knowledge, permit the system builder to associate it with the appropriate performance information, and support the explanation facility's access via those associations.

Most expert system building tools lack the ability to conveniently represent explanatory information or to associate such information with that used for task performance. This lack is largely responsible for the inadequate explanations of the resultant systems: without supporting tools, system builders are unlikely to improve explanation competence. Although many expert system tools possess the basic capability to record some form of execution trace, the limited recorded information and inadequate manipulation and presentation of that information neutralizes the value of even this restricted explanation ability. Lewis and Lynch[45] report that typical "explanation traces tend to be bulky and involved—even for simple rule sets...—and unsophisticated users sometimes find them difficult to interpret." Traces typically consist of the names of rules listed in the order they fired. The links between such firings, database assertions, the goals of the system, and the conditions responsible for the activations are not presented nor maintained.

In a very limited sense, the working memory of the system shows the result of the reasoning process and some intermediate conclusions and assertions. However, the only information recorded is the fact of each assertion. When the assertion was made and what caused the system to make it are not recorded. Furthermore, information about any assertions that were made and subsequently retracted is lost. Some systems allow the use of debugging tools provided by the support environment. The information from such tools is typically only available by interrupting the expert system during execution and then only includes procedure calls and environment parameters active at the moment. Basic expert system building tools provide insufficient facilities for recording and manipulating the information necessary for competent explanation.

The main problem with attempting to retrofit explanation capability into an existing expert system building tool is the lack of constructs for representing the explanatory information and for attaching such information to other knowledge in the knowledge base. For example, if the system must be able to justify the application of a particular heuristic to a class of generic problems, that justificatory information must be available—whether it comes from a deep model of the mechanism, a corpus of knowledge in the field, or directly from a domain expert's experience. If a deep model of the domain exists, the expert system building tool must provide the necessary representation vehicle for encoding that model and for the explanation facility to access that model in ways that facilitate explanation generation. If a deep model is not available then mechanisms for recording and attaching surface-level explanations need to be provided.

SAL uses *explanation structures* and *multilevel association* to support the recording and attachment of explanatory information. The term *explanatation structures* refers to available constructs for representing explanatory information—that is, for encoding and recording information that is incorporated into the knowledge base for its explanatory function. *Multilevel association* refers to the integration of that information with the task-oriented knowledge. The association accounts for the organization, storage, and retrieval of explanatory information.

Generalized explanation structures for recording and organizing explanatory information provide a robust foundation for explanation. Some of the information is provided directly by the system builder while other information is generated by the system itself, utilizing inferenced generation. To be useful, explanatory information must be associated with the relevant task-performance knowledge, in the former case by explicit direction of the system builder and in the latter as determined by the course of the system's reasoning. Thus the recording of explanatory information is both static—where explanations and justifications are being encoded with the task knowledge itself—and dynamic—in terms of the active state of reasoning and the explanation dialogue. That is, information may be statically determined at system building time or dynamically determined during execution or during explanation.

The specific types of explanatory information recorded and used by SAL are shown in Fig. 33. Event information relates to activities performed by the inference engine. The other information is concerned with knowledge structures in the domain knowledge base or working memory. The distinction between the categories of EVENT and KNOWLEDGE STRUCTURE is for convenience of discussion only. For example, the exemplification of the

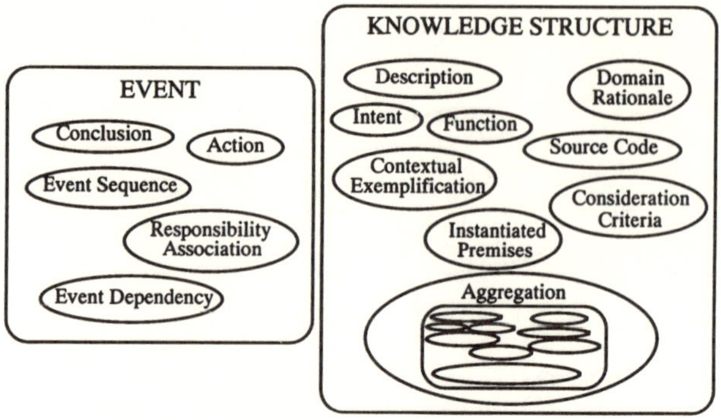

FIGURE 33. Types of explanatory information.

application of a structure involves both that structure and the event with which it was involved.

The EVENT itself is simply what activity occurred, be it a *conclusion* such as an assertion to working memory or an *action* such as the invocation of a procedure. *Responsibility association* indicates what particular knowledge structure was responsible for the event, i.e., whose execution/activation resulted in the given event. *Event sequence* refers to the order in which events occurred. *Event dependency* records which events are responsible for the occurence of others. The *sequence* and *dependency* information link the overall reasoning chain and also provide information as to what caused given structures to be contemplated.

Associated with each KNOWLEDGE STRUCTURE is a *description* of that structure itself which gives the general picture. The *intent, function,* and *domain rationale* are precisely as described in the preceding section. *Contextual exemplification* is an example of application of the structure, a generic entity, to the specific case at hand—that is, how the structure fits into the context of the current consultation at the time it was invoked. *Source code* is the knowledge structure expressed in the original format in which it was presented by the system builder. As described earlier in this chapter, this is the most accurate description of a structure's operation considered out of the active context and also provides key information about the representation of knowledge when viewed in combination with the function and intent of the structure. The *consideration criteria* indicate under what conditions the structure is appropriate to activate. Examples include the "condition" part of a "condition/action pair," the "switch" of a "case statement," and the "if" clause of an "if–then" control structure. The *instantiated premises* are the particular data that satisfied the consideration criteria resulting in the activation of the structure. Finally, *aggregation* information links the structure to the aggregate structure of which it is a part. That aggegrate, itself a knowledge structure, would also have the same information—namely, description, intent, function, etc.—describing how the structures function collectively.

5.2.4. Explanation Attachment. Explanation attachment refers to the association of explanatory information with task-performance knowledge. Standard expert system building tools do not provide the constructs necessary to attach explanatory information. New language forms are needed to provide links between task-oriented and explanatory information. For example, *explanation forms* added to the ROSIE grammar provide places to hold explanatory information. The generic forms record both static information and instructions to the system regarding the inferences necessary to generate dynamic explanations. The syntactic structure of the most general explanation form is

in general, *explanatory-information*;
  for example, *explanatory-information*;
  for example, *explanatory-information...*

where *explanatory-information* is any arbitrary ROSIE pattern. An arbitrary number (zero or more) of "for example" clauses are allowed. Details of the explanation attachement mechanisms added to ROSIE are described in Ref. 79.

Another class of explanation forms are those associated with input and output. Explanatory information can be associated with user prompts and with menu-selectable responses. In SAL these attachments are used to tell the user why information is being requested, to elaborate on the meaning or intent of a question, to provide assistance in understanding how a question should be aswered, and to explicate the meanings of possibly unfamiliar or misunderstood terms.

The levels at which explanations can be attached in the knowledge base affect the types of constructs that the system can explain and the system's ability to produce explanations in varying amounts of detail. *Multilevel association* allows system builders to attach explanatory information at arbitrary levels in the knowledge-base framework. The attachment syntax for SAL varies with the level of association but is usually indicated by preceding the explanation form with the conjunction "because."

The basic requirements of Step 2, the information resource aspect of the methodology, are that there be appropriate constructs and techniques available for the encoding, recording, and attachment of explanatory information. This explicitly recognizes that there is information that is not necessary nor even useful for the task performance part of the system. The information must be associable with the relevant portions of the knowledge base and interpretable by the inference engine for recording that information along with a historical trace of the chain of reasoning and for dynamically constructing explanatory text. SAL employs new structures and mechanisms for recording, organizing, retrieving, and manipulating the necessary explanatory information.

5.3. Performance–Explanation Integration

The remaining aspect of the methodology governs the generation and presentation of explanations. This covers what the explainer does with the selected information to formulate an explanation and the nature of the interaction by which the explanation is communicated to the user. Considerations of structure, language, and media are all relevant. Discourse knowledge may be brought into play to facilitate the transfer of the desired information. Assuming that the explanatory information is available, the explainer must filter that information to derive the key pieces that will best serve the user's needs. This may require knowledge representation structures that relate the information in ways that support the explanation task with little bearing on the performance program.

A self-revealing system satisfies a user's need for explanatory information by navigating through the history structures, working memory, and the static knowledge base; selecting the appropriate information; reformulating it for presentation; and displaying the result. For example, selecting the "Why?" item from the explanation-option menu posts an "explain" goal for the item found in the history list by following the SUPERGOAL link of the current structure. Summary and tutorial functions are handled similarly by posting a "summarize" goal.

The final step of REVEAL concerns integrating the behaviors necessary for a "self-revealing" system with those responsible for problem solving. The explanation process consists of the subtasks shown in Fig. 34. Explanation requires support mechanisms to (1) recognize and identify a need for explanation, (2) select information to satisfy that need and reformulate this information with respect to active context, and (3) display the resulting reformulation. Integrating these with basic task performance enhances explanatory power and provides a cohesive user interface.

The information resources previously described provide the basic knowl-

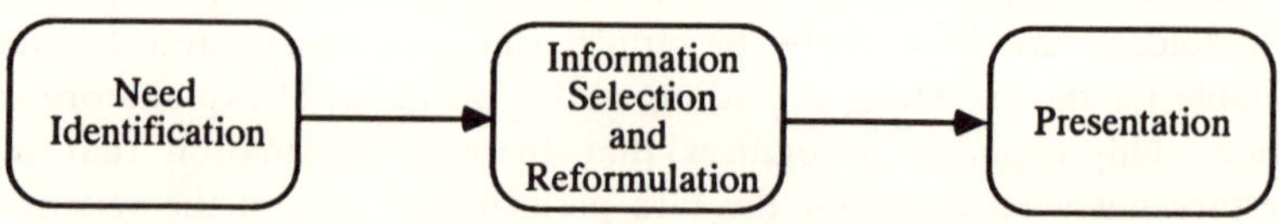

FIGURE 34. Explanation tasks.

ledge that users require to understand the operation and underlying rationale of a system. Unaided users must attempt to navigate their own way through this mass of data. Additional assistance from the system is necessary to select only the appropriate information and present it coherently. REVEAL addresses these concerns by specifying techniques and mechanisms for generating comprehensible explanations, that is, for presenting explanatory information lucidly to users.

Need identification refers to the explainer's task of determining the user's goals or requirements. The characterizations of users' needs presented earlier provide suitable bases for recognizing explanation needs. Expectations derived from user and interaction models (described in this section) aid the interpretation of user input in the active context.

Information selection involves mapping determined needs to information resources. It covers identifying the information to be transferred that will satisfy the established goal. For surface-level explanation, the linking or mapping between explanation need and explanatory information is generally explicit, established by the system builders *a priori*. Where a deeper model is available, more complex analysis is necessary to derive the appropriate linkage.

Information reformulation refers to the task of massaging the selected information into a form best suited to satisfying the identified need. This step involves filtering information on the basis of the active context. A primary consideration is the knowledge believed already in possession of the user. For example, SAL's user model for domain experts presumes that they are familiar with the general meanings of domain-specific terms. Working memory can retain information about the explanation dialogue so that repetitve sequences are avoided—e.g., different interpretations and responses may be given to subsequent entries of a physically identical request.

Presentation deals with the display of explanations to users and interaction between them and the system. Considerations of human cognition and the nature of discourse affect the ability of an explainer to transfer information to a listener. Generic window, menu, selection, and text-generation techniques may be used to address presentation concerns, including focus of attention, context awareness, expectation cognizance, and presentation of options.

These tasks are not monolithic. That is, each is not a single event or activity but is realized by a conglomeration of events and activities. The ingredient events occur at various times in the life-cycle of the system, from initial design through consultation. For example, determining the information most appropriate for answering a question such as "What is cyanosis?" may be done at system building time, effected by hard-coding relevant responses. Alternatively, the system may perform these activities *on-the-fly* at the time that the question is asked, using the active state of the knowledge base to fine-tune what information is most relevant and how best to communicate it to the user.

5.3.1. User and Interaction Models. The key to effective communication is matching the content and structure of exchanged messages with the intended goals. Good explanations satisfy the needs of the user. An explainer must take into account the knowledge and beliefs held by the user and the state of the current dialogue. Information selection, preparation, and presentation can be adjusted to achieve an adequate and efficient transfer of knowledge. A *user model* describes a user's characteristics, experience, and knowledge. An *interaction model* accounts for dialogue conventions and the current state of the communication in progress. In this section I discuss the features of user and interaction models that facilitate addressing explanation goals in an effective manner.

User models account for the differences among system users, including their goals, needs, knowledge of the domain and the problem, expert systems experience, etc. Figure 35 categorizes the features of a user model that affect the success of an explanation. A model of the user's knowledge and beliefs helps the system interpret explanation requests and determine what information to communicate. Knowledge of communication impediments and facilitators helps the explainer avoid ineffective presentation techniques such as display clutter. They also enable the exploitation of efficacious methods such as pop-up windows.

Different types of users typically require different kinds of explanation. An explainer must be capable of tailoring its explanations to the varying needs and characteristics of its users.[42, 47, 48, 71] For example, if the user is known to be well acquainted with the domain, the generated explanations can omit elaborations of domain concepts that would be necessary for less knowledgeable users. The user model provides the means for discriminating the appropriate aspects of explanation for the current user.

Users typically have multiple needs that change during the course of an

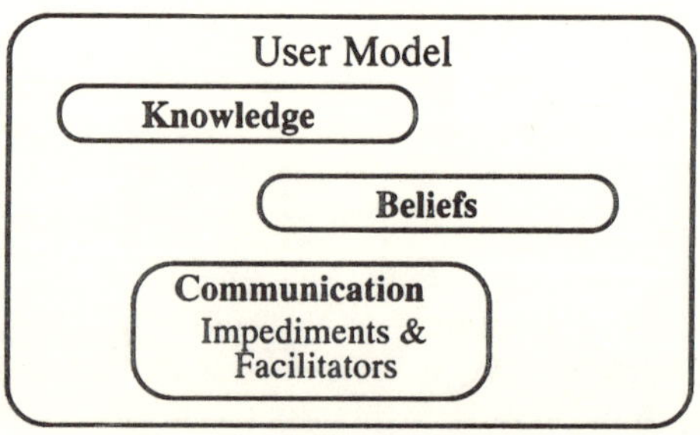

Figure 35. User model features.

> Is Gerald Potter's condition progressive?
> *Elaborate*
> It x-rays or lab tests have changed over time, showing a deterioration of the plaintiff's medical condition you should answer YES. If not, you should answer NO.

⋮

> Did Gerald Potter work on board ships?
> *Why?*
> If the plaintiff worked on board ships the severity of exposure is greater since it occurred in a confined space.

FIGURE 36. Elaborate and why–Match to user's needs (SAL).

interactive session with the system. For example, in response to very similar requests for data a user may at one point need to have a term explained and at another point wish to understand the implication of the reasoning being pursued. Figure 36 demonstrates this phenomenon in a dialogue between SAL and an end user. The user model must be dynamic within a consultation, responding to user feedback.

Effective explanation requires information about how human–computer communication works. The nature of discourse and the current state of a user–system dialogue are contained in an *interaction model.* Interaction models are rarely explicit. Discourse knowledge in SAL is integrated throughout the knowledge base. While this is sufficient for competent explanatory performance, an explicit model would facilitate the development of more robust and flexible techniques for human–computer interaction.*

A model of cooperative communication is shown in Fig. 37 (after Ref. 87). The communicative goals are established as a result of identifying the comprehender's needs to be fulfilled, such as being aware of the system's interpretation of a given response. The message realized via the communication vehicle (e.g., text or images) will be interpreted with respect to the comprehender's knowledge and thus must provide the necessary cues for that party to infer the desired information. In particular, the explainer must combine the intended goals with knowledge of what is being discussed and what the comprehender already knows. Winograd points out that "the design of an utterance depends critically on the producer's expectations that the comprehender will make use of knowledge and intelligent reasoning to find interpretations and fill in information that is not explicit."[87] In essence, the effectiveness of an explainer depends on the ability to exploit the

* A substantial foundation for further work on interaction models may be found in the research on *Dialogue-Games,*[44, 46] investigations of the *structure of conversations,*[5] and research on *discourse* and *focus.*[28–30]

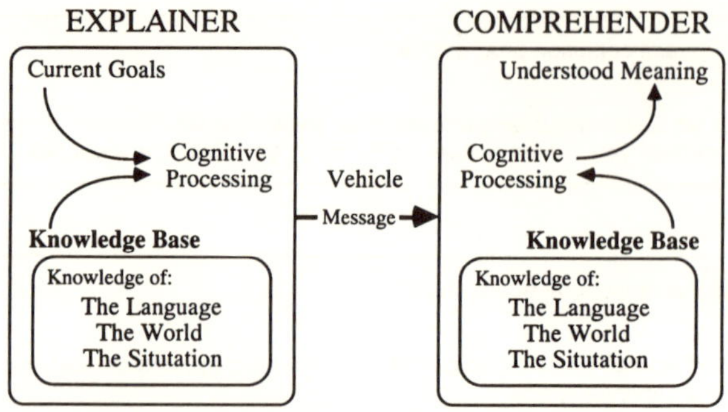

FIGURE 37. Model of cooperative communication.

knowledge possessed by the user to facilitate transfer of the desired information.

Interaction and user models drive an expectation-based explanation process in SAL. The use of expectations in processing is an effective way of representing active context.[3, 4, 20, 24, 57, 58, 61] Expectations derived from the models help identify the needs to be addressed by aiding the interpretation of user input as described in the following section.

5.3.2. Need Identification/Request Recognition. In addition to their assessment value, the characterizations of users' needs provide guidance in accomplishing the task of recognizing explanation requests. They suggest the design of support tools to facilitate explicitization and identification of users' requests. Need recognition is primarily a function of context, specifically of the particular user and events preceding the request. Requests may be made *explicitly* or *implicitly* by the user or may be *system initiated.*

Explicit request. The user may request an explanation via any input medium, such as typing in a question, making a menu selection, or pushing the "help key." Such action serves to indicate that an unfulfilled need exists and that the explainer is expected to provide information that satisfies the need. The request must be interpreted with respect to the current context because the same realization may have very different interpretations. For example, a generic "help" request should be interpreted as referring to an immediately preceding event. Similarly, the question "How is x-ray evidence used?" has a different interpretation if initiated by an informant, a programmer, or an end-user.

Implicit request. Sometimes what the user requests explicitly is not what the user wishes to know or not all that is desired to be known. Upon interpreting the user's explicit request in the active context, the system may determine that an unstated implicit goal is most appropriate to address. Alternatively, the request may be implicit in the interpretation of a non-requesting action taken by the user. The understanding of utterances as acts

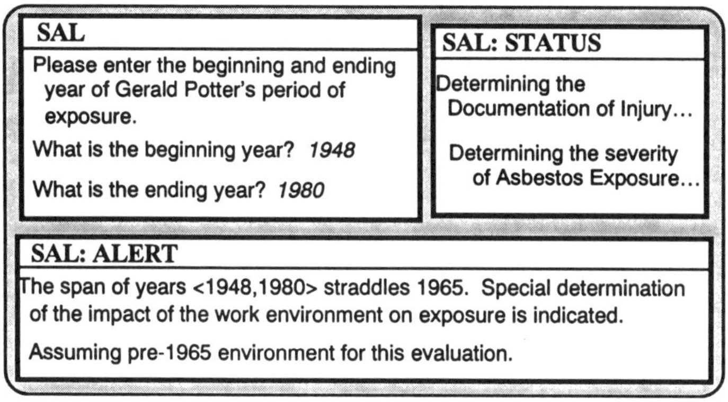

FIGURE 38. Status and alert windows (SAL).

rather than as representations has been explored under the rubric of *speech acts.*[1, 27, 63, 60]* The task of interpreting user communication as explanation requests follows the work on the theory of *indirect speech acts*[17, 34, 64] and on *whimperatives.*[26, 59]

System initiated. The conclusion that an explanation is called for may in fact not be in response to a request at all but determined independently by the explainer. An example of this kind of system-initiated explanation is a status message that informs the user of the current state of the consultation, as shown in Fig. 38. This may establish context for interpreting future information display, assure the user that the system is not waiting for a response or other input, or simply indicate that the system has not hung itself up in an infinite processing loop. SAL often employs selection menus that present possible responses. Without these, the user would have to ask for that information directly or, worse, may not realize that some response is possible or appropriate and may not know enough to request additional help. This covers the case where the user does not know that he or she does not know something and therefore does not realize that a clarification request should be made.

SAL employs a simple model of user–system interaction that accounts for the differing needs of different user types. This metaknowledge is used to initiate explanation and interpret requests. SAL pops-up a *status window*, an *alert window*, *selection menus*, and *elaboration windows* to inform users about current context and expectations. The function of the status window was described previously. An alert window is used to inform the user of anomalies, discontinuities, and limitations detected by SAL. Figure 38 shows SAL's use of an alert window to inform the user that the current case data

* The importance of speech act theory for language understanding by computer is discussed in depth in Ref. 86.

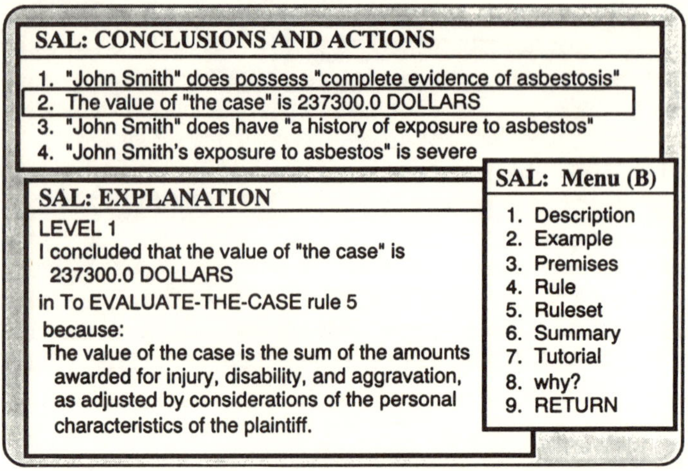

FIGURE 39. Menu of explanation options (SAL).

appear to fall outside of its expertise. By providing users with more knowledge about their possible actions and the interpretations that the system will apply to their input, they are in a better position to make their particular needs known and to interact with the system effectively.

Expectations derived from a user model help identify needs by aiding the interpretation of user input. Thus, when a domain expert asks why the system is requesting a particular datum the reply typically addresses a conceptual knowledge viewpoint. In contrast, when a programmer asks the same question, the reply may be performance related. SAL also uses expectations in generating menus of anticipated responses to its queries. Such menus inform the user of possible responses and facilitate their entry. For example, as shown in Fig. 39, SAL presents a number of explanation options made easily accessible to the user by popping-up a window with a mouse-sensitive menu.

5.3.3. Information Selection and Reformulation. A need exists for new approaches for selecting information to convey and reformulating such information to generate an explanation. A major flaw in current expert system explanation is swamping users with excessive amounts of information at too low a level of detail. Expert system building tools typically provide minimal explanation facility consisting of an execution trace, sometimes supplemented by a simple knowledge-base editor. (More often than not, a basic text editor serves the latter function.) These mechanisms suffer from three fundamental problems: (1) only the most detailed level is exposited, (2) all information is uniformly presented, and (3) nothing more than a trivial organization is supported.

A trace of execution does not provide for aggregation or summarization, but displays only the lowest level of detail. It does not support the selection of information on the basis of explanation needs. Every rule firing that occurred remains in the trace and is displayed, regardless of whether the effect was for interface, control, general problem solving, or domain reasoning. The events are presented only in the trivial organization of order by occurrence, not by related intent or other semantically relevant relationships.

Analogous problems exist with knowledge base editors: they display only the lowest level of detail without paraphrase or summarization; selection on the basis of content is not supported; and the typical organization structure is simply linearly sequential as entered by the user. An effective explainer must be capable of adjusting the level of aggregation and of editing the amount of information to match as nearly as possible those that best satisfy the needs of the user.

Distinguished inference selection. Systems need to be selective about what information they choose to record and explain. Using a technique we call *distinguished inference selection,* knowledge structures and events can be incorporated or left out of the reasoning history, generated explanations, or both. A primary use of distinguished inference selection in SAL is to tailor explanation for different user types by selecting different information. Figure 40 compares the inference information selected by SAL for different user types. The "generic" column is the total number of database assertions independent of the explanation facility. Database assertions are the only recorded trace information in a generic expert system tool—actions are not

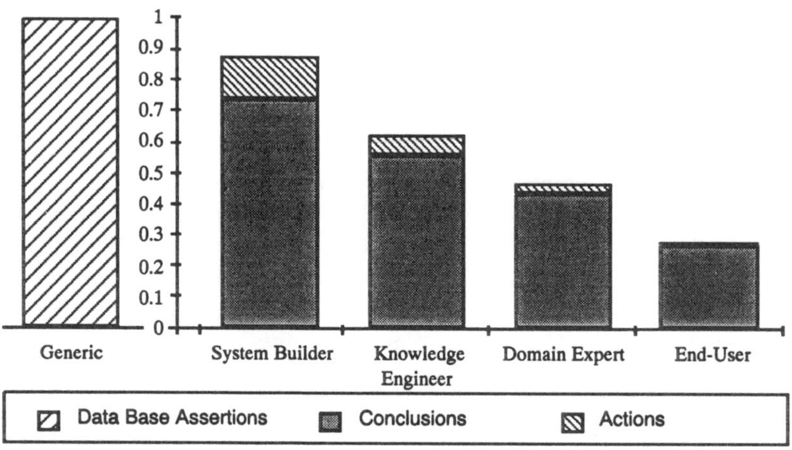

FIGURE 40. Distinguished inference selection (SAL): Comparison of information selected for different types of users.

recorded. This figure is used to normalize the remaining statistics. For a typical SAL session, there are approximately 100 database assertions. Notice that roughly one-third of the quantity of information selected for system builders is selected for end users. Also, actions comprise a larger portion of the selected information for system builders than any other user type.

A key to controlling the quantity and detail of generated explanations is the ability to choose only relevant structures and activities for inclusion. Distinguished inference selection mandates marking various structures for incorporation in the explanation trace of reasoning. In general, expert system building tools do not permit control over the inferences that are to be traced and explained.*

The selection of information may be performed entirely by system builders at the time the system is constructed or may be performed by the system itself—another application of inferenced generation. SAL allows every explainable construct to be either selected or rejected for inclusion in the explanation history or for presentation during the consultation session. Different user types with diverse explanation needs are thus accommodated. The basic method for selecting a construct is by the use of one of the special syntactic forms for explanation attachment. Any construct suitable for inclusion in the reasoning history that includes a special syntactic form is automatically traced and integrated into the list. Nontagged constructs are simply ignored by the explanation mechanisms. It is not possible to request an explanation of a nontagged item, and nontagged items will not appear in the chain of reasoning used for explanation.

Dynamic granularity. The needs of the user may be more dynamic than can be accounted for by simple *a priori* tagging; for example, the user may wish to see explanation at a very high level of aggregation one moment, then later decide that deeper inspection is necessary. As another example, the user may wish to have "domain naïve" explanations for one part of the consultation, and "domain expert" explanations for another. To account for this, dynamic flags in the user model may be set and checked, and the explanations generated accordingly via inferenced generation.

The appropriate choice of quantity and level of information used in explanation shifts during the course of a consultation session. A user may change levels spontaneously or in response to explanations. For example, a user who initially desires a general explanation may find that the given explanation provides a heightened understanding, which now requires a

* An exception, the RuleMaster system,[23, 55, 56] allows modules to be declared as "SILENT" or "MUTE." If declared SILENT, the system suppresses any explanation that may be generated during the execution of the module (including calls to other modules, regardless of their declaration). If declared MUTE, no explanation is generated for the module or any module that it calls.

I did conclude that "John Smith" did retire for reasons-of-health, in To decide a plaintiff did retire for reasons-of-health, rule 3, because

Settlement value will include an amount for disability if the plaintiff retired early because of asbestosis.

Why?

I needed the conclusion that "John Smith" did retire for reasons-of-health in order to conclude that "John Smith" is disabled, in To decide a plaintiff is disabled, rule 3.

Why?

I needed the conclusion that "John Smith" is disabled in order to conclude that the disability-compensation of the case is 75000 DOLLARS, in To generate a disability-compensation of a case, rule 5.

FIGURE 41. Pursuing super-goals (SAL).

deeper level of explanation. Alternatively, upon seeing a low-level description of a process step, a user may feel that a higher-level description of the process is needed. This is often true when looking at specific conclusions or actions of a system: once the particular low-level event that resulted in the conclusion has been identified, the user often wishes to know what higher-level goal of the system (supergoal) was being pursued that required the lower-level action. This is shown in the user–SAL dialogue of Fig. 41. Each "Why?" request results in an explanation of the next-higher-level goal.

Context reconciliation. Determining an appropriate explanation depends on the following contextual aspects: User Characterization, Active State of Reasoning, and Active State of Explanation Dialogue. *User characterization* reflects the model of the target of explanation, including such factors as domain familiarity, system usage goals, and technological expertise. These are maintained by the user model described earlier. The *active state of reasoning* accounts for the status of the reasoning component at the time the explainable activity occurs. This may be "real time," as when a question is asked during the course of a consultation, or "historical perspective," as in a post-consultation explanation session. Since the nature of explanation is like that of any other dialogue or discourse, explanation also depends on the *active state of explanation dialogue.* This accounts for the immediate context of what has recently transacted between explainer and user and is embodied in the interaction model described earlier in this chapter. Such knowledge permits recognizing that a certain presentation would be redundant with information just shown. In a slightly different context, the same presentation could be relevant as establishing a framework for the user to understand other information.

Selection/reformulation dimensions. There are four key information selection and reformulation dimensions developed here: quantity/brevity, complexity/simplicity, specificity/generality, and concreteness/abstraction. A good explanation, appropriate for the user and interaction, selects only relevant information and presents it at the appropriate level of detail and within the proper context. Each dimension represents a continuum, with gradations between the end points presented.

Quantity/brevity: Good explanations present just the right amount of information. Too little information forces the user to make additional requests to achieve the desired level of understanding. Worse yet, if presented with an inappropriate amount of information the user may misunderstand the answer and thus come to believe something that is false. On the other hand, if too much information is presented, the user may be overwhelmed and must either struggle to make sense out of it or simply give up and not achieve the desired comprehension. Figure 42 compares these two situations as they affect a simple trace of rule firings in SAL. Each screen image shows seven of the traced conclusions and actions. Note that only one of the seven items shown in the nonselective case appears in the selective case. The items displayed in the selective case are higher level, thus the items visible at one time encompass a greater part of the consultation. Particular subsets of the knowledge and rule firings are of interest to specific types of users—they should not all be shown to all users. Presenting excess information that is of no interest to the user makes it difficult for that user to sift through and determine what is relevant and what is not.

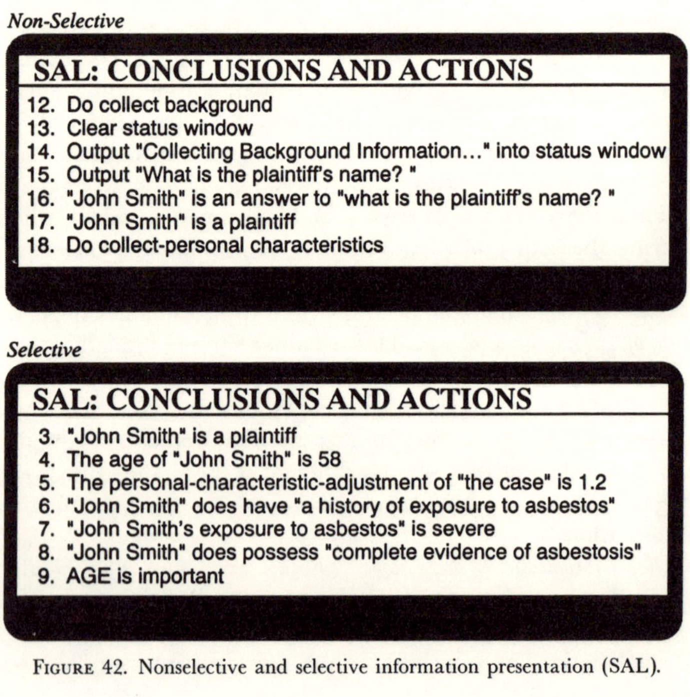

FIGURE 42. Nonselective and selective information presentation (SAL).

Complexity/simplicity: The explainer must decide the appropriate level of complexity (simplicity) to meet the needs of the user. Typically, end users do not need nor will they benefit by a deep, complex explanation of a behavior. Rather, they desire a more surface-level understanding. This situation is directly analogous to consultation with a human specialist where a client does not want an explanation framed in terms of first principles, but at a higher, more easily understood, level. On the other hand, domain experts actively probing the knowledge base to determine representational correctness or depth of domain knowledge may wish to view information at a fine granularity to expose intimate complexities. Similarly, system builders often need to see various levels of detail, such as how data are being processed, how conflicts are resolved, or how bits are being transferred from one place to another. For example, a knowledge engineer may be satisfied with a description that says that the picture stored along with text will be displayed simultaneously to an end-user, but a programmer may also wish to see how the picture is stored and whether or not multiple copies of the same image are stored redundantly or if pointers are used instead.

Specificity/generality: The specificity (generality) of information concerns the level of aggregation chosen for presentation. It is often appropriate to provide an overview explanation of a behavior that neglects detail in order to provide the big picture. This understanding allows users to focus requests for more specific explanation on areas of special interest. It is sometimes fitting to show information in the finest detail—for example, when it is important to find out exactly how a given assertion to the database came about or to see exactly how a given fact was used. Precisely which rule fired is often relevant to finding out why a conclusion was posted to the knowledge base, as shown in the lower screen of Fig. 43. Often one merely needs to know more generally about how a given conclusion is made, such as an overview about the collection of rules involved in making a decision like that shown in the upper screen of Fig. 43. In a rule-based system, explanations may be appropriate that only concern one level of detail: terms, individual clauses of a rule, individual rules, collections of rules, or larger aggregations.

Concreteness/abstraction: The level of abstraction is another dimension of information selectivity. Here the concern is whether the explanation should be general and discuss the abstract notion/case, or framed in concrete terms with respect to the current consultation. Often, the most desirable explanation concerns the way the system approaches a given type of problem rather than the way a particular problem happened to be solved this time through. Even end users trying to understand a specific conclusion may be better served by a general description of the process in the abstract than by seeing the particular instantiation of that process as followed for the

General

> I did conclude that "John Smith" is disabled in To decide a plaintiff is disabled, rule 3
>
> *Summary*
>
> This set of rules determines whether or not the plaintiff is disabled. Factors taken into account include whether or not the plaintiff is working a reduced schedule or has reduced job status, whether or not the plaintiff's activities are restricted, and the plaintiff's credibility.

Specific

```
I did conclude that "John Smith" is disabled in  To decide a plaintiff is disabled, rule 3
Rule
IF ( ((the plaintiff is not working-full-time or the plaintiff's job-status is reduced)
   and  (the plaintiff is not retired or the plaintiff did retire for reasons-of-health)
   and  (the plaintiff's pulmonary-function-test does show "serious restrictive disease"
          or (the plaintiff's pulmonary-function-test does show "severe restrictive disease"))
   or  (the plaintiff's activities are restricted and the plaintiff's activities are not
          slightly-restricted and the plaintiff's credibility is not low) )
Conclude that the plaintiff is disabled.
```

FIGURE 43. General and specific explanations (SAL).

consultation at hand. Here again, these goals/needs vary with user type and during a given consultation session as well. Most expert system-building tools do not provide aggregation or overview capabilities. SAL provides several levels of abstraction in explanation as illustrated previously by the pursuit of supergoals.

5.3.4. Presentation. This aspect of explanation concerns the nature of interaction between system and user—the actual explanation dialogue. These issues involve the structure of explanation, the media and protocols used for input and output, and the framing of explanatory information.

The presentation of explanatory information can aid or hinder the goal of satisfying the user's need. Additional burdens can result by cluttered screens or confusing usage protocols. Key elements include establishing and exploiting the pertinent *context*, and directing the user's *focus* to relevant aspects. These two factors are discussed here.

An explainer cognizant of the *active context* can use this information to help interpret explanation requests and other user actions, and to direct the construction of explanations. For example, the system may recognize a request for terminological explication as actually requesting assistance in making the connection between what is being asked for and the current line of reasoning being pursued.

A critical aspect of explanation is establishing the *relevant context* within which the explanation is to be understood. The communication of concepts depends on the ability for the listener to disambiguate possible interpretations of what is presented. Sometimes the appropriate context may be established by a dialogue cue, such as an introductory remark, or it may be effected by using a specifically designated window. An explanation method must account for recognizing and exploiting the active context and establishing the relevant context where it differs from the current one.

The issue of *focus* involves cueing the listener to the relevant parts of a communication exchange. In a dynamic display, for example, both context and focus can be addressed by using windows for different explanation functions (such as status posting or elaborating the set of possible responses to questions) and having them pop up and hide at appropriate times. SAL exploits the use of windows, menus, font-style variation, and dynamically generated text to address context and focus considerations.

Communication vehicles. Appropriate selection and use of a communication medium can ease the burden of understanding and overcome communication impediments. For example, a system that anticipates a user's next request can minimize the work required to initiate that requests by making it a mouse-selectable menu item or the default next action. Expression media suitable for use with contemporary expert system building tools are shown in Table 7. *Item selection menus* present items to be operated on or explained. *Function selection menus* present the types of explanations or types of needs to be addressed. The remaining items are self-explanatory.

Since different information is best communicated via different media an explanation facility should allow system builders a range of possible I/O methods. For example, a graphical trace is appropriate for a reasoning chain or taxonomic information whereas a definition or information request may best be presented via a textual description.

We have seen how the needs of users can be addressed by incorporating knowledge-based explanation techniques into an expert system structure. Need identification, information selection and reformulation, and presentation—the activities that contribute to the realization of a self-revealing system—are accomplished by exploiting the knowledge-based technology inherent in the expert system building tool.

TABLE 7
Communication Media

Display	Input
Formalized text	Item selection menu
Natural language text	Function selection menu
Pictorial images	Item mouse-hold
Pop-up menu	Typed request/response
Pop-up window	Function key

6. Conclusions and Future Work

Improving the understandability of a system addresses many of the basic problems with knowledge transfer (both explanation and knowledge acquisition). The groundwork must be laid by clearly defining the objective goals of explanation. Next, we explicitly gather the necessary information resources. Finally, by applying the full power of KBS technology to the task (inferenced generation) we can provide effective explanation and thus achieve greater understandability.

REVEAL is a concrete method for producing software systems that explain their own structure and behavior, i.e., *self-revealing* systems. Knowledge-based systems are the primary target, since the *need* for understandability as well as the difficulty in achieving it are greatest for such systems. The incorporation of more extensive explanatory capabilities provides much-needed assistance for defining the limits, sources, and utilization of knowledge and aids knowledge acquisition.

The method is applicable independently of the particular support environment, with the caveat that the system must include provisions for integrating explanatory information. It treats the problem generally, and is not just a technical solution limited to an instance of the problem. Furthermore, the method addresses deeper reasoning and rationales that are otherwise lacking or "compiled out" of most expert systems.

Existing expert systems all provide very similar explanation facilities despite their vast differences in the domain addressed and implementation characteristics. MYCIN demonstrated that a limited explanation facility was feasible based on the task-performance system itself and an execution trace. A trace capability and canned text constitute the vast majority of current explanation facilities. Many researchers have noted that these unabridged surface-level traces are not sufficient to meet the explanation needs of users. We have presented other types of explanation and detailed specific technical features necessary to support them.

The main departure of this work from previous work is the application of knowledge-based technology to knowledge transfer. *Inferenced generation* enables the system itself to generate explanations by the same techniques and mechanisms it uses to effect task-performance reasoning. This allows dynamic creation of explanatory displays and provides for flexible user and interaction models. Support for inferenced generation is provided by introducing explanation structures, multilevel association, distinguished inference selection, and understandable code.

REVEAL is predicated on addressing the explanation *needs* of users. Before a solution is attempted, the goals and intent are explicated and considered. We have detailed several dimensions along which users' needs may by characterized. This codification is also useful for assessing explanation competence, and provides a foundation for automatic recognition and identification of explanation requests.

Explanation involves communicating information to achieve new understanding. Satisfactory explanation often requires a different organization of task-performance information and knowledge which is not necessary for system operation. We treat explanatory knowledge as a resource. This leads us to consider the types of knowledge involved, means for incorporating them into the knowledge base, and methods for accessing and manipulating them. Because most expert system shells do not provide for recording and associating explanatory information, it is generally infeasible to retrofit a fully capable explanation facility onto an existing system. Nonetheless, many of the aspects of REVEAL can be applied to even the most simple expert system building tool, with associated improvement in the understandability of developed systems.

Explanation is just one aspect of the use of self-knowledge. Waterman predicted that self-knowledge would allow expert systems to "create the rationale behind individual rules" and "tailor their explanations to fit the requirements of their audience."[76] REVEAL provides the tools to accomplish these functions now. Self-knowledge is the key to building systems capable of changing their own internal structure through rule correction, knowledge-base reorganization, and system reconfiguration.[43, 76]

This work has focused primarily on the explanation aspect of knowledge transfer as it forms the necessary foundation. There are additional ways we can assist knowledge acquisition in particular. Specific difficulties with knowledge acquisition include *using the appropriate representation* for conceptual knowledge, *avoiding inconsistent representations* of knowledge, and *filling-in missing knowledge.*

Mechanisms that can support KBS developers and help mitigate the knowledge acquisition problem include extending the self-revealing software paradigm to include *similarity recognition* (a form of pattern matching) for the knowledge represented in the system, *gap identification*, and *fringe exploration.* Similarity recognition (for example, using a constraint-based pattern matcher or classifier) can be used to explore generalization, specification, explosion, and collapse of existing concepts. This can help suggest similar representations for similar concepts and aid in the selection of appropriate representations. By directly employing existing concepts, a technique we call *conceptual building blocks*, the user can be helped to avoid inconsistent representations of the same concept. Inconsistent representation is a problem where information may be entered redundantly (multiple representations for the same concept) or with disconnected concepts where the linkage between two high-level concepts is broken since the linking concept is represented differently in each case.

Gap identification involves uncovering missing knowledge and requires a paradigm for "knowing what you do not know." Here the system must be able to identify potential gaps in the knowledge base, either missing concepts or relationships among them. Again, the utility of commonality recognition in a conceptual knowledge base is key. Fringe exploration is a knowledge acquisition strategy for expanding the knowledge base by pushing the frontier of encoded knowledge. Each of these areas are subjects for further research.

Appendix: SAL—An Expert System for Asbestos Litigation

System for Asbestos Litigation (SAL) is an expert system that models the legal reasoning underlying settlement of product liability claims arising from asbestos exposure. SAL elicits the facts of a particular case to which it applies expertise and problem-solving techniques, extracted from attorneys and claims adjusters, to produce a dollar value for the associated claim. The primary goal in developing SAL was to aid the development of a methodology for knowledge-based system explanation that is suitable for application to similar domains and the diverse uses of explanation. Other goals included the extraction and formalization of expertise required for case evaluation in this domain and obtaining a deeper understanding of the process of case evaluation and its relation to legal decision making. SAL additionally serves as demonstration model for a vehicle to assist the courts in speeding asbestos claim settlements.

Most of the concepts and features of REVEAL are realized in the SAL system. Notably, SAL utilizes inferenced generation. It employs its own reasoning procedures, methods, and mechanisms for constructing the explanations it presents. SAL can access its entire world state at the instant an explainable event occurs and apply its reasoning engine's power to generate an appropriate, customized explanation. The full knowledge base and complete history of events are also available when an explanation is requested and displayed.

SAL demonstrates the application of the REVEAL methodology for producing self-revealing systems to a knowledge engineering language that supports complex control processes and a real-world complex problem domain. SAL is implemented in ROSIE,[21, 38, 68] a rule-based knowledge engineering language with an English-like syntax. The basic ROSIE system was extended to include enhanced explanation capability for this research.[79]

The user communicates with SAL via a collection of pop-up windows and programmable menus. The windows, which appear only when needed, focus attention, establish context, and effectively organize displayed information. The menu items are mouse selectable and have associated explanatory text that is displayed in an "Elaboration" window upon demand. This feature has proven especially useful for dealing with the medical information needed by SAL.

Legal Domain

Building an expert system for the legal domain illustrates the issues involved in addressing real-world problems. The domain is inherently complex and lacks well-defined and formalized expertise. Legal concepts and practices are largely intuitive. The book knowledge that is available for some subdisciplines consists primarly of examples of applied expertise, not explication of underlying deeper models. Although some domain knowledge is quite explicit, e.g., laws and legal concepts, these are expressed in natural language in a way that exploits its richness of meaning, interpretation, and ambiguity. These laws and concepts are not readily translated into a constrained computer language. The specific difficulties of applying artificial intelligence to the legal domain are explored in Refs. 53 and 80.

Unlike the legal area, there are other domains that have reasonably clear and well-understood models of the underlying mechanisms involved. For example, expert systems applied to electronics use knowledge about circuits and their operation for prediction and explanation.[36, 37] The difficulty in locating similar deep knowledge in the legal area further inhibits the development of competent explanation facilities. Deeper explanations of legal rules invoke complex, ill-defined concepts (e.g., deterrence, fairness), and resorting to different justifications can produce different rules. For example, laws should be equitable and efficient, but some efficient rules might be inequitable. The problem is especially acute because in this area, as in other complex problem areas, it is important for the system not only to explain how it reached a conclusion but also to provide a convincing argument that the method used is valid.

Evaluating Asbestos Product Liability Claims

The implementation of SAL confronts the difficulty found in scaling artificial intelligence solutions from narrow applications to real-world problems. It addresses a complex, poorly structured, sociolegal problem domain—asbestos product-liability case litigation and settlement. Over 21,000,000 exposures to asbestos have occurred and that asbestosis and other lung-related diseases are involved in more than 20,000 unresolved asbestos product-liability cases.[35] Waterman and Peterson[54, 82] point out that the need for a means to analyze and evaluate individual cases within an overall framework that is adaptable to changing facts and law suggests the application of expert system technology.

The development of SAL identified areas of agreement among experts including important features that influence their evaluation of cases. It provides a model of case evaluation and a framework for further study of legal decision making. The case evaluation schema developed for LDS (Legal Decisionmaking System)[81, 82] forms the basis of SAL's case evaluation model for asbestos product-liability claims. LDS was an exploration vehicle which demonstrated the feasibility of rule-based modeling for legal decision making.[83] Its depth of domain coverage was quite limited and it did not address explanation other than by the use of canned text and by employing ROSIE as the development language.

Case Evaluation Schema

SAL's case evaluation schema is shown in Table 8. The elements of LOSS, LIABILITY, RESPONSIBILITY, and CHARACTERISTICS included in SAL are only part of those used by lawyers in evaluating

TABLE 8
SAL Case Evaluation Schema

$$\text{Value} = f\begin{pmatrix}\text{LOSS}\\ \text{LIABILITY}\\ \text{RESPONSIBILITY}\\ \text{CHARACTERISTICS}\end{pmatrix}$$

Category	Description	SAL
LOSS	Special damages resulting from direct losses such as medical expenses and lost income, plus general damages resulting from less direct losses such as injury trauma, fear, and faculty loss.	The medical damages due to asbestos exposure, based on symptoms, x-rays, other laboratory results, and disability. Lost income, based on medical damages, current work status, and prior income.
LIABILITY	Estimates of the probability of establishing liability against the defendant.	The plaintiff's history of exposure to asbestos.
RESPONSIBILITY	An estimate of the proportion of responsibility assigned to the plaintiff for his or her own carelessness.	The proportion of responsibility assigned to the plaintiff for smoking, knowledge of danger, etc.
CHARACTERISTICS	The numerical adjustment for subjective factors such as litigants' characteristics, lawyers' skill, and case jurisdiction.	The adjustment for personal characteristics such as the plaintiff's age.

asbestos cases. A full asbestos model would need, for example, additional rules about whether defendants had been shown to have sold asbestos products used by a plaintiff. Characteristics would include rules about credibility and the sympathetic or unsympathetic nature of a plaintiff.

The schema also organizes the rule sets and rules in SAL. Each component requires a large number of rules, the majority being located in the LOSS component of the schema. SAL contains approximately 400 rules that represent expert legal knowledge. These rules were obtained from lengthy interviews with asbestos litigation experts associated with the United States District Court in Cleveland, Ohio, including attorneys engaged in actual asbestos litigation cases. Their expertise, both legal and medical, was initially encoded as sets of informal rules. After enough knowledge to determine a useful and coherent organization had been collected and analyzed, the expertise was formally encoded in ROSIE (extended for explanation as described in Ref. 79).

The experts expressed their knowledge at the surface representation level. SAL uses only this surface knowledge and expertise because, as with many expert system task domains, explicating a mechanism/causal model for this domain is a difficult, expensive undertaking and would only slightly improve SAL's case evaluation performance. However, such a model would provide deeper levels of reasoning which could extend SAL's explanation capabilities. An unresolved issue is: What should be modeled at deeper levels for this domain?[(53)]

References

1. J.L. Austin, *How to Do Things with Words*, Oxford University Press, Oxford, 1962.
2. A. Barr and E.A. Feigenbaum, *The Handbook of Artificial Intelligence*, William Kaufmann, 1981.
3. L. Birnbaum and M. Selfridge, Problems in conceptual analysis of natural language, No. 168, Computer Science Department, Yale University, 1979.
4. L. Birnbaum and M. Selfridge, Conceptual analysis of natural language, in *Inside Computer Understanding: Five Programs Plus Miniatures*, Schank and Riesbeck, Eds, Lawrence Erlbaum Associates, Hillsdale, New Jersey, 1981, pp. 318–353.
5. G.P. Brown, A framework for processing dialog. MIT/LCS/TR-182, Massachusetts Institute of Technology Laboratory for Computer Science, 1977.
6. J.S. Brown, R.R. Burton, and A.G. Bell, SOPHIE: A step toward creating a reactive learning environment, *Int. J. Man-Machine Stud.* **7**, 675–696, 1975.
7. B.G. Buchanan, D. Barstow, R. Bechtal, J. Benett, W. Clancey, C. Kulikowski, T. Mitchell, and D.A. Waterman, Constructing an expert system, in *Building Expert Systems*, Hayes-Roth, Waterman, and Lenat Eds., Addison-Wesley, Reading, Massachusetts, 1983, pp. 127–167.
8. W.J. Clancey, The epistemology of a rule-based expert system—A framework for explanation. *Artif. Intell.* **20**, 215–251, 1983.
9. W.J. Clancey, Classification problem solving, Proceedings of AAAI-84, pp. 49–55, 1984.
10. W.J. Clancey and R. Letsinger, NEOMYCIN: Reconfiguring a rule-based expert system

for application to teaching, Proceedings of Seventh International Joint Conference on Artificial Intelligence, pp. 829–836, 1981.

11. R.E. Cullingford, M.W. Krueger, M. Selfridge, and M.A. Bienkowski, Automated explanations as a component of a computer-aided design system, *IEEE Trans. Sys.*, Man, *Cyb.* **SMC-12**, 168–181, 1982.
12. B. D'Ambrosio, Building expert systems with M.1, *BYTE* **10**(6), 371–375, 1985.
13. R. Davis, Applications of meta–level knowledge to the construction, maintenance and use of large knowledge bases, STAN-CS-76-552, Stanford Artificial Intelligence Laboratory, Stanford University, 1976.
14. R. Davis, Use of meta–level knowledge in the construction and maintenance of large knowledge bases, Ph.D. thesis, Computer Science Department, Stanford University, 1976.
15. R. Davis, Interactive transfer of expertise: Acquisition of new inference rules, *Artif. Intell.* **12**, 121–157, 1979.
16. R. Davis and B.G. Buchanan, Meta-level knowledge: Overview and applications, Proceedings of IJCAI-5, pp. 920–927, 1977.
17. A. Davison, Indirect speech acts and what to do with them, in *Syntax and Semantics*; Vol. 3: *Speech Acts*, Cole and Morgan Eds., Academic Press, New York, 1975, pp. 143–185.
18. G.A. Drastal, Experiments with RULE WRITER, a tool for building expert systems, Proceedings of Seventeenth Hawaii International Conference on System Sciences (HICSS). pp. 167–173, 1984.
19. R.O. Duda, P.E. Hart, N.J. Nilsson, P. Barrett, J.G. Gaschnig, K. Konolige, R. Reboh, and J. Slocum. Development of the PROSPECTOR consultation system for mineral exploration. Stanford Research Institute, 1978.
20. M.G. Dyer, *In-Depth Understanding: A Computer Model of Integrated Processing for Narrative Comprehension*, MIT Press, Cambridge, Massachusetts, 1983.
21. J. Fain, D. Gorlin, F. Hayes-Roth, S. Rosenschein, H. Sowizral, and D.A. Waterman, The rosie language reference manual. N-1647-ARPA, The RAND Corporation, 1981.
22. P.K. Fink, J.C. Lusth, and J.W. Duran, A general expert system design for diagnostic problem solving, *IEEE Trans. Pattern Anal. Machine Intell.* **PAMI-7**, 553–560, 1985.
23. B. Finkle, Tapping into the knowledge power of expert systems, *Comput. Design* March 15, 1986.
24. A.V. Gershman, Knowledge-based parsing, Ph.D. thesis, Computer Science Department, Yale University, 1979.
25. G.A. Gorry, Computer-assisten clinical decision making, *Meth. Inf. Med.* **12**, 45–51, 1973.
26. G.M. Green, How to get people to do things with words: The whimperative question, in *Syntax and Semantics*, Vol. 3: *Speech Acts*, Cole and Morgan, Eds., Academic Press, New York, 1975, pp. 107–141.
27. P. Grice, Logic and conversation, in *Syntax and Semanctics*, Vol. 3: *Speech Acts*, Cole and Morgan, Eds., Academic Press, New York, 1975, pp. 41–58.
28. B.J. Grosz, Discourse, in *Understanding Spoken Language.* Walker, Ed., Elsevier North-Holland, New York, 1978, pp. 229–234.
29. B.J. Grosz, Discourse analysis, in *Understanding Spoken Language*, Walker, Ed., Elsevier North-Holland, New York, 1978, pp. 235–268.
30. B.J. Grosz, Focus spaces: A representation of the focus of attention of a dialog, in *Understanding Spoken Language*, Walker, Ed., Elsevier North-Holland, New York, 1978, pp. 269–285.
31. D.W. Hasling, W.J. Clancey, and G. Rennels, Strategic explanations for a diagnostic consultation system, STAN-CS-83-996, Department of Computer Science, Stanford University, 1983.
32. F. Hayes-Roth, D.A. Waterman, and D.B. Lenat, An overview of expert systems, in *Building Expert Systems*, Hayes-Roth, Waterman, and Lenat, Eds., Addison-Wesley, Reading, Massachusetts, 1983, pp. 3–29.
33. G.G. Hendrix, KLAUS: A system for managing information and computational resources, No. 230, SRI International, 1980.
34. J. Heringer, Some grammatical correlates of felicity conditions and presuppositions, No. 11, Department of Linguistics, Ohio State University, 1972.

35. J.S. Kakalik, P.A. Ebener, W.L.F. Felstiner, and M.G. Shanley, Costs of asbestos litigation. R-3042-ICJ, The RAND Corporation, 1983.
36. V.E. Kelly, The CRITTER system: Automated critiquing of digital circuit design, LCSR-TR-55, Laboratory for Computer Science Research, Rutgers University, 1984.
37. V.E. Kelly and L.I. Steinberg, The CRITTER system: Analyzing digital circuits by propagating behaviors and specifications, Proceedings of The National Conference on Artificial Intelligence, American Association for Artificial Intelligence, pp. 284–289, 1982.
38. J.R. Kipps, B.A. Florman, and H.A. Sowizral, The new rosie® reference manual and user's guide, R-3448-DARPA/RC, The RAND Corporation, 1987.
39. B. Kuipers, Qualitative simulation of mechanisms, MIT/LCS/TM-274, Massachusetts Institute of Technology Laboratory for Computer Science, 1985.
40. A. Leal and J. Pearl, An interactive program for conversational elicitation of decision structures, *IEEE Trans. Sys. Man Cyb.* **SMC-7**, 368–376, 1977.
41. W.G. Lehnert, Human and computational question answering, *Cog. Sc.* **1**, 47–73, 1977.
42. W.G. Lehnert, *The Process of Question Answering*, Lawrence Erlbaum Associates, Hillsdale, New Jersey, 1978.
43. D.B. Lenat, R. Davis, J. Doyle, M. Genesereth, I. Goldstein, and H. Schrobe, Reasoning about reasoning, in *Building Expert Systems*, Hayes-Roth, Waterman, and Lenat, Eds., Addison-Wesley, Reading, Massachusetts, 1983, pp. 219–239.
44. J.A. Levin and J.A. Moore, Dialog-games: Metacommunications structures for natural language interaction, *Cog. Sc.* **1**, 395–420, 1977.
45. J.W. Lewis and F.S. Lynch, GETREE: A knowledge management tool, Proceedings of Trends and Applications conference on Automating Intelligent Behavior, IEEE Computer Scociety, pp. 273–277, 1983.
46. W.C. Mann, J.A. Moore, and J.A. Levin, A comprehensive model for human dialogue, Proceedings of The Fifth International Joint Conference on Artificial Intelligence (IJCAI-77), 1977.
47. K.R. Mc Keown, Generating natural language text in response to questions about database structure, Ph.D. thesis, The Moore School of Electricial Engineering, University of Pennsylvania, 1982.
48. K.R. Mc Keown, M. Wish, and K. Matthews, Tailoring explanations for the user, CUCS-172-85, Department of Computer Science, Columbia University, 1985.
49. R. Neches, W.R. Swartout, and J. Moore, Enhanced maintenance and explanation of expert systems through explicit models of their development, Proceedings of IEEE Workshop on Principles of Knowledge-Bases Systems, IEEE Computer Society Press, pp. 173–183, 1984.
50. R. Neches, W.R. Swartout, and J. Moore, Explainable (and maintainable) expert systems, Proceedings of Ninth International Joint Conference on Artificial Intelligence (IJCAI-85), pp. 382–389, 1985.
51. D.M. Neves and J.R. Anderson, Knowledge compilation: Mechanisms for the automatization of cognitive skills, in *Cognitive Skills and Their Acquisition*, Anderson, Ed., Lawrence Erlbaum Associates, Hillsdale, New Jersey, 1981, pp. 57–84.
52. J. Paul, Self-revealing software: A method for producing understandable knowledge-based systems, Ph.D. thesis, University of California, Los Angeles, 1988.
53. J. Paul, D.A. Waterman, and M.A. Peterson, SAL: An expert system for evaluating asbestos claims, Proceedings of The First Australian Artificial Intelligence Congress, Computerworld EXPO, 1986.
54. M.A. Peterson, New tools for reducing civil litigation expenses, R-3013-ICJ, The RAND Corporation, 1983.
55. Radian Corporation, *RuleMaster Reference Manual*, 1986.
56. Radian Corporation, *RuleMaster Tutorial Manual*, 1986.
57. C.K. Riesbeck, Conceptual analysis, in *Conceptual Information Processing*, Schank, Ed., North-Holland, Amsterdam, 1975.
58. C.K. Riesbeck, An expectation-driven production system for natural language understanding, in *Pattern-Directed Inference Systems*, Waterman and Hayes-Roth, Eds., Academic Press, New York, 1978.

59. J. Sadock, Whimperatives, in *Studies Presented to Robert B. Lees by His Students*, Sadock and Venek, Eds., Linguistic Research, Inc., 1970, pp. 223–238.
60. R.C. Schank, Questions and thought, RR-385, Department of Computer Science, Yale University, 1985.
61. R.C. Schank and C.K. Riesbeck, *Inside Computer Understanding: Five Programs Plus Miniatures*, Lawrence Erlbaum Associates, Hillsdale, New Jersey, 1981.
62. A.C. Scott, W.J. Clansey, R. Davis, and E.H. Shortliffe, Explanation capabilities of production-based consultation systems, *Am. J. Comput. Ling.* 1–50, 1977.
63. J.R. Searle, *Speech Acts: An Essay in the Philosophy of Language*, Cambridge University Press, 1969.
64. J.R. Searle, Indirect speech acts, in *Syntax and Semantics*, Vol. 3: *Speech Acts*, Cole and Morgan, Eds., Academic Press, New York, 1975, pp. 59–82.
65. E.H. Shortliffe, An artificial intelligence program to advise physicians regarding antimicrobial therapy, *Comp. Biomed. Res.* **6**, 544–560, 1973.
66. E.H. Shortliffe, *Computer-Based Medical Consultations: MYCIN*, Artificial Intelligence Series, Elsevier North-Holland, New York, 1976.
67. E.H. Shortliffe, Explanation capabilities for medical consultation systems (tutorial), Proceedings of AAMSI Congress '84, pp. 193–197, 1984.
68. H.A. Sowizral and J.R. Kipps, rosie: A programming environment for expert systems, R-3246-ARPA, The RAND Corporation, 1985.
69. D.W. Stampe, Meaning and truth in the theory of speech acts, in *Syntax and Semantics*, Vol. 3: Speech Acts, Cole and Morgan, Eds., Academic Press, New York, 1975, pp. 1–39.
70. W.R. Swartout, A digitalis therapy advisor with explanations, Master's thesis, Massachusetts Institute of Technology, 1977.
71. W.R. Swartout, Producing explanations and justifications of expert consulting programs, Ph.D. thesis, Massachusetts Institute of Technology, 1981.
72. W.R. Swartout, XPLAIN: A system for creating and explaining expert consulting programs. *Artif. Intell.* **21**, 285–325, 1983.
73. M. Van Horn, *Understanding Expert Systems*, Bantam Books, Toronto, 1986.
74. W. van Melle, E.H. Shortliffe, and B.G. Buchanan, EMYCIN: A knowledge engineer's tool for constructing rule-based expert systems, in *Rule-Based Expert Systems*, Buchanan and Shortliffe Eds., Addison-Wesley, Reading, Massachusetts, 1984, pp. 302–328.
75. A. Walker, Automatic generation of explanations of results from knowledge bases, RJ-3481, IBM Research Library, 1982.
76. D.A. Waterman, *A Guide to Expert Systems*, Addison-Wesley, Reading, Massachusetts, 1986.
77. D.A. Waterman, How do expert systems differ from conventional programs? *Expert Sys.* **3**(1), 16–19, 1986.
78. D.A. Waterman, R.H. Anderson, F. Hayes-Roth, P. Klahr, G. Martins, and S.J. Rosenschein, Design of a rule-oriented system for implementing expertise, N-1158-1-ARPA, The RAND Corporation, 1979.
79. D.A. Waterman, J. Paul, B. Florman, and J.R. Kipps, An explanation facility for the rosie knowledge engineering language, R-3406-ARPA/RC, The RAND Corporation, 1986.
80. D.A. Waterman, J. Paul, and M.A. Peterson, Expert systems for legal decision making, in *Applications of Expert Systems*, Quinlan, Ed., Addison-Wesley, Sydney, 1987, pp. 23–47.
81. D.A. Waterman and M.A. Peterson, Rule-based models of legal expertise, Proceedings of First Annual National Conference on Artificial Intelligence, 1980.
82. D.A. Waterman and M.A. Peterson, Models of legal decisionmaking, R-2717-ICJ, The RAND Corporation, 1981.
83. D.A. Waterman and M.A. Peterson, Evaluating civil claims: An expert systems approach, *Expert Sys.* **1**, 65–76, 1984.
84. J.L. Weiner, The structure of natural explanation: Theory and application, SP-4035, System Development Corporation, 1979.
85. J.L. Weiner, BLAH, A system which explains its reasoning, *Artif. Intell.* **15**, 19–48, 1980.
86. T. Winograd, What does it mean to understand language? *Cog. Sc.* **4**, 209–241, 1980.
87. T. Winograd, *Language as a Cognitive Process*, Vol. I: *Syntax*, Addison-Wesley, Reading, Massachusetts, 1983.

II

DISPLAY AND DESIGN

VIDEO
AN EXPERT CONSULTANT FOR VISUAL DESIGN

J.W. Roach

1. Introduction

Behavioral experimenters have hypothesized that fundamental principles of human communication can be discovered to help designers of computer systems create a better human–computer communications interface. A large number of experiments have contributed to the growth of a corpus of such principles. Compilations of principles have been published, yet computer systems continue to have poorly designed interfaces because

1. The principles are not widely known; and
2. The principles cannot be readily applied.

Discovering principles is insufficient; the actual utility of a discovery must be validated and a method for incorporating the ideas into a design process must be made operational. The construction of automated consultants that can advise system builders who configure human–computer interfaces would make the knowledge of interaction principles much more readily usable. Formalization of the principles for inclusion in an automated consultant would also lead to an understanding of the principles at a different level: either an operational embodiment of a principle can be found or the "principle" cannot be considered a truly scientific statement about human–computer interaction.

We here discuss the construction of intelligent coaches/consultants that can help dialogue authors realize the full potential of a dialogue by configuring that dialogue according to principles discovered in behavioral tests. A consultant could either critique a dialogue specification and offer

J.W. ROACH • Department of Computer Science, Virginia Polytechnic Institute and State University, Blacksburg, Virginia 24061.

suggestions for improvement, or it could help design the dialogue from specifications. A partial list of the consultant systems needed includes visual design, command language design, voice interface designe, manual entry code (joystick, touch panel, mouse, etc.) design, coordination of input/output modes, and system personality design. The major emphasis in this paper is an explanation of a visual design consulting system.

People consider visual design to be an art. Yet there are some very definite principles of design that can be captured in a computer program. In addition to principles of design, standard layouts provide a stylized means for conveying information based on the principles; the general visual design consultant would use layouts to provide good designs without having to create its own formats.

Ultimately, we want to provide an expert consultation program to help dialogue authors create better visual interfaces. This chapter contains a progress report on the construction of VIDEO.

Work first began on this system over seven years ago. The failure of our attempts to build an initial system led to much more successfull work and to the design of a new kind of knowledge-bases system. We shall discuss the nature of the consultation problem for visual design, the architecture of the system, and the knowledge base that it needs.

2. *A "True Consultant" for Visual Design*

The critical part of the visual design consultant is the building of a domain description for the user's problem. The question and answer session gathers structural information about the domain, but a less structured period must also be allowed in which the dialogue author specifies connections and dependencies between conceptual objects in the problem domain. Answers to the questions and the specified dependencies determine the kind of layout, color combinations, and format of the actual textual and pictorial information.

In the years following the inception of this project, we cast about for a good, general purpose interviewing tool. We discovered that an interview tool that let a client create a problem description was not enough. A whole series of additional representations are needed to translate the client's problem description into a rendered image. The general architecture of this gradually evolving knowledge-based system we habe dubbed a "true consultant" because the set of problems to be solved is not totally defined at the time when the system is initially configured. The current generation of VIDEO has made some progress toward construction of such a true consultant.

Violating principles of good visual design causes people to struggle to encode the information being presented, so perception is partially blocked. Programmers who have designed interfaces in the past have not considered beauty to be a design goal. Rather, they have produced software in minimum time to save money, a false economy since products with a beautiful display and good interface continue to be valued over time. The hit or miss production of good displays for interfaces will only stop if we begin systematically using good design principles. Graphic design theory and practice and human factors experimentation give us the operational principles for designing aesthetically pleasing visual displays.

3. *First Attempt to Build a System*

A critical aspect of the consultation problem centers about determining the nature of the task that the end user is performing. Human beings who serve as visual design experts have access to a vast amount of real world knowledge and can easily listen to a general purpose description of a task and extract the relevant details. VIDEO cannot and does not have such a general understanding capability. We tried to create a system that led the dialogue author through a structured question/answer dialogue. The questions elicited from the dialogue author characteristics of the task that affected the display.

We hypothesize that information can best be communicated by combining pictorial information with text. Displaying text is known to be difficult, but at least the parameters that can be adjusted are well understood; pictorial data are not so well circumscribed. VIDEO cannot store all possible shapes or icons that a dialogue author might wish to use. Consequently, we planned to supply a package that allowed the dialogue author to create shapes appropriate to the problem. A graphics editor was to be used for this purpose.

The dialogue author must be able to describe shapes, and the editor must have some internal representation of shapes that is complete and easy to manipulate. The consultation system must also be able to use these descriptions to reason about shapes, arranging them in balanced and pleasing compositions as well as creating images of two- and three-dimensional scenes. Ideally, the consultant should have knowledge about shapes, size, and distance, and general pictorial knowledge that people take to be common sense.

The result of a consultation will be a display that can be validated for its communicative effectiveness. Automatic generation of code that could drive a display and be interfaced to an application program can be achieved by generating specifications in a declarative language such as PROLOG.[6]

3.1. Determining the Content of the Image

The first design of our consultant emphasized certain concept categories: pictorial versus alphanumeric versus graphical data, tense versus relaxed situations, whether there were alarm conditions, the required attention span, the nature of the user's environment, and more specific concepts about pictorial data such as whether the data represented are three dimensional or two dimensional. Information about these concepts was elicited from the dialogue author in a question-and-answer session. The questions amount to an information-gathering routine similar to the debriefing procedures used in medical diagnostic expert consultants.

Since the graphic design expert cannot understand and store all possible images and symbols for all possible applications, the dialogue author must be able to make and alter images and symbols for objects. The author needs a graphics editor that is simple to learn and tolerant of mistakes. A pattern-matching natural language command system is a practical interface to the editor because the domain of discourse is restricted. Pattern-matching systems are robust and tolerant of incomplete or irregular input. Their main weakness for this application, however, is a tendency to lose details.

The parser must handle indirect commands masked as statements: "The red box should be on the other side." There should be a naming facility, allowing the dialogue author to refer to sets of images by natural names, like "stop sign." The system should resolve complex noun phrases specifying objects in the domain, ellipses, and pronoun references. The editor must also supply default values for things common sense would control. For example, the dialogue author should be able to create a box without specifying its color or other attributes at first. It is hard to verbalize some graphics operations, so coordinated input from an analog device of some kind—mouse, joystick, or digitizer pad—must be available.

3.2. Pilot Experiment: Building the First System

Here we describe the considerations that made up the first generation version of VIDEO. The dialogue author designing a display had two terminals available—a normal terminal to communicate with the visual design consultant and a graphics screen on which the display being created could be shown. Figure 1 shows the basic structure of the interaction sequence where the expert system tried to configure the design. The dialogue author consulted with VIDEO, which then built a description of the display. The dialogue author examines the results and edits it as desired. VIDEO was to examine each editing change to make sure that visual design principles are not violated.

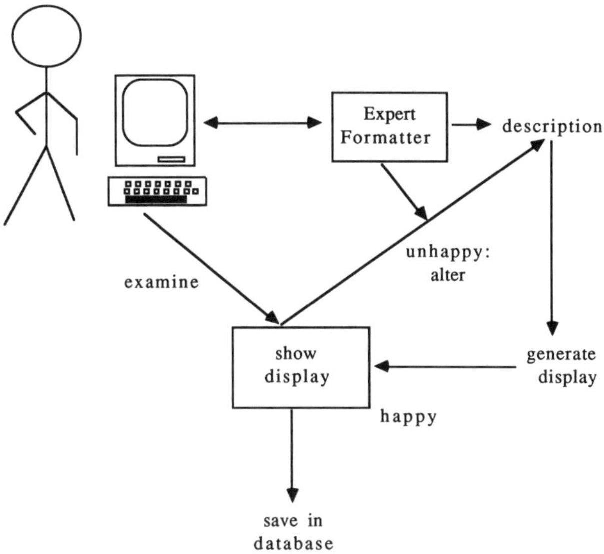

FIGURE 1. Interaction configuration of the initial VIDEO system.

In addition to determining the overall structure of the system, we designed a tutorial module that explains the principles of design and the layouts to novice designers. The tutorial explains the connection between the principles and questions being asked the dialogue author, shows the various components of design with the current graphics device (what shapes, lines, shading, etc. are available), and gives full examples of visual designs. The tutorial also serves as a help file for a dialogue author when using the system. For example, if the dialogue author wants to know why the system is asking a particular question, the part of the tutorial explaining that question can be invoked.

The work reported here is implemented in PROLOG, a production system language employed in artificial language applications. The language was interfaced to an early version of the Dialogue Management System (DMS),[12] a multiprogramming environment designed to support the creation and testing of human–computer interfaces. DMS supported high-level input/output services for the DEC VT100 terminal, the Hewlett-Packard 2621 terminal, and the DEC GIGI color graphics terminal.

The capabilities and usefulness of a visual design consultant system were demonstrated on a small scale with a design aid program written to help an application programmer employ a few of the principles and guidelines from the Williges and Williges report. The aid helped the application programmer create a menu-selection interface for a given application program. The programmer used the aid to design a hierarchy of menus, each with a title, instructions, and choices. Some choices lead to submenus, while others lead to some specific action performed by the application software. The programmer could designate certain choices as more important, or more dangerous, and have those choices indented or enlarged. The programmer can also control the colors employed for each choice. The aid automatically formatted each menu, taking into consideration the relative sized of normal choices, emphasized choices, titles, and instructions. It also prevented the use of too many choices in one menu and provided advice on splitting menus and ordering choices within a menu. Menu editing capabilities were built into the aid. A VT100 terminal was used by the application programmer to communicate with the menu design aid. A GIGI terminal was used to display the menus.

The first generation VIDEO system led us to understand the consultation situation for graphic arts design. Its ability to define problems and map the problem definition into knowledge about graphic arts was very minimal. Our work concentrated instead on understanding graphic arts principles and rendering an image given a rendering plan.[9, 11, 12] While useful, this work certainly failed to fulfill the goals set forth for VIDEO.

4. *CHARTMAKER, A Second Generation Visual Design Consultant*

In this section we describe a system to define a client's problem and then map that problem definition into background knowledge about graphic arts. We have initially restricted the graphic arts knowledge to a particularly well-understood domain, chartmaking. Our previous work had concentrated on graphic arts knowledge and on rendering images, and we felt that we had reached some reasonable understanding of that problem. The simplification adopted here, therefore, did not seem unduly restrictive. As we shall see, the problem definition phase was not affected by our simplification. Certain aspects of mapping the problem definition into background knowledge, however, were not entirely general and further work will be needed.

Since conventional expert system representation methods do not provide a viable framework for problem definition, we began investigating interviewing tools that would help build problem models. The impetus for this investigation was not to find the optimal user interface, but, more importantly, to find a mechanism able to capture knowledge in a problem definition setting just as a human consultant might.

The basic tool of interviewing is, of course, questioning. Clearly, one must ask questions to gain information, and this is what both experts and conventional expert systems do. The fundamental difference lies in the type of questions asked. As stated before, the typical human consultant allows more freedom than do the standard "yes/no" or "select one from the following" questions of conventional expert systems. The human expert seeks richer content by requiring more detailed answers in an effort to build a mental model of the problem. This mental model is vital to a system in the design domain because the expert needs a solid understanding of the client's viewpoint in order to design what the client wants. If the expert is to accomplish this, he must permit the client to describe the situation in his own terminology. To be successful in the design domain, the consultant must be able to "step into the client's shoes" and to see the problem from the client's perspective. Because conventional expert systems have focused on domains that are completely well defined prior to any consultation, they require less information and work with a less complex model, typically something like a decision tree.

A human consultant interviews a client using unrestricted natural language. We cannot rely on such a procedure and therefore must seek powerful yet well-defined tools for building problem models. Industrial engineering offers just such a modeling tool: system dynamics modeling (SDM). System dynamics provides a model of an objective reality using directed graphs that link relevant concepts by arrows. The arrows in the graph indicate a causal influence from one concept to another, and a link is usually accompanied by a plus (+) or minus (−) sign indicating direct and inverse influences respectively.[3] The aim of the problem model in the automated consultant is to represent the client's reality, so system dynamics provides a viable framework for problem representation in the design domain. In system dynamics, the user specifies both the concepts of the model and their influences on one another (see Fig. 2). System dynamics provides a strong generic basis for representation. This is evidenced by some of its actual applications: modeling a magazine publisher's business concerns, situational political influences,[10] and inventory tracking and control.[3] DYNAMO, a system modeling program, employs system dynamics in order to simulate influence in causal models.[10] DYNAMO, however, focuses on simulation while our work concentrates on using the structure of system dynamics as a framework for knowledge representation. The generic qualities of system dynamics make it the obvious choice for representing knowledge in the design domain. In order to exploit the capabilities of system dynamics, the automated design consultant uses the basic influence model as a cognitive map of the user's beliefs. The system and the user can build a useful and informative model of the user's problem by incorporating both the user's terminology and perspective, just as a human expert does.[5]

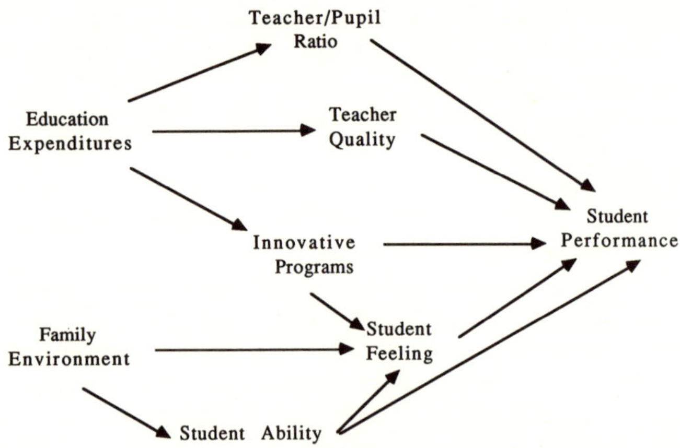

FIGURE 2. A sample influence diagram indicating causal factors on a student's performance in school.

4.1. Applying Background Knowledge to the Client's Problem

Once the consultant has let the client define a problem, background knowledge must be applied to create the design. We have divided background design knowledge into two parts: a theory of design and knowledge about graphics, that is intially about charts. We use a theory of design derived from a theoretical work in this area[1, 4, 15] that calls for a sequenced set of representations. We synthesized rather simple graphical knowledge about charting from Refs 13 and 14.

4.1.1. A Framework for Design Problems. The design problem framework for a true consultant requires more than a single representational structure. The system dynamics model forms only one of a series of internal representations that eventually lead to a rendered chart. Actually, the theory of a design consultant presented here requires a series of three internal representations, each associated with a distinct phase of problem interpretation: a situation model, an intent model, and a rendering model. This "pipeline" of representations is a major departure from traditional expert systems. Instead of forcing all aspects of a consultation into a single structure, the structure is designed to suit the nature of each particular aspect of a given problem. For the design domain, the entire consultation can be broken down into three basic areas: the client's conception of the situation, the client's primary interest (goal), and the model for the rendered chart. Through the pipeline, each ensuing model is constructed using information derived from the previous model and from background information already known about the domain.

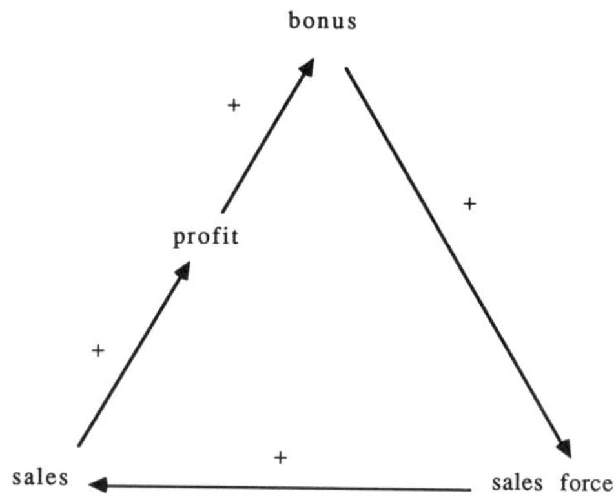

FIGURE 3. A systems dynamic model indicating how increased sales result in improved bonuses to the sales force.

The situation model is the user's interpretation of the situation. The user describes the problem by constructing an "influence" diagram (Fig. 3) consisting of concepts and influence links. The influence links represent the casual interaction of the concepts in the given problem. Another fundamental difference between CHARTMAKER and conventional expert systems lies in the interview phase. Instead of the system asking the user predefined answer-type questions, the user describes the problem in his own terminology and constructs the problem as he sees it—not as an instance of some predefined classification system.

Once the complete situation model is constructed, the user is asked to identify the important or key concepts in the model. A key concept reflects a basic issue that the presenter wishes to convey through use of the chart. After the client identifies the key issues, CHARTMAKER reduces the situation model by focusing on the key concepts and their causal effects on one another. The system then produces a new, simpler model of the problem. This model consists only of the key attributes and their derived (transitive closure) influence links (Fig. 4). This reduced model represents the user's intent in the presentation. For example, the user's intent could be to show

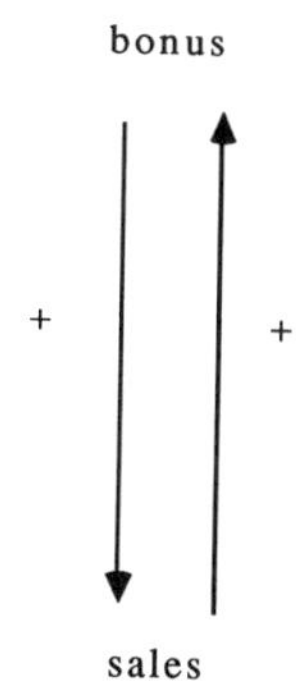

FIGURE 4. Simplified version of Fig. 3 showing sample key nodes and their connecting arcs formed by transitive closure.

a relationship between two key quantities, and this intent is portrayed by a set of reduced representations. This intent model is the second in the representation pipeline.

The reduction phase represents a new approach to goal elicitation. Instead of asking the user directly what type of chart he wants, the system derives this information from the user's problem description. Indirect goal elicitation is vital in the design domain. For example, when a client requests a human factors specialist to design a computer screen display, the specialist does not ask the client "What type of screen layout do you want?" because such a question eliminates the usefulness of the specialist. The consultant typically asks questions like, "What information do you need to see most often?" or "What information do you look for first?" The consultant uses these questions to construct a mental model of the client's situation. From this mental model, the consultant applies background knowledge in graphic arts and human factors to the basisc structure of the problem in order to create a screen design. In essence, the client cannot explicitly describe the type of screen desired but can describe the situation that dictates the design, and the consultant uses this situational description in determining the client's "goal."

4.1.2. Charting Knowledge. From the goal model the automated consultant creates a rendering model. The rendering model (Fig. 5) contains chart-specific information, such as axis labels, interval spacing, and even the particular type of chart to be constructed. In constructing the rendering model, the system relies on background charting knowledge. For example, even when the system knows that it is plotting a trend-type chart, it must still decide between a line chart and a column chart. This decision usually

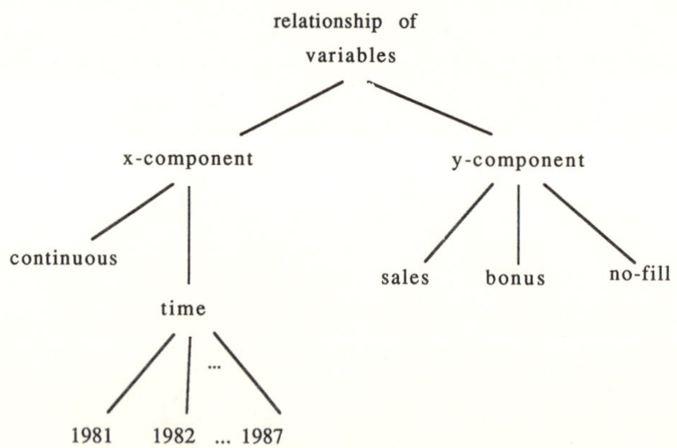

FIGURE 5. Internal representation of a plan automatically generated by VIDEO for rendering a chart.

depends on the number of sampling intervals, and if only a few intervals are available, the automated consultant will select a column chart.

4.2. Comparison with Other Charting Systems

In constructing a system dynamics model, the goal is to define a user's problem as thoroughly as possible before attempting to render a design. During the planning process, or problem definition phase, the problem is first defined by the user in terms of the necessary concepts and their associated influences. Instead of requiring the user to specify "pie" or "bar" as in Lotus-123 and Microsoft Chart, or even to select a functional description like "trend" or "comparison of variables," the system determines the appropriate chart from the nature of the data and the user's actual objective in the presentation. This is achieved by drawing inferences from the situation model. The user's implied objective may not necessarily be as simplistic as "to show a pie chart," but may be something more abstract and less delarative, such as "to show the importance of increased sales on employee bonuses." From this high-level description and the user's indication of key concepts, the automated consultant determines the actual purpose of the presentation without actually asking the user to state the goal. In essence, the user may not always be able to do this succinctly, but he can always describe his situation—as he would to a human graphic design consultant.

Many modern commercial graphics packages allow the user to render the chart of choice quickly, but the charting domain is substantially more complicated than "plot bar chart" or "plot pie chart." Amazingly, the most advanced and innovative systems currently available—Microsoft Works, Jass, Cricket Graph, and Lotus 123—require the user to specify explicitly the type of chart desired,[(2, 7, 8)] with no regard for the appropriateness of the chart to the data or the presentation. The emphasis of these and other "state-of-the-art" graphics programs is on improving the visual aspects of the chosen graph—using clever graphic design techniques like shading and coloration to enhance the aesthetic appeal of the chart. While these programs may create appealing visual displays, the actual display may be totally inappropriate for the charted data, the intended audience, and the uninitiated presenter. The absence of progress in user interfaces for these programs provides added impetus to use charting as the target domain for the expert system reported here.

The CHARTMAKER system that we have built uses a design theory to map from a user-described problem into a set of chart types (i.e., a classification system for charts) which forms only a small part of an expert's knowledge about charts. The freely described problem model, the goal model, the pipeline of descriptions, and the knowledge needed to map from one description to the next create a new model of expertise, the "true consultant." In this model, we have separated many different areas of an expert's knowledge that were either combined together or neglected in the rule-based paradigm of expertise.

5. Discussion

Now that we have described an initial working system, CHARTMAKER, we obviously wnat to determine the generality of the metodology outlined here. The framework for constructing CHARTMAKER is quite general; the major difficulty in adapting it to other domains comes in mapping a problem description into design knowledge. Several difficulties currently present us from realizing this mapping: we are considering several different alternative design theories, and without a definite design theory we are unsure of how to proceed with a mapping. At this time, however, systems utilizing the CHARTMAKER framework for realizing an airline reservation agent and a carrier air traffic control screen have been figured out. We believe that these systems can be programmed and can also be generalized.

6. Conclusions

A true visual design consultant for improving human–computer interaction cannot be realized as a normal knowledge-based system. We have devised a new knowledge-based system framework and identified the knowledge sources needed for implementing VIDEO. A full-blown system will not be possible for a long time; however, realizations of the consultant for a few specific problems will be available in the near future.

References

1. M. Asimow, *Introduction to Design*, Prentice-Hall, Englewood Cliffs, New Jersey, 1962.
2. C. Cox and D. Caldeway, Software reviews, *Computer Language*, Miller Freeman Communications, San Francisco, 1987, pp. 93–121.
3. R. Coyle, *Management System Dynamics*, Wiley, New York, 1977.
4. N. Cross, *The Automated Architect*, Pion Limited, London, 1977.
5. C. Eden, S. Jones, and D. Sims, *Messing About in Problems*, Pergamon Press, New York, 1983.
6. R. Kowalski, *Logic for Problem Solving*, North-Holland, New York, 1979.
7. B. Meng, Get to the point, *Macworld* April, 137–143, 1988.
8. S. Reed, The graphics challenge, *Personal Computing*, Hayden, Totowa, New Jersey, 1988, pp. 155–167.
9. S. Reilly and J. Roach, Improved visual design for graphics display, *IEEE Comput. Graph. Appl.* **4**(2), 42–51, 1984.

10. G. Richardson and A. Pugh III, *Introduction to System Dynamics with DYNAMO*, MIT Press, Cambridge, Massachusetts, 1981.
11. J. Roach, J. Pittman, S. Reilly, and J. Savarese, A visual design consultant, *Proceedings of the IEEE 1982 International Conference on Cybernetics and Society*, Seattle, Washington, October 1982, pp. 194–198.
12. J. Roach, S. Reilly, and J. Tatem, The dialogue author, intelligent tools: An automated advisor for visual design, in *Human–Computer Dialogue Design*, R. Ehrich and R. Williges, Eds., Elsevier Science Publishers, Amsterdam, 1986, pp. 165–177.
13. M. Spear, *Practical Charting Techniques*, Mc Graw-Hill, New York, 1969.
14. E. Tufte, *Visual Display of Quantitative Information*, Graphics Press, Chesire, Connecticut, 1983.
15. J.W. Wade, *Architecture, Problems, and Purposes, Architectural Design as a Basic Problem-Solving Process*, John Wiley, New York, 1977.

PARALLEL COORDINATES
A TOOL FOR VISUALIZING MULTIVARIATE RELATIONS

ALFRED INSELBERG AND BERNARD DIMSDALE

1. Introduction and Basic Results in Two Dimensions

The visualization of multivariate data and relations has attracted considerable attention in recent years (see, for example, Refs. 1–6, 9, 11, 13, 22, 23, 28, 29, 37, 39, 40, 42–47; this being only a partial bibliography), being of interest in many fields. Based on a multidimensional system of *parallel coordinates*, the development of the methodology described here began in 1978. Preliminary results on some representations and (paper and pencil) constructions for N-dimensional lines and hyperplanes appeared in 1981.[30] The results were extended in Refs. 31, 35 and 36. Interest in this method grew in recent applications to robotics,[15, 21] statistics,[14, 28, 48] computational geometry,[34] and other areas (see Refs. 17, 19, 24, 27, 33, 41). In conjunction with the design of the new air traffic control system an algorithm for early conflict detection and resolution* was derived using the results on lines in P^N, the projective N-space.[35, 36]

The approach presented here bears a superficial similarity with *nomography*[10] as well as a method of *graphical computation* in Ref. 25, a *profiles algorithm* in Ref. 26, and a special case of (1 and 1 plot in two dimensions) the *M* and *N plots* in Ref. 18 (the last four references as quoted in Ref. 7). Yet it differs in that parallel coordinates are used to obtain graphical *representations* of multidimensional *relations rather than just finite point sets* with

* Patent pending.

ALFRED INSELBERG • IBM Scientific Center, Santa Monica, California 90404 and Department of Computer Science, University of Southern California, Los Angeles, California 90089-0782. BERNARD DIMSDALE • IBM Scientific Center, Santa Monica, California 90404.

the systematic study of the mathematical structure of the representations leading to algorithms for *synthetic* constructions. It is a system for doing and *visualizing* analytic and synthetic multidimensional geometry.

On the plane with xy Cartesian coordinates, and starting on the y axis, N copies of the real line, labeled x_1, x_2, ..., x_N, are placed equidistant and perpendicular to the x axis. They are the axes of the *parallel coordinate* system for Euclidean N-dimensional space R^N all having the same positive orientation as the y axis—see Fig. 1. A point C with coordinates $(c_1, c_2, ..., c_N)$ is represented by the polygonal line whose N vertices are at $(i-1, c_i)$ on the x_i axis for $i = 1, ..., N$. In effect, a 1–1 correspondence between points in R^N and planar polygonal lines with vertices on x_1, x_2, ..., x_N is established. A *convex* hypersurface in R^N is represented by the envelope (see Ref. 8), taken one variable at a time, of the family of polygonal lines representing all points on the hypersuface. The key idea is that the description of a higher-dimensional object is captured, to a considerable extent, in the *two-dimensional representation* of the envelope of the polygonal lines representing its points. In short, a nonprojective mapping between N-dimensional and two-dimensional sets for *any* N is established. This will become clear as the discussion proceeds starting from line representation immediately below.

Points are denoted by capitals and lines (or arcs of curves) by lower-case letters, respectively. In parallel coordinates, the corresponding symbols are shown with a bar superscript (i.e., $\bar{l}$ represents the line l, $\bar{P}$ represents the point P, etc.).

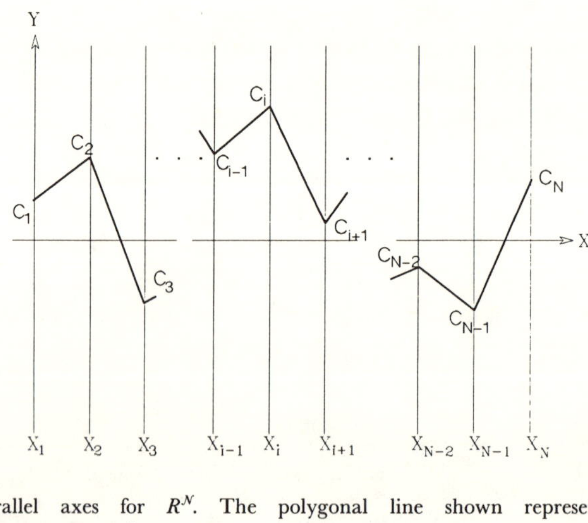

FIGURE 1. Parallel axes for R^N. The polygonal line shown represents the point $C = (c_1, ..., c_{i-1}, c_i, c_{i+1}, ..., c_N)$.

2. *The Fundamental Point* $\leftarrow\rightarrow$ *Line Duality and Curves in* R^2

Points on the plane are represented by segments between the x_1 and x_2 axis and, in fact, by the *line* containing the segment. In Fig. 2, the distance between the x_1 and x_2 axes is d. The line

$$l: x_2 = mx_1 + b, \qquad m < \infty \tag{1}$$

is a collection of points A. They are represented by the infinite collection of lines $\bar{A}$ on the xy plane which when $m \neq 1$ intersect at the *point*

$$\bar{l}: \left(\frac{d}{1-m}, \frac{b}{1-m}\right) \tag{2}$$

given with respect to the xy Cartesian coordinates. That is, the line l is represented by the point $\bar{l}$. The reason for representing a point P by the *whole* line $\bar{P}$, rather than just the *segment* between the parallel axes, is that $\bar{l}$ may lie outside the strip between the axes. For lines with $m = 1$, we consider

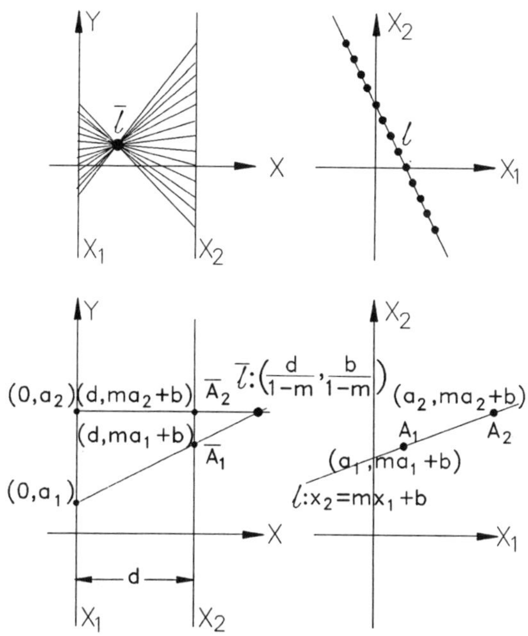

FIGURE 2. In the plane parallel coordinates induce quality. Top part shows the family of segments, representing points on a given line, *intersecting* at a point. Lower part shows the duality line $l \leftarrow\rightarrow \bar{l}$ point in general.

xy and x_1 and x_2 as two copies of the *projective plane*,[16] so that the line l corresponds to the *ideal point* $\bar{l}$ with tangent direction (i.e., slope) b/d (that is a point at infinity in the *specified direction*). Conversely, in the x_1x_2 projective plane the *ideal point* with slope m is mapped into the *vertical line* at $x=d/(1-m)$ of the *xy* projective plane. Hence, we have a *duality* between points and lines of the projective plane. This duality as expressed by means of homogeneous coordinates is a linear transformation, called *correlation* in projective geometry, between the line coordinates $[m, -1, b]$ of l and the point coordinates $(d, b, 1-m)$ of $\bar{l}$. Specifically, the correlation

$$C_A: l \longleftrightarrow \bar{l}, \qquad k(\bar{l}) = A[l] \tag{3}$$

where $[l]$ and $(\bar{l})$, the line and point (homogeneous) coordinates, respectively, are taken as *column vectors*, k is a proportionality constant, and

$$A=\begin{bmatrix} 0 & -d & 0 \\ 0 & 0 & 1 \\ -1 & -1 & 0 \end{bmatrix} l, \qquad A^{-1}=\begin{bmatrix} d^{-1} & 0 & -1 \\ -d^{-1} & 0 & 0 \\ 0 & 1 & 0 \end{bmatrix} l \tag{4}$$

maps lines of the x_1x_2 plane into *points* of the *xy* plane. This shows that the computation involved in going to and from parallel coordinates is minimal.

The practical problem of plotting points $\bar{l}$ representing lines with slope close to 1 can be solved by a simple stratagem. For a line $\bar{l}$ with *positive* slope m a reflection about the x_2 axis (i.e., $x_1 \to -x_1$ yields a line l' with *negative* slope $-m$. Consequently, the point $\bar{l}'$ lies between the x_1 and x_2 parallel axes. For any constructions and computations performed with $\bar{l}'$ the result must again be reflected to obtain the actual value of x_1 for the line l.

Consider now a curve in Cartesian coordinates as the collection of its

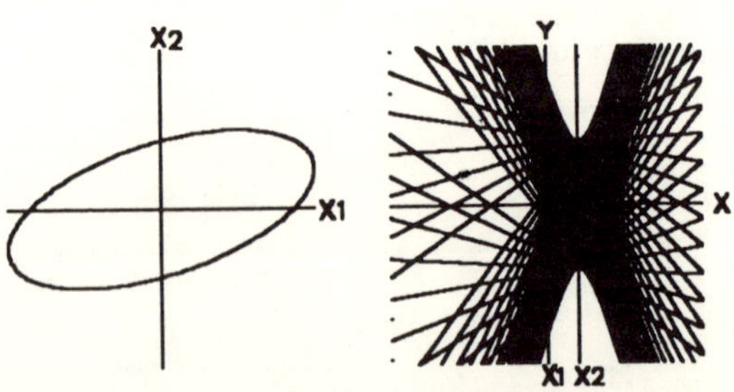

FIGURE 3. A (point) curve is mapped into a (line) curve formed by the tangents at each one of its points. Here an ellipse is mapped into a hyperbola. In general, conics are *always* mapped to conics.

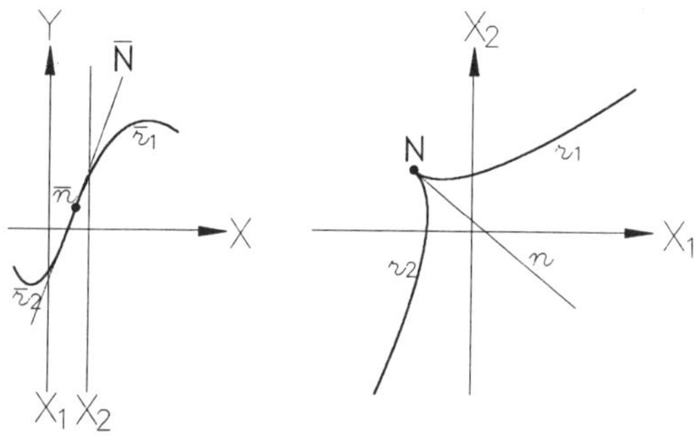

FIGURE 4. Cusps map into reflection points.

points (let us call it "*point curve*"). By means of the correlation C_A the image of this collection of points is a collection of lines which can be considered as tangents to another curve ("*line curve*"—see Fig. 3). On the plane conics map into conics but this property is more general. Sections of a double cone whose base is *any* bounded convex set, let us call them *gconics*, map into other such sections corresponding to bounded and unbounded convex sets as well *hstars* (generalized hyperbolas). This yields a new duality between bounded and unbounded convex sets and hstars as well as a duality between convex merge (convex union) and intersection. Based on these results efficient new algorithms for convex hull construction and the convex merge and intersection of convex sets were derived (see Ref. 34).

For nonconvex curves there is a striking duality between cusps and inflection points as shown in Figs. 4 and 5 which is *independent of the orientation.* Since it is computationally advantageous to test if a curve has a cusp rather than an inflection point, the duality is useful in the detection of inflection points in such applications as CAD/CAM and pattern recognition in general.

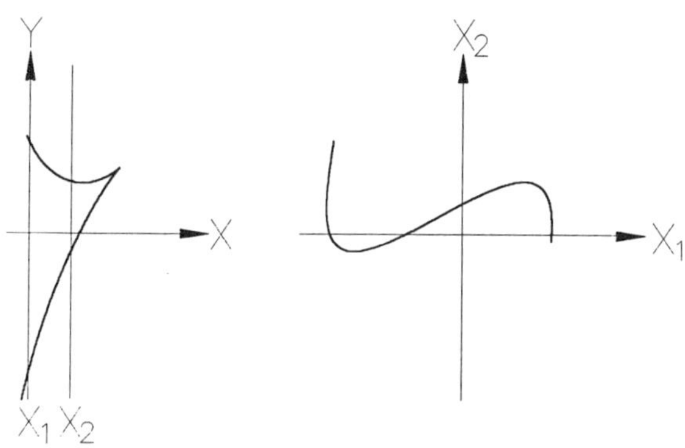

FIGURE 5. Inflection points map into cusps.

3. Extensions to Higher Dimensions

3.1. Multidimensional Lines

One of our objectives is to represent multivariate relations *geometrically* so that pencil and paper construction of points on the line is possible. Other objectives include the calculation of closest points on a pair of lines; and the development of a foundation for investigating the geometry of hyperplanes. Consider now a line l in P^N described by

$$l_{i,i+1}: x_{i+1} = m_i x_i + b_i, \qquad i = 1, ..., N-1, \qquad m_i \neq 0 \tag{5}$$

In the $x_i x_{i+1}$ plane the relation labeled $l_{i,i+1}$ is a line and by the correlation C_A translated appropriately it is represented by the *point*.

$$\bar{l}_{i,i+1}: \left(i - 1 + \frac{1}{1-m_i}, \frac{b_i}{1-m_i}\right) \tag{6}$$

There are $N-1$ such independent relations in the given set of equations, ergo the line l is represented by the corresponding $N-1$ points. For example, in Fig. 6 we see several points on a line interval in R^{10}. It is clear from the diagram how a point can be constructed on the line, for any given initial value of one of the variables. It is also clear how, given the equations or the coordinates of their equivalent points in parallel coordinates, points on the line can be calculated.

In the given set of equations the constraint $m_i \neq 0$ requires some atten-

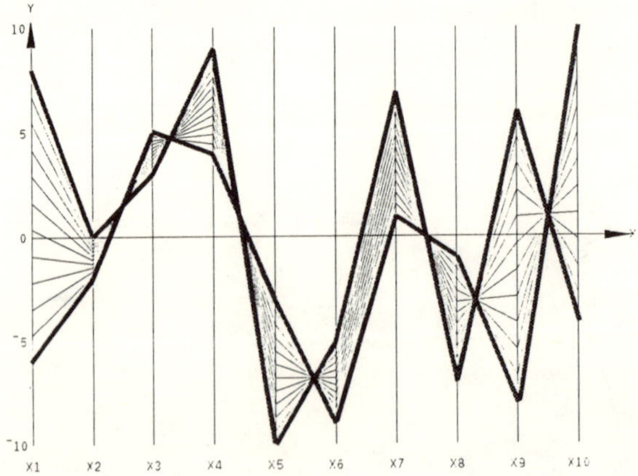

FIGURE 6. Interval on a line in R^{10}. Heavier polygonal lines indicate the end points.

tion. Note that if some intermediate $m_i = 0$ then the variable x_{i+1} has a fixed (nonvarying) value. There is no reason why a straight line should not have such a property. However, there is a deficiency in the procedure outlined above. For, starting with a value of x_1 the values of the variables up through x_{i+1} can be constructed (calculated) without difficulty, but it is not possible to continue. Conversely, starting at, for example, x_N it is not possible to construct (compute) values of the x's before x_{i+1}. One solution to a problem such as this is, for example, to replace the linear relation between x_{i+2} and x_{i+1} by a linear relation between x_{i+2} and x_i, thus bypassing the fixed value x_{i+1}. Note that there are still $N-1$ equations, hence $N-1$ points defining the line in parallel coordinates. Obviously there are many ways to represent a line algebraically by sets of linear equations each equation of which specifies a relation between two variables. Using graph theoretic methods it can be shown[35] that the distinct number of such systems is N^{N-1}, each of which are represented in parallel coordinates by exactly $N-1$ points.

Another difficulty has to do with graphic display of the $N-1$ points that define a line. For any slope $m_i = 1$ the corresponding point becomes an ideal point and can be represented as a direction. For slopes sufficiently near 1 the corresponding point is simply off scale. This problem can be resolved by modifying the definition of the representative point as follows. If $m_i > 0$, replace m_i by $m_i' = -m_i$. It can be shown that if this is done with appropriate attention to detail, points on the original line can be constructed (computed) by (1) constructing a pseudopoint with the modified representation, and (2) changing the signs of components in a manner consistent with the order in which the signs of slopes are changed. It can also be shown that the modified representative points are bounded in a manner greatly simplifying the graphical representation. For example $\bar{l}_{i,i+1}$ becomes $\bar{l}'_{i,i+1}$ whose x-coordinate is bounded by $(i-1, i)$ and whose y coordinate is absolutely bounded by $|b_i|$.

Another procedure for definition of a line in P^N is to select two distinct points P_1, P_2. The set of points on the line is then

$$P = \lambda P_1 + (1-\lambda) P_2 \tag{7}$$

Given two lines, or possibly segments each defined by a pair of end points, it is not difficult to calculate the shortest distance between the two lines, and the respective points of closest approach on each line. Also it is equally easy to solve this problem with the constraint that some one variable be the same on both lines. In particular the points of closest approach of two timed trajectories require of course that the time be the same for both trajectories. By the same token, if the lines intersect, their common point can be found

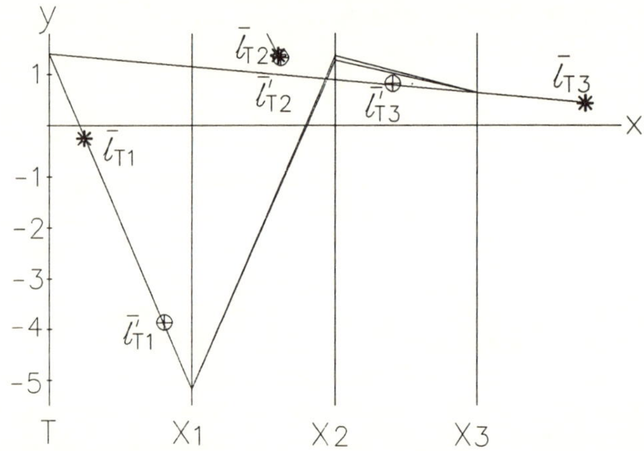

FIGURE 7. Nearly intersecting lines. Two points, one on each line, are found that are almost coincident.

and shown by an easy construction. It turns out that this construction can be used to discover *proximity* when the two lines are nearly intersecting or better yet when their minimum distance is less than some specified bound. This can be computed very efficiently *without need for computing the actual distance.* The construction is shown in Fig. 7 (for details see Ref. 36).

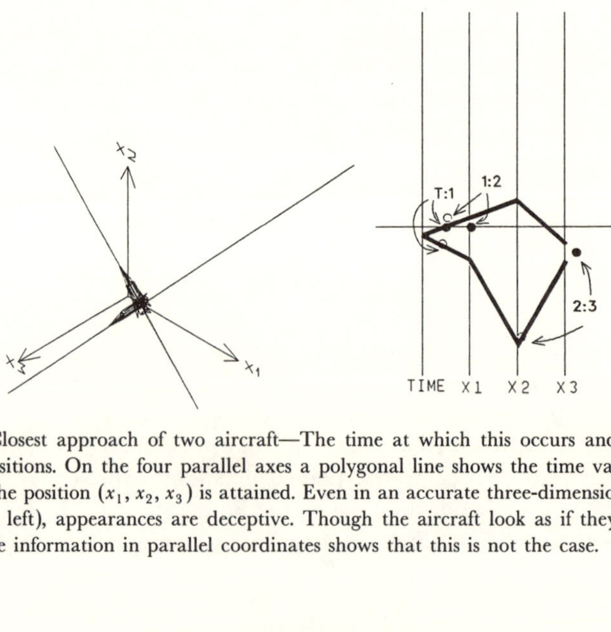

FIGURE 8. Closest approach of two aircraft—The time at which this occurs and their corresponding positions. On the four parallel axes a polygonal line shows the time value on the T axis, when the position (x_1, x_2, x_3) is attained. Even in an accurate three-dimensional isometric view (above left), appearances are deceptive. Though the aircraft look as if they are close to colliding, the information in parallel coordinates shows that this is not the case.

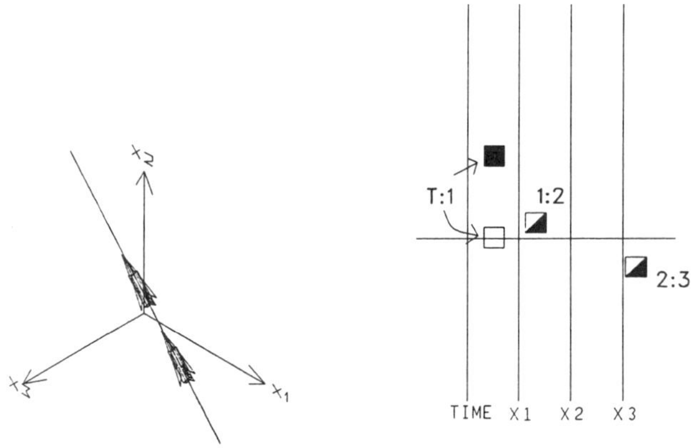

FIGURE 9. Two aircraft flying the same path with the same velocity. Note that the 1:2, 2:3 points (indicated by boxes) [these are the $\bar{T}$s given by Eq. (5)] are shared, indicating that the paths in three dimensions are the same. When that occurs, the leftmost T:1 corresponds to the greater speed. Here the airplanes have the same velocity since the two T:1 points have the same horizontal position.

There is an important application to *air traffic control** based on this representation. The trajectory of an aircraft is a function of time T and three space variables, say x_1, x_2, x_3. A straight line trajectory for constant vector velocity can be represented by three stationary points say T:1, 1:2, 2:3 as shown in Figs. 8 and 9. The time axis can be thought of as a "clock." At any given time T, the position of the aircraft is found by selecting the value of T on the T axis. From this representation lines that are "nearly" intersecting in *time and space* can be visualized as in Fig. 7. Such a representation also clarifies the problem of understanding collision as the intersection in space *and* time (i.e., both aircraft must be at the same position at the same time, a four-dimensional intersection) as contrasted to passing through the same position but at different times. An example is shown in Fig. 10.

Based on this methodology an algorithm for *conflict detection and resolution* was derived and tested on some complex scenarios provided by the FAA in conjuction with the design of the new air traffic control system.[38] By way of illustration a resolution not involving altitude changes (i.e., in two dimensions) for one of these scenarios (scenario 8) is presented. The initial position and constant velocity of the six aircraft involved is shown in Fig. 11. As can be seen from Fig. 12, several conflicts arise. An expert air traffic controller resolved the conflicts with four (4) aircraft remaining at the same altitude and changing the altitude of two. It is usually more desirable to avoid

* Patent pending.

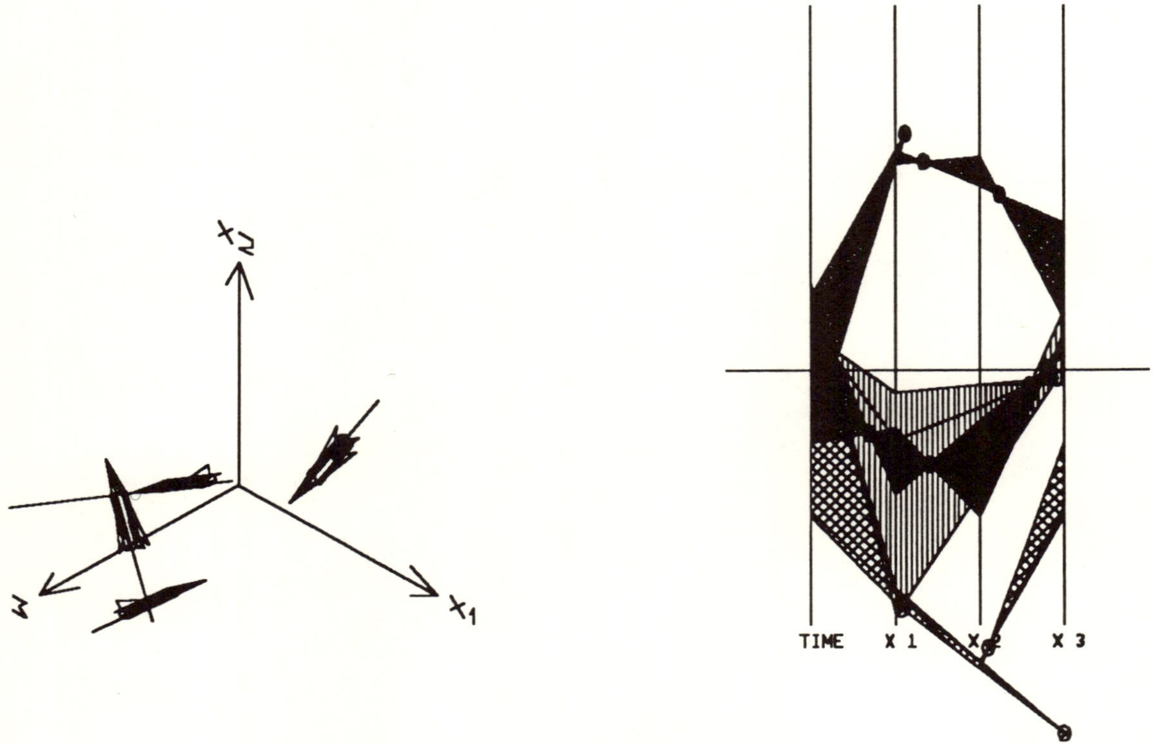

FIGURE 10. Aircraft flying straight line paths for specified time periods (segments). Here three paths intersect yet two aircraft actually collide. Such a situation is impossible to discern in three dimensions, though it is easily found (both algorithmically and visually) in parallel coordinates.

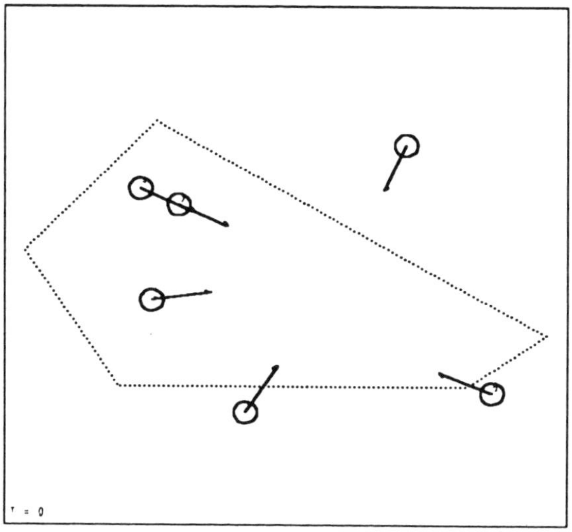

FIGURE 11. Six aircraft flying at the same altitude. Initial positions ($T = 0$ sec) and circles centered at each aircraft with radius 2.5 nm (separation standard) are shown to scale. Arrows represent the velocity vectors.

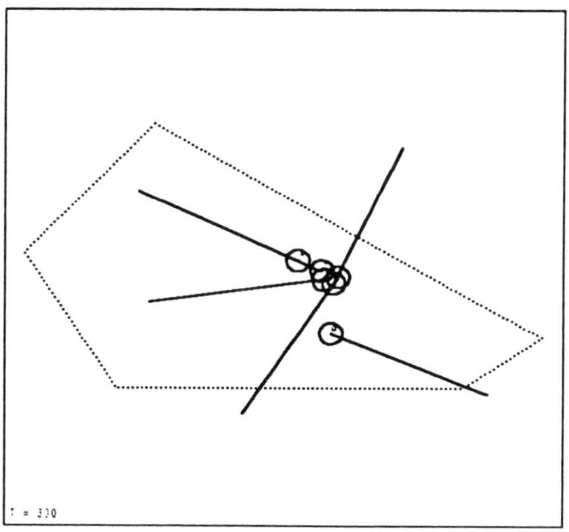

FIGURE 12. Conflicts among the six aircraft. A conflict occurs when the separation between any two aircraft is less than 5 nm (i.e., two circles intersect). Here several conflicts occur within the first 5 min. The time elapsed in seconds is indicated in the lower left-hand corner.

altitude changes. Our algorithm was able to resolve all conflicts without altitude changes. There are two key steps.

The first involves the computation of certain parallelograms which enclose the circular safety zones centered on each aircraft. For example, in Fig. 13 the parallelograms with respect to aircraft 4 and the remaining aircraft are shown. There may be none, one, or two adjacent parallelograms with each circle, and they are computed from the relative velocity of that aircraft with respect to aircraft 4. Any aircraft flying with the same velocity as 4 in order to avoid the circles *must avoid entering the parallelograms.* The second step involves the representation of lines in parallel coordinates by points so that the trajectory information for each circle is represented by intervals on the same line. Aircraft 4, in this case, is in conflict with another aircraft if and only if the points representing the trajectory of 4 are interior to the intervals associated with that aircraft. The closest conflict-free path for 4, parallel to the original trajectory, is represented by one of the end points of the union of these intervals. Unfortunately, it is not possible to provide a short exposition on this interesting application, and a proper one would take us too far afield. The interested reader is referred to two papers currently in preparation.[12, 32]

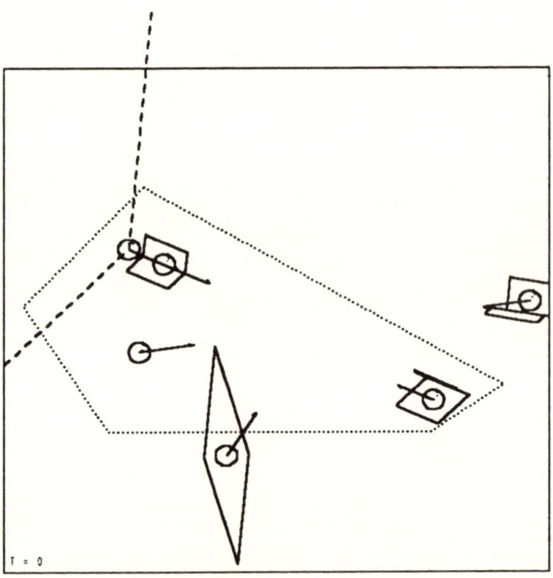

FIGURE 13. Conflict parallelograms. Here potential conflicts between aircraft 4 (where the two dashed lines intersect) and remaining aircraft are analyzed. An aircraft with the same velocity as 4 in order to avoid entering a circle must necessarily avoid the parallelogram(s) associated with that circle.

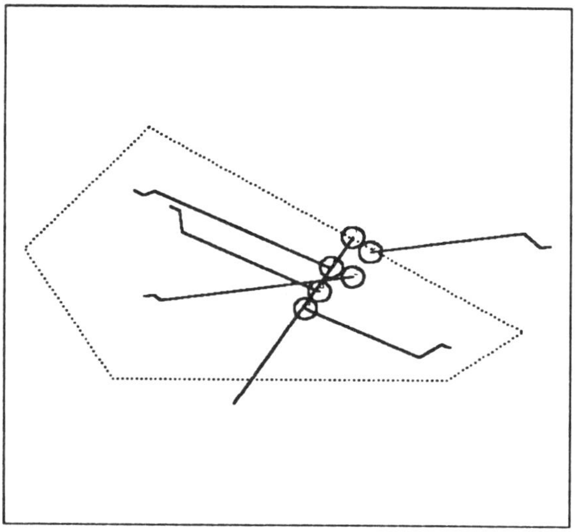

FIGURE 14. Three pairs of tangent circles.

Let us see how this works out. To resolve the conflicts, at $T=0$ five aircraft start maneuvers which eventually lead each of them to a new course parallel, and within 5 nm of the original one, where they continue flying at their previous constant velocity. This is shown in Fig. 14. There and also in Fig. 15 are seen instances of tangent circles. This is because the algorithm minimally disturbs the aircraft from their original paths. The

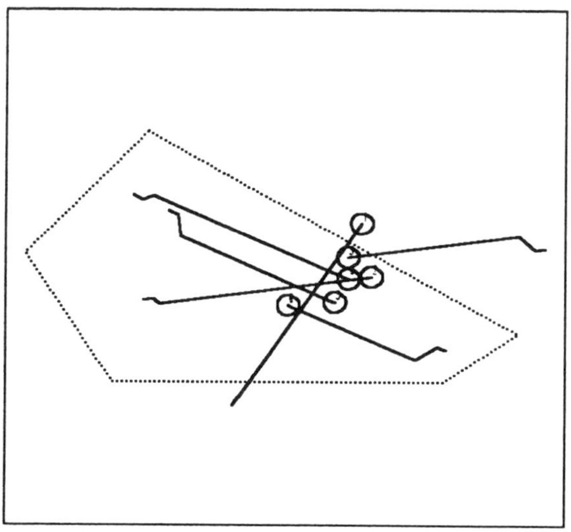

FIGURE 15. Triple tangency.

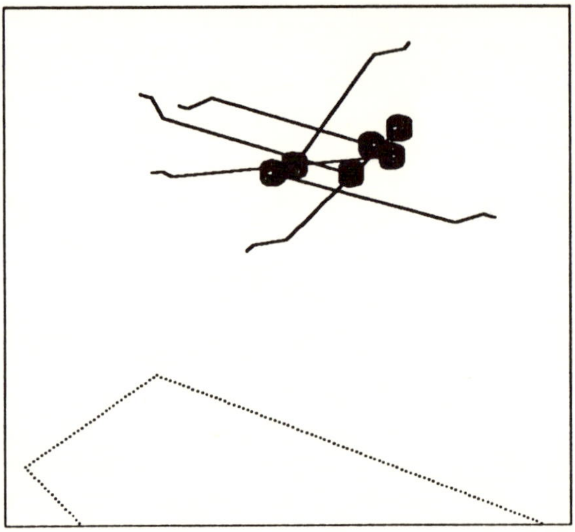

FIGURE 16. Resolution in three dimensions. Aircraft are at different altitudes and protected space is cylindrical.

method applies to aircraft at various altitudes as shown in Fig. 16 and for protected spaces which can be any convex shape, not just cylindrical. The generalization of this approach to more general motion planning in robotics is currently being investigated.[21]

3.2. *Hyperplanes*

Since a line in two dimensions is represented by a point whose x coordinate is $1/(1-m)$, where for convenience we set $d=1$, the set of parallel lines with slope m is represented by a vertical line at $X=1/(1-m)$ in parallel coordinates. In N dimensions a set of parallel lines transforms into $N-1$ vertical lines. This is the basis for the first of two representations of hyperplanes. Namely, since a hyperplane in N dimensions is a linear $N-1$-dimensional object, therefore, it can be represented by a system of $N-1$ parallel coordinates x_i, $i=1, 2, \ldots, N-1$. The parallel axes represent the family of lines parallel to each of the coordinate axes (i.e., $x_i=\text{const}$). Actually, a set of $N-1$ axes represent a whole *family* of parallel hyperplanes. A specific hyperplane of the family is identified by choosing a particular point on it and including the corresponding polygonal line in the representation. A specific example for three dimensions shown in Fig. 17—see Ref. 30 for a more extensive discussion.

Given a point and a hyperplane, it is possible from this kind of representation to decide whether the point is "above," "below" or on the hyperplane and hyperplanes so represented can be intersected with lines or

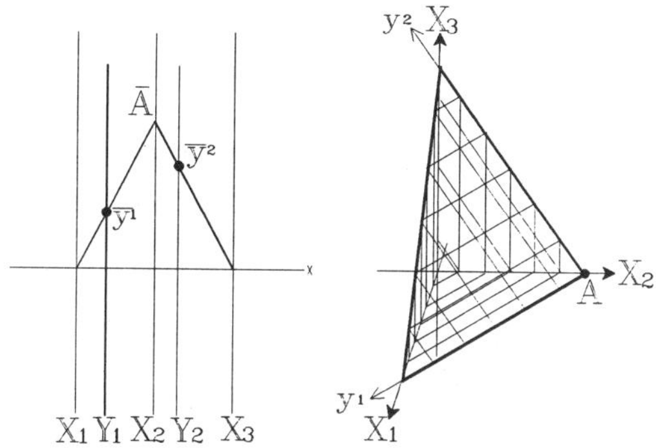

FIGURE 17. In P^3 planes are represented by two vertical lines and a polygonal line $\bar{A}$. This generalizes to $\mathcal{N}$-dimensions where hyperplanes are represented by $\mathcal{N}-1$ parallel lines and a polygonal line representing one of its points.

other hyperplanes. These and other properties have important implications for areas like linear programming and convex-hull construction in $R^{\mathcal{N}}$.

There has been considerable interest for this methodology in the area of statistics known as exploratory data analysis. In Fig. 18 a set of about 2000 eight-dimensional points is shown. They are actual measurements of an

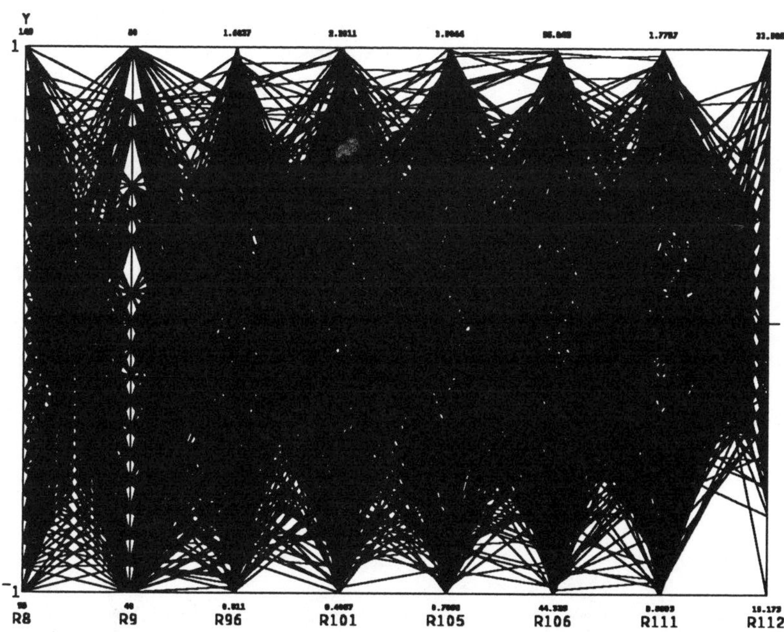

FIGURE 18. Industrial data. Note pattern between the variables R111 and R112.

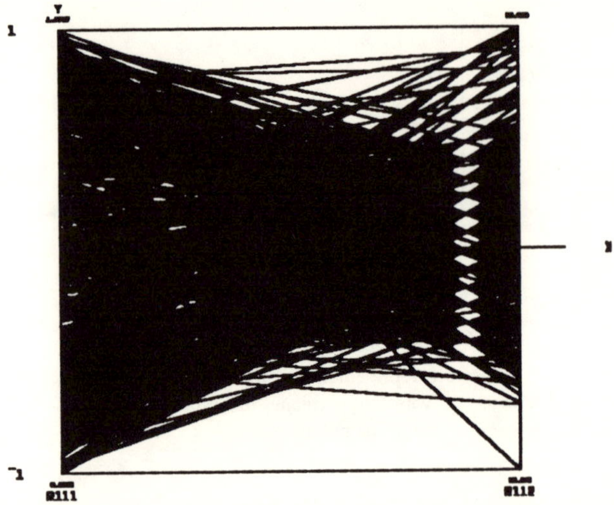

FIGURE 19. Enlarged R111–R112 portion of previous plot.

industrial process involving a number of variables, not all of which are shown here. The pattern in between the R111 and R112 axes resembles one of the vertical lines in the hyperplane representation. This is confirmed subsequently in Figs. 19 and 20. That a linear relation would be found in industrial data by mere inspection was surprising to us.

A feel for the power of the representation can be gained from Fig. 21, from which, with a bit of practice, the vertices, edges and faces, and their interrelationship, of the hypercube can be recognized.

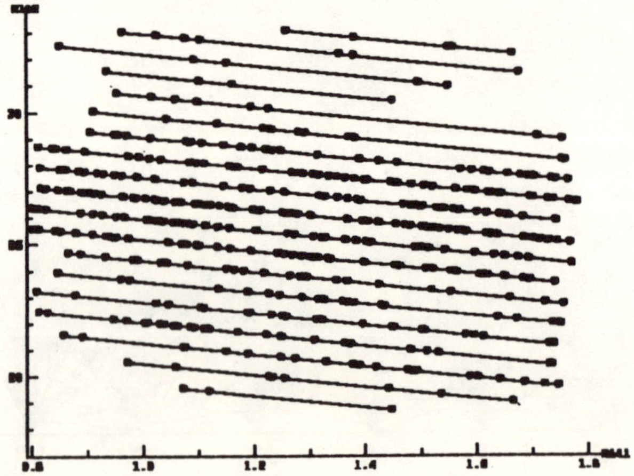

FIGURE 20. R111 versus R112 linear relation between these two and another parameter.

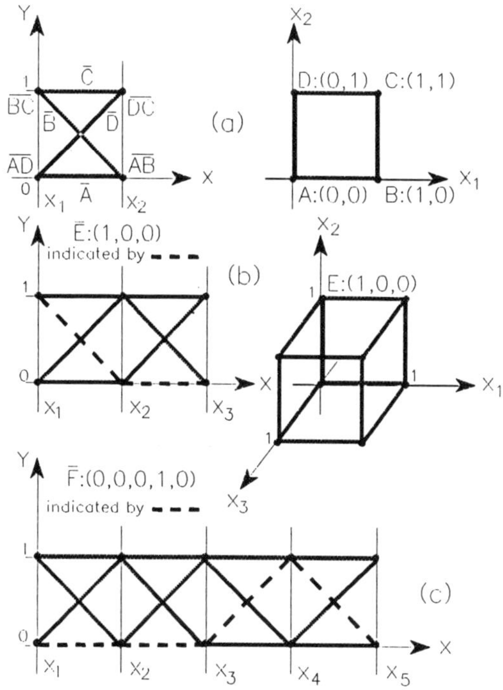

FIGURE 21. Hypercube representation in parallel coordinates. Graph of (a) square, (b) cube in three dimensions, and (c) cube in five dimensions, all having unit side.

There is an alternate elegant hyperplane representation due to Eickemeyer.[20] To motivate it let us look at a set of three-dimensional points (consider at first only the portion in the first three parallel axes) shown in Fig. 22, which happen to be on a plane π and are not equally spaced. So the vertical lines revealing coplanarity of the previous representation are not seen. In fact, no discernible "structure" seems evident. Now let us form lines on the plane π from pairs of these points. These lines are represented by a pair of planar points indexed with two subscripts (i.e., two out of the three *collinear* points $\bar{l}_{12}$, $\bar{l}_{23}$, $\bar{l}_{13}$) obtained from (6). The collection of lines through the $\bar{l}$ points for every line l on any plane turns out to form a pencil of lines at a point $\bar{\pi}_{123}$ as shown in Fig. 23 (i.e., all the lines intersect at the *same* point). In fact this turns out to be a necessary and sufficient condition for coplanarity! Since a planar point has two independent coordinates, more information is needed to identify the *specific* plane. This may be done by taking another copy of the three parallel coordinate axes as shown on the right-hand side of Fig. 22 (hence the six axes seen) and represent the original points on π with respect to these

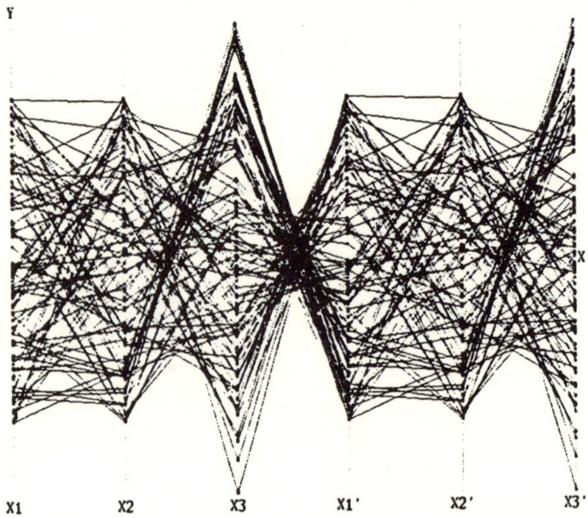

FIGURE 22. On the first three axes a set of coplanar points is shown. The points are not equally spaced with respect to any of the variables.

axes which effectively have a different horizontal distance between them than the original set, repeating the process and getting another pencil of lines as shown in Fig. 24 at another point say $\bar{\pi}_{231'}$ which turns out to have the same y coordinate as $\bar{\pi}_{123}$ so the coordinates of the two points have enough

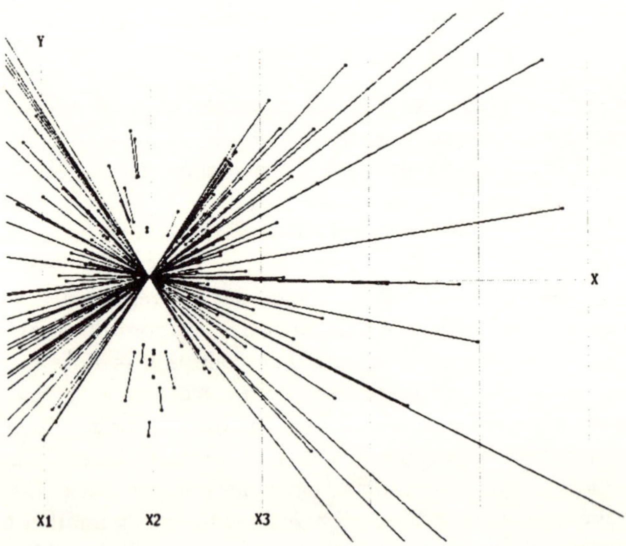

FIGURE 23. Coplanarity. The two planar points representing lines on three-dimensional plane form a pencil of lines.

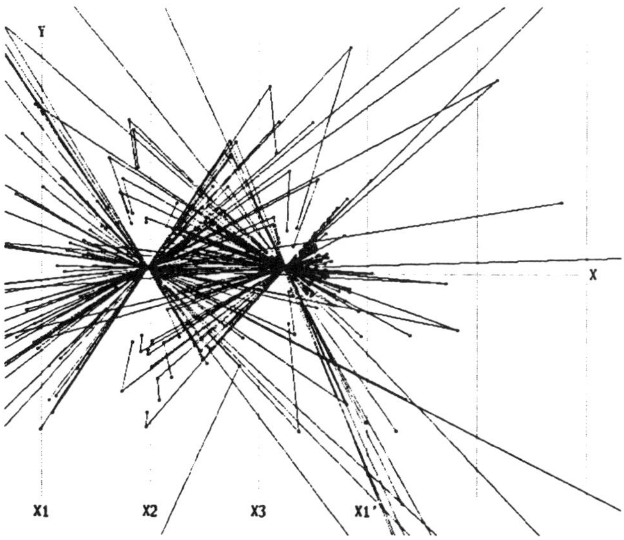

FIGURE 24. A plane in three dimensions is represented by two points.

information between them to identify the plane π that they represent. It is rather remarkable that a plane in three dimensions, a two-dimensional object technically called *2-flat*, can be represented in terms of two points which are obtained from the lines (one-dimensional objects calles *1-flats*) on the plane. Each such line is in terms represented by *two* points. To distinguish the situation where points represent lines we label them with two indices and points representing planes are labeled with *three* indices. Altogether then, in order to discover coplanarity of *points* it is recommended that *lines* be formed by appropriately sampling the points by pairs. Then coplanarity can be found *independently of the coordinate system and VIEWPOINT of the observer*, which does not enter in this representation.

It is not difficult to think of situations where discovering coplanarity is of interest, i.e., computer vision, statistics, CAD/CAM, etc.. In such cases one would expect the data to have errors, so it is important to see if this representation is numerically stable with "noisy" data. So a set of points that is within "20% of coplanarity" in the sense that given

$$\pi, \pi_n: c_1x_1 + c_2x_2 + c_3x_3 = c + c_n \qquad (8)$$

points on the nominal plane π are found with c_n while points on the "noisy" plane π_n are computed by allowing the variation $|c_n| = \geqslant .2|c|$ from the nominal as shown in Fig. 25. Repeating the process by forming lines, again two clusters of points are clearly seen in Fig. 26. According to some *best fit*

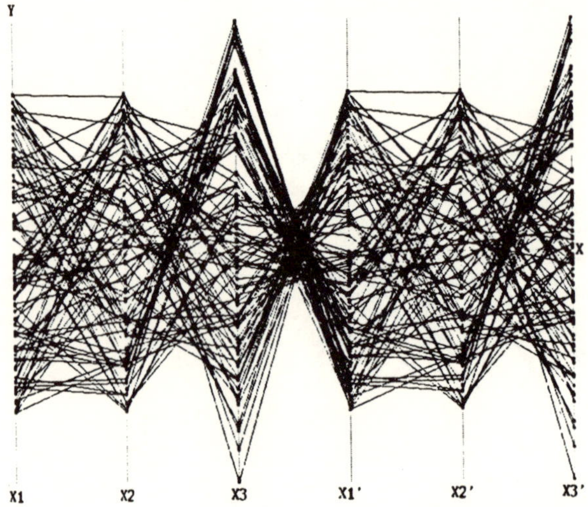

FIGURE 25. Points on a "noisy" plane in three dimensions on left three axes.

criterion a pair of points can be chosen to represent the plane. Then the points in the clusters show the variance.

This representation, as well as the numerical stability, turns out to be completely general. As an example consider the six-dimensional points with the same kind of variation from some nominal hyperplane shown in the first

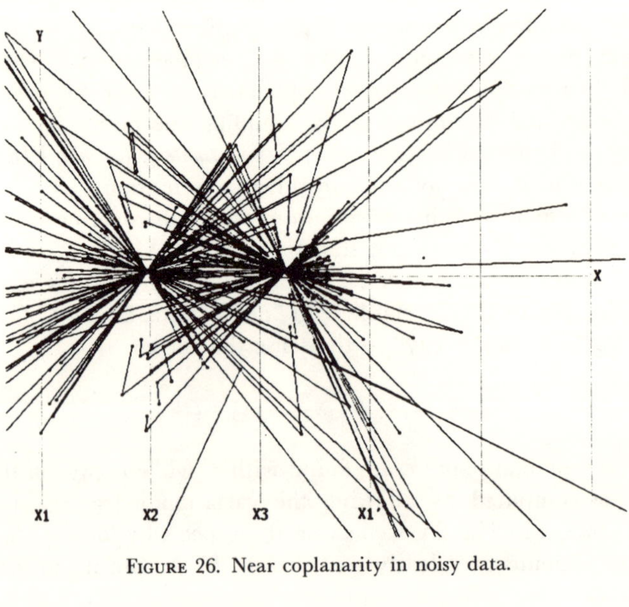

FIGURE 26. Near coplanarity in noisy data.

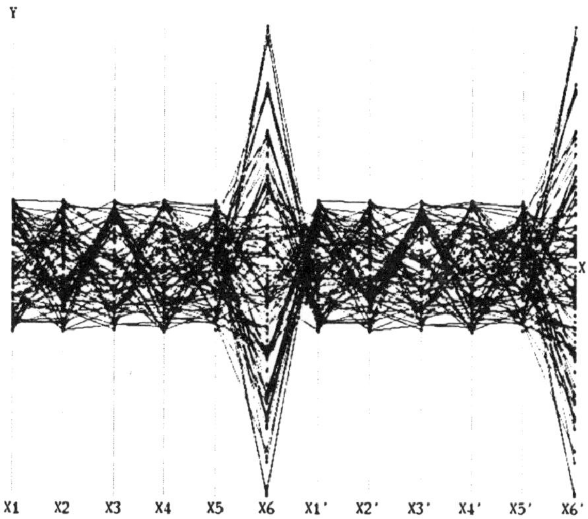

FIGURE 27. Points (0-flats) on a noisy hyperplane in six dimensions.

six axes in Fig. 27. Repeating the process as is shown in Figs. 28, 29, 30 by joining successive pairs of points representing linear objects increasing dimensionality, the near coplanarity is discovered from the pencil of lines formed by the points representing the four-dimensional hyperplanes (4-flats) contained in the hyperplane as shown in Fig. 31. Repeating this process for the successive axes permutations 234561′, 34561′2′, 4561′2′3′,

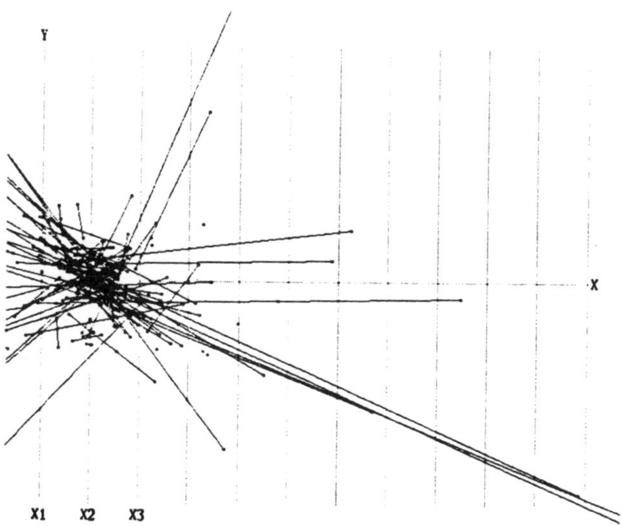

FIGURE 28. Portions of lines (1-flats) formed from the previous points. No structure is apparent.

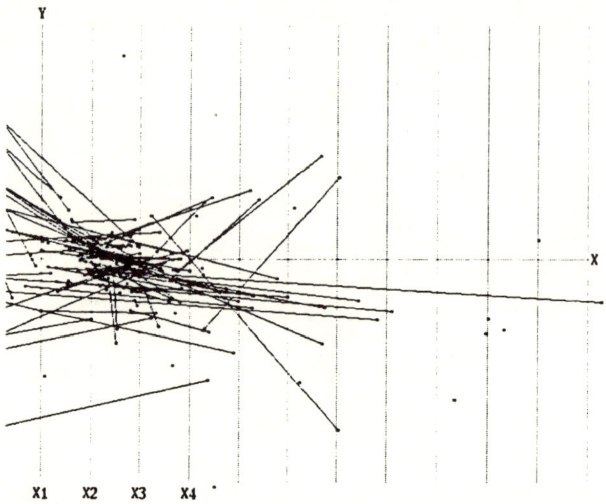

FIGURE 29. Portions of planes (2-flats) formed from the previous lines. Again no structure is seen.

561′2′3′4′ yields four more such points. The position of these five points, labeled by the six indices corresponding to their axes permutation, completely identifies the nominal hyperplane. In general, in N dimensions an (N-1)-flat is represented by N-1 pencils of lines (i.e., points) representing (N-2)-flats, each point labeled with N indices.

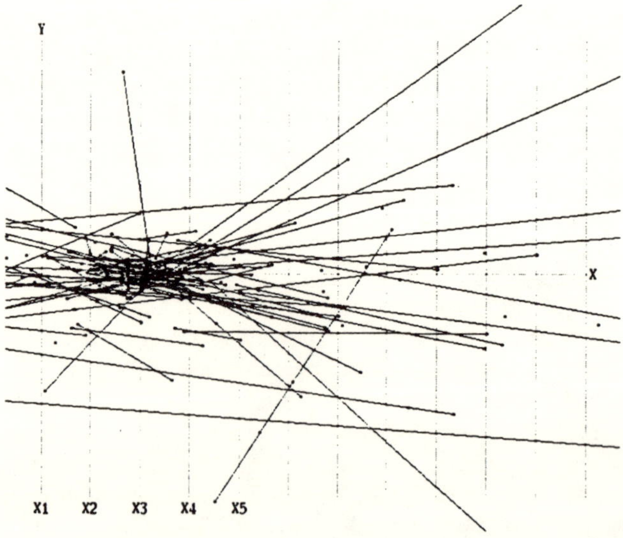

FIGURE 30. Portions of 3-Flats formed from the previous 2-flats. Still no structure.

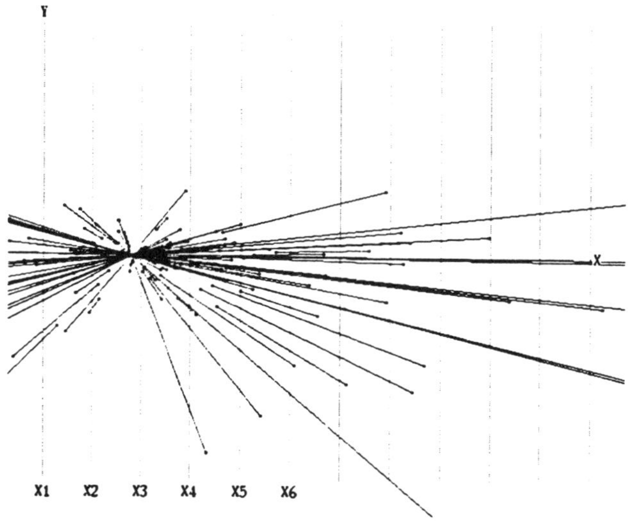

FIGURE 31. Portions of 4-flats formed from the previous 3-flats. The points are very near to a hyperplane (5-flat) in six dimensions.

Points belonging to several planes may also be "separated" in this way. Starting with 100 points in three dimensions the process is repeated and 100 lines are formed by sampling pairs of points as shown in Fig. 32. The positions where lines intersect, as well as the number of intersections at that position, is recorded (Fig. 33), with only two positions having more than two

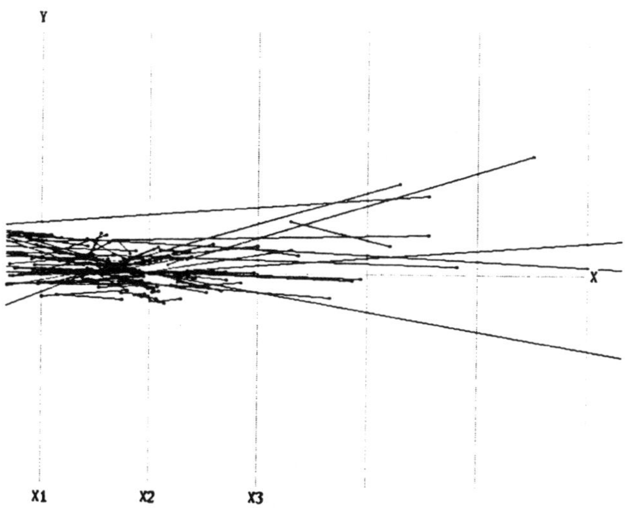

FIGURE 32. Portion of lines formed from a point set.

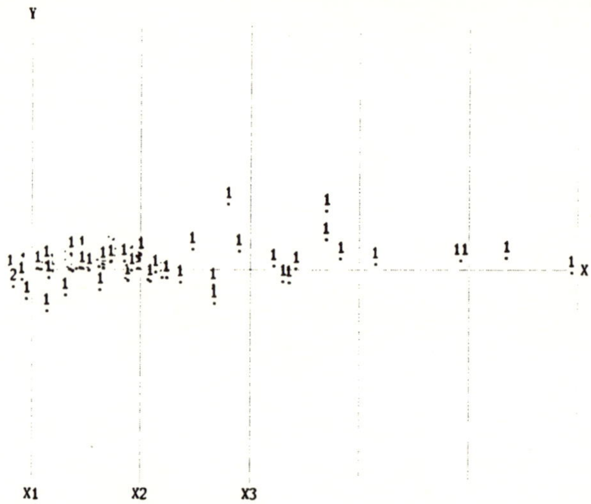

FIGURE 33. Histogram showing number of intersections per position.

"hits" (Fig. 34), indicating that the original point set was on *two* different planes. This separation method works also in higer dimensions—see Fig. 35, where the 4-flats are formed from a set of 500 points in five dimensions. The number of intersections per position is shown in Fig. 36, with two positions having more than one (Fig. 37), showing that the points are on two

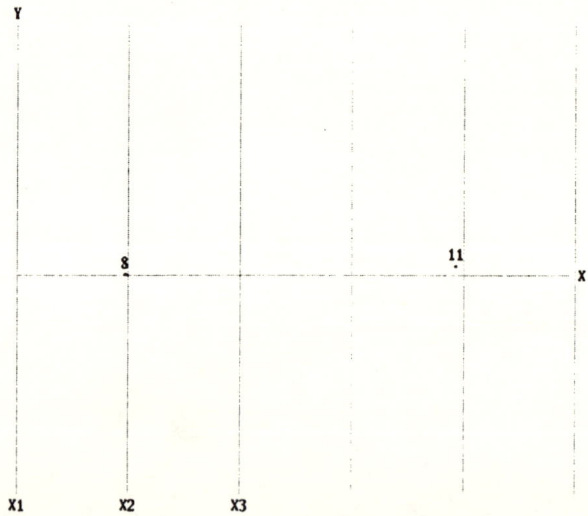

FIGURE 34. Only two points have more than one intersection. This shows that the original points belonged to two different planes. To completely identify the planes this process may be repeated with another axes permutation to obtain two more points.

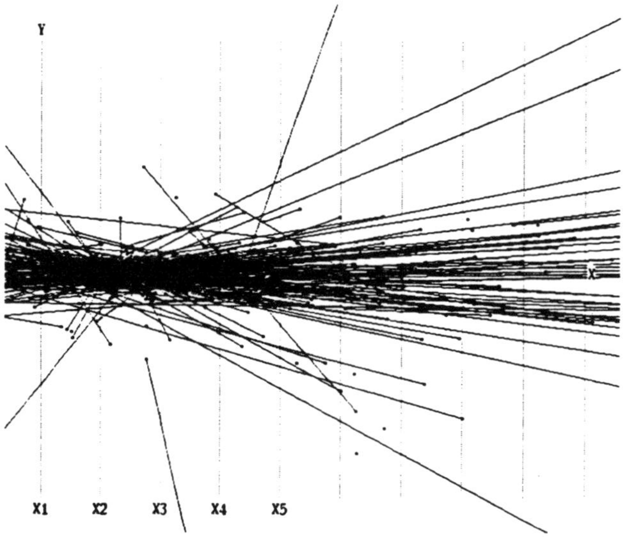

FIGURE 35. Portions of 4-flats formed from a set of points in five dimensions.

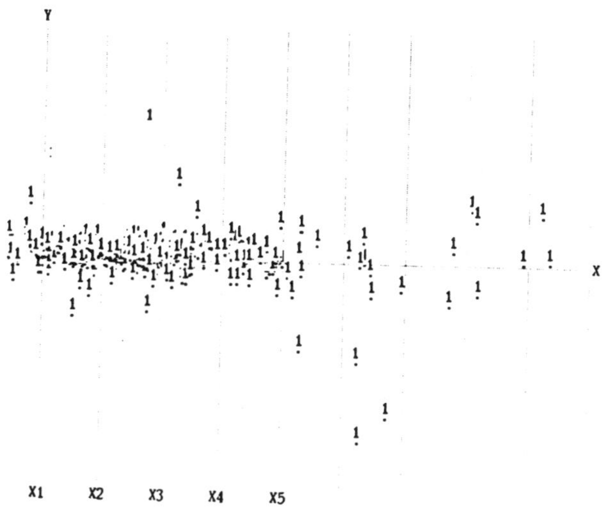

FIGURE 36. Number of intersections per position.

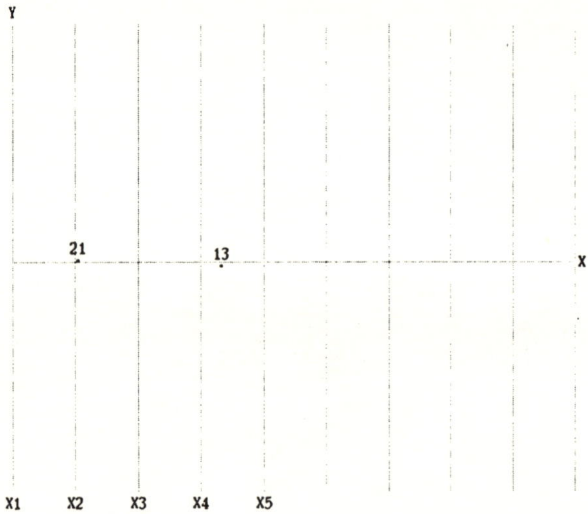

FIGURE 37. Two "hits" with more than one intersection. Points are on two hyperplanes.

separated hyperplanes. In general a *p-flat* in N-space, specified by the $(N\text{-}p)$ linearly independent equations,

$$\pi_{i_1 \ldots i_{(p+1)}}: \sum_{k=1}^{p+1} c_{i_k} x_{i_k} = c_0 \tag{9}$$

$1 \leqslant i_k \leqslant N$, $i_j \neq i_r$, $j \neq r$, is represented by the $(N-p)p$ points, given in homogeneous coordinates, by

$$\bar{\pi}_{i_1^* \ldots i_{(p+1)}^*}: \left(\sum_{k=1}^{p+1} d_{i_k^*} c_{i_k}, c_0, \sum_{k=1}^{p+1} c_{i_k} \right) \tag{10}$$

where the axes $x_1, \ldots, x_N$ are followed by the axes $x_{1'}, \ldots, x_{N\eta}$, d_i is the distance from the y axis to the x_i axis and

$$i_k^* = \begin{cases} i_k, & j \leqslant k \\ i'_k, & j > k \end{cases} \tag{11}$$

This is a very recent result and its ramifications are just being explored. The identification of several hyperplanes and in fact the subject of geometric modeling of polyhedra in N-dimensions, with the very exciting prospect of visualizing polyhedra in any dimensions in terms of connected indexed points, is now open to investigation.

3.3. Finding Interior Points to Certain Hypersurfaces in P^N

The problem of finding interior points to hypersurfaces is important in applications like process control. Here an algorithm for finding and *visualizing* such points is given. In the interest of brevity the algorithm is discussed concurrently with the representation of some of the hypersurfaces to which it applies. The point is constructed by succesively defining its coordinates in some order. The following steps are required:

1. Initially find the *range* of the variable whose value is going to be selected.
2. Select a value of the variable in this range. This reduces the dimensionality of the point selection problem.
3. Find the range of the next variable for the reduced convex hypersurface and select a value in that range.
4. Repeat the previous step until the last variable is selected.

The convexity and boundedness of the convex hypersurface guarantee that in each case the range is an interval. The success of the algorithm hinges on finding the intervals efficiently.

As we will see, for bounded convex hypersurfaces the intervals are found by taking the projections of sections of the hypersurface. The boundary of these sections is transformed into the parallel coordinates resulting in the stepwise display of the point selection process. Such a display provides

1. Information on the *local* properties of the hypersurface; and
2. *Proximity* information for the selected point—i.e., a measure of distance of the point from the boundary.

These points are illustrated subsequently when the algorithm is applied to some specific kinds of hypersurfaces—for example, the algorithm for certain nonconvex hypersurfaces where the available ranges are a *finite union of intervals* at every stage of the point selection process.

3.4. The Hyperellipsoids

Let A be a positive definite symmetric matrix and consider the function

$$F(X) = (X, 1)\begin{pmatrix} A & e^T \\ e & \alpha \end{pmatrix}\begin{pmatrix} X^T \\ 1 \end{pmatrix}$$

where X is a row vector $(X_1, X_2, \ldots, X_N)$, e is a row vector, and alpha is a constant. The equation

$$F(X) = 0$$

defines a hyperellipsoid with real points X provided that

$$e(A^{-1})^T - \alpha > 0$$

A point is interior, on, or exterior to the surface according as $F(X)$ is less than, equal to, or greater than 0. Given a point X it is straightforward to determine its location relative to the hyperellipsoid. The object here is to find, component by component, an interior point.

There are two kinds of processes required for this purpose. Fixing the value of one of the variables produces a planar section of the hyperellipsoid which is itself a hyperellipsoid of dimension one less than the original; this necessitates the computation of a new array for the resulting hyperellipsoid. The other requirement is to calculate the available range of the remaining free variables by taking the projection of the reduced hyperellipsoid on a coordinate plane. Now let

$$A = \begin{pmatrix} A_1 & B^T \\ B & C \end{pmatrix}, \qquad e = (e_1, e_2)$$

where A_1 consists of the first two by two submatrix of A, C is A with the first two rows and colums deleted, B and B^T account for the remainder of A, e_1 is the first two elements of e, and e_2 is the remainder of e. It is relatively straightforward to prove the following.

THEOREM. Let $\tilde{X} = (X_1, X_2)$. The projection of $F(X)$ on the X_1, X_2 plane is the ellipse

$$f(\tilde{X}) = 0 = (\tilde{X}, 1)\begin{pmatrix} \bar{A} & \bar{e}^T \\ \bar{e} & a \end{pmatrix}\begin{pmatrix} \tilde{X}^T \\ 1 \end{pmatrix}$$

and

$$\bar{A} = A - B^T C^{-1} B$$
$$\bar{e} = e_1 - e_2 C^{-1} B$$
$$\bar{\alpha} = \alpha - e_2 C^{-1} e_2^T$$

Given the projection, the maximum ranges of X_1, and X_2 are easily calculated. The available range for X_2, given X_1, is also easily calculated.

The hyperbola that is the transform of the ellipse into parallel coordinates is essential for display of the selection process. For one thing, the available range of X_1 is the interval on the X_1 axis between the two branches of the hyperbola, and similarly for X_2. Furthermore, the *range* of

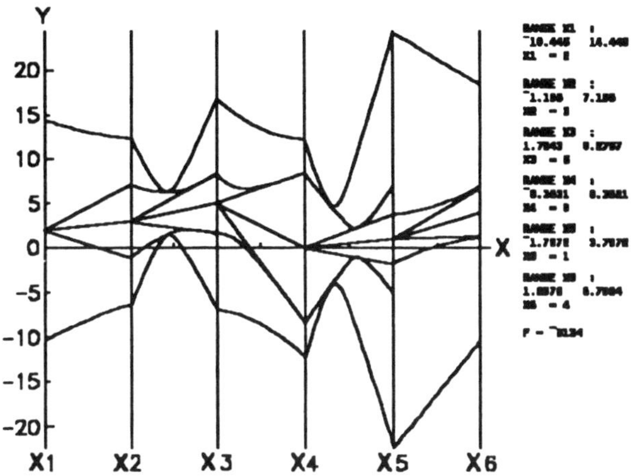

FIGURE 38. The polygonal line represents the point found interior to the hyperellipsoid.

X_2, once X_1 is selected, is *exhibited* by drawing tangents to the hyperbola from the point on the X_1 axis specified by the chosen value of X_1. As the process is iterated through all the variables, the entire sequence of selections and ranges is presented in a parallel coordinate display. Furthermore the hyperbolas themselves provide indications as to the sensitivity of various choices. For an example of the display, see Fig. 38. In that picture the outer hyperbolic segments are the transforms of the original hyperellipsoid projections on the coordinate planes. The inner segments come from the successive slices of the hyperellipsoid and show the successive ranges. Note that a point is exterior to the hyperellipsoid if and only if the intermediate envelopes, representing successive slices of the hypersurface, are crossed by the polygon, is interior if it nowhere touches, and is on the surface otherwise. The computational complexity is $O(\mathcal{N}^3)$ (see Ref. 19).

3.5. *The Hyperellipse*

The procedure for hyperellipsoids calls for repeated projections on coordinate planes, at a considerable cost in computational complexity. The obvious suggestion is to define an object whose projections do not change as the iteration proceeds.

Consider a collection of hypercylinders whose projections on coordinate planes are ellipses. The intersection of such a set of hypercylinders, if it exists, is clearly a convex object. With some attention to detail such objects can be defined so that they exist, are bounded, and have a property called *consistency* which, briefly, means that a valid value for *any* variable is consistent with a valid value for *any other* variable. A notable property of this kind of surface is that the computational complexity is $O(N)$, since the surface is defined by $N-1$ ellipses, whence $N-1$ quadratic equations must be solved (see Ref. 19).

Though necessarily brief, we hope to have conveyed a notion of a new geometrical tool for visualizing and analyzing multivariate processes. Parallel coordinates have a "built-in mechanism" for generalizing "lower-dimensional" intuition and results without any intrinsic limit on the dimensionality.

3.6. Intelligent Process Control

3.6.1. Representation of the Domain. The following points are worth noting at the outset of this section:

1. Controlling a process requires considerable human intelligence. The proposed system, or any other, is called *intelligent* if it can replace—to a considerable extent—the human operator for this task. Or if an operator is present, the system can render *intelligent help* by choosing the *critical* parameters—at any given time—with explanations on why they are critical. Further, there is the ability to do *what if scenarios* and see what options there exist for getting out of "tight spots."
2. It is known that any *dynamical* system with N parameters can be considered as a *static* system in $N+1$ parameters: the original N parameters *and time.* This applies also to processes.

A general convex hypersurface can be represented by computing the *envelope* of the collection of polygonal lines representing its points as defined in the beginning. For multidimensional polyhedra such a representation is *unique*; otherwise the description of the hypersurface needs to be retained in addition to its parallel coordinate representation. The representation of a convex hypersurface in R^{20} is shown in Fig. 39. It corresponds to a particular (nonlinear) relation which for our purpose describes a *process* involving 20 parameters labeled $X_1, \ldots, x_{20}$. An N-tuple is a *feasible point for the process* $\Leftrightarrow$ it is *interior to the hypersurface.* This is the *geometrical equivalent* of the statement that a particular combination of parameter values satisfies all the constraints imposed on the process. When a mathematical description for the process is not available, an approximation (a "*best fit*" according to some criterion) for the envelope can be obtained from available data. Geometrically, the relation between adjacent axes is indicated by the shape of the envelope, and here, due to the hypersurface's convexity, is an hstar. To economize on display space only the strip between the axes is shown.

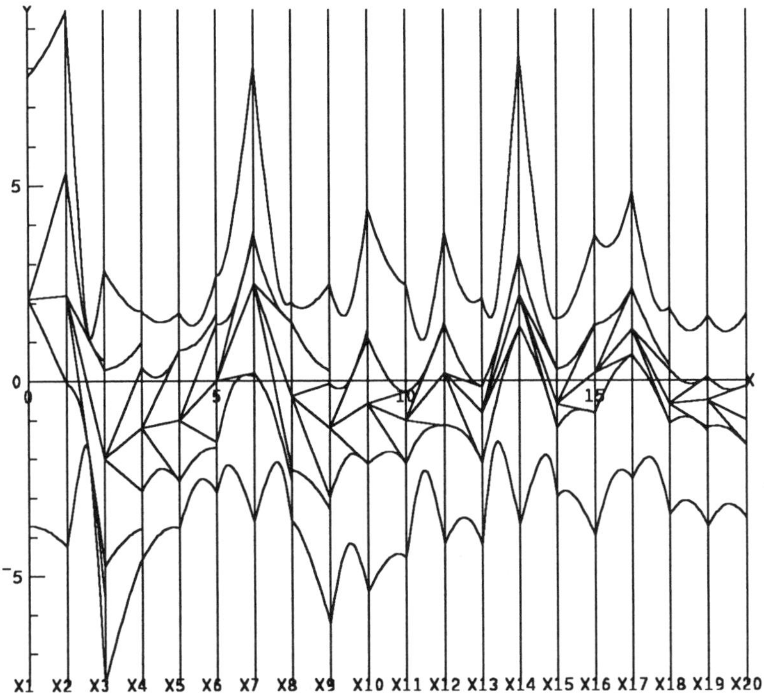

FIGURE 39. Finding a feasible point for a process represented by a hypersurface.

Further, the envelope's shape clearly depends on the particular order in which the variables appear (as axes labels). As we will see, rather than a problem, this provides an opportunity for choosing *preferred orderings of the variables*, subject to certain criteria.

3.6.2. The Rule. Controlling the process is equivalent geometrically to staying in the interior of the hypersurface. We apply the *algorithm* for doing so. The interval on the x_i axis between the upper and lower portions of the envelope indicates the range of pararmeter x_i. When a value for $x_1 = a_1$ is chosen in the available range of X_1 the number of variables in the relation is reduced by one (i.e., from 20 to 19 variables in our example). Again with reference to Fig. 39 from the fixed value of x_1 *tangents* to the upper and lower portions of the envelope are drawn (here the points of tangency are not seen since they lie outside the strip between the x_1 and x_2 axes). The envelope of the 19-dimensional hypersurface is then obtained from the (description) of the original hypersurface with the value $x_1 = a_1$. Again for display space economy only the strip between x_2 and x_3 of the *new* envelope is shown and is always *contained* between the upper and lower portions of the original envelope. The *restricted available range* of x_2, due to the constraint of fixing the value of one of the variables, is the interval on the x_2 axis between the upper and lower tangents. Proceeding, a value $x_2 = a_2$ is chosen within the *current available range of* x_2 reducing the dimensionality by one (e.g., 18) and the whole process is repeated. A polygonal line is shown which always lies between the intermediary envelopes. It represents an interior point to the hypersurface and all interior points can be found in this way. So this provides us with a way of finding and *displaying feasible points for the process.* Also the algorithm is very fast and suitable for real time operation.

3.6.3. The Display as an Instrument Panel. When the value of a variable is fixed, the *complete* envelope of the 19-dimensional surface may be drawn showing at a glance not only the current value of any of the parameters (in this case x_1) but, unlike a conventional "instrument panel," the *current available range of all the parameters*; this is due to the interrelationship among them. There are other salient features of the display. Note that for the parameters x_{13}, x_{14}, x_{15} the available ranges are the narrowest. This indicates that the feasible point is *closest to the boundary* with respect to these *critical* parameters. Hence, where there is need for rapid decisions in view of a lot of information (e.g., as is piloting very fast aircraft), significant information reduction can be achieved by controlling the critical parameters whose *priority is decided by their proximity to the boundary.* Time permitting, the remaining variables can also be adjusted. This also brings out the possibility of displaying or highlighting only the information desired (possibly specific subsystems) at any given time. When a point is not feasible the display shows the corresponding polygonal line crossing some of the intermediate envelopes. It is then evident *which parameters must be changed and by how much* in order to bring the polygonal line inside the envelope and the process in control. In short, this instrument panel provides a variety of information which an operator would somehow need to know (at least implicitly derive or guess based on his experience) in order to control the process.

3.6.4. Order, Learning, and Explanations. Ideally, one would like to have a display where the *relation between any pair of variables, say* x_i *and* x_j is apparent. There are $N(N-1)/2$ such pairs for N variables, which suggests that $O(N^2)$ displays with different orderings may be needed. Fortunately this is not the case. It turns out[48] that for $N = 2n$ exactly n "well-chosen" displays suffice while for $N = 2n + 1$, $n + 1$ displays suffice. This is due to the fact that a complete graph of N vertices can be obtained as a union of that many "properly chosen" spanning trees. For example, for $N = 6$ the permutations 126354, 231465, 342516 circularly contain every possible pair (independent of the order) of *adjacent* subscripts from 1 to 6. Concerning the ordering of the variables, an advantageous ordering is according to sensitivity. In fact, it is possible to implement a dynamic ordering where the order at each step is decided according to some rule, like for example, the variable displayed and fixed next is the one with the smallest available range.

When an *approximation* of the hypersurface is obtained from data, the representation can be altered in view of new information. When a new feasible point is found that is not obtained via the algorithm from the approximate representation of the hypersurface, using the approximation technique again would *enlarge* the hypersurface in some region. On the other hand an *incorrect* feasible point obtained from the approximation would yield a *restriction* of the hypersurface. So a continuously more accurate representation of the process can be obtained as a result of successive trials. Hence, the system has the ability to *learn.*

At any stage of the algorithm an *explanation of why* a specific value of a variable results in feasible or infeasible point is available in terms of whether the polygonal line intersects the intermediate envelopes. Also, having arrived at a partial choice of parameter values it is possible to conduct "*what if scenarios*"—an important look-ahead feature—before actually fixing the value of the remaining parameters. What is proposed then is a new technique for process control which also suggests a *generalization of the concept of expert system* to include those whose *knowledge representation* is a hypersurface together with a rule for "navigating" in the interior.

ACKNOWLEDGMENTS

It is a pleasure to acknowledge the valuable suggestions and help received from A. Hurwitz, J. Eickemeyer, G. Wong, R. Addison, M. Boz, P. Fiorini, T. Chomut, H. Lee, and L. Tobias. The authors are very grateful to two anonymous referees for their significant suggestions and corrections. This work was supported in part by NASA grant No. NCC 2-587 through the USC affiliation of the first author.

References

1. E. Anderson, A semigraphical method for the analysis of complex problems, *Proc. Natl. Acad. Sci. USA* **13**, 923–927, 1957.
2. D.F. Andrews, Plots of high-dimensional data, *Biometrics* **29**, 125–136, 1972.
3. D. Asimov, The grand tour: A tool for viewing multidimensional data, *SIAM J. Sci Stat. Comp* **6**, 128–143, 1985.
4. T.F. Banchoff and C.M. Strauss, Real time computer graphics analysis of figures in four-space, in Ref. 9.
5. V. Barnett, Ed., *Interpreting Multivariate Data*, Wiley, New York, 1981.
6. J. Bertin, *Semiologie Graphique*, Gauthier-Villars, Paris, 1967.
7. M. Bolorfoush and E.J. Wegman, On some graphical representations of multivariate data, in *Proc. of Interface '88*, American Statistical Association, Alexandria, Virginia, 1988.

8. V.G. Boltyanskii, *Envelopes*, Translated from the Russian by R.B. Brown, Pergamon Press, New York, 1964.
9. D. Brissom, Ed., *Hypergraphics: Visualizing Complex Relationships in Art, Science and Technology*, American Association for the Advancement of Science, Westview Press, Boulder, Colorado, 1979.
10. O.S. Brodetsky, *A First Course in Nomography* G. Bell & Sons, London, 1949 (first published 1920).
11. J.M. Chambers, W.S. Cleveland, B. Kleiner, and P.A. Tukey, *Graphical Methods Data Analysis*, Duxbury Press, Boston, 1983.
12. Y. Chen, M. Hsieh, A. Inselberg, and H. Lee, Constrained planar conflict resolution for air traffic control, *Proc. 2nd Canadian Conf. on Complex Geometry*, University of Ottawa Press, Ottawa, 1990, pp. 160–163.
13. H. Chernoff, The use of faces to represent points in k-dimensional space graphically, *J. Am. Stat. Assoc.* **68**, 361–368, 1973.
14. T. Chomut, Exploratory data analysis using parallel coordinates, M. Sc. thesis, UCLA Department of Computer Science, IBM LA Science Center Report, 1987-2811, 1987.
15. S.M. Cohan and D.C.H. Yang, Mobility analysis of planar four-bar mechanisms through the parallel coordinate system, *J. Mech. Mach. Theory* **21**, 63–71, 1986.
16. H.S.M. Coxeter, *Projective Geometry*, Univ. of Toronto Press, Toronto, 1974.
17. A. Desai and L.C. Walters, Graphical presentations of Data Envelopment Analysis: Management implications from parallel axes representations, *Decision Sci.* **2**(22), 335–353, 1991.
18. P. Diaconis and J. Friedman, M and N plots, in *Recent Advances in Statistics*, Academic Press, New York, 1983.
19. B. Dimsdale, Operating point selection for multivariate systems, IBM LA Science Center Report No. ZZ20-6249, 1981.
20. J.S. Eickemeyer, Visualizing p-flats in N-space using parallel coordinates (unpublished).
21. P. Fiorini and A. Inselberg, Configuration Space Representation on Parallel Coordinates, *Proc. IEEE Conf. Robotics Autom., Scottsdale, Arizona*, 1215–1219, 1989.
22. B. Flury and H. Riedwyl, Graphical representation of multivariate data by means of asymmetrical faces, *J. Am. Stat. Assoc.* **76**, 757–765, 1981.
23. J. Friedman and J. Tukey, A projection pursuit algorithm for exploratory data analysis, *IEEE Trans. Comput.* **C-23**, 881–890, 1974.
24. C. Gennings, K.L. Dawson, W.H. Carter Jr., and R.H. Myers, Interpreting plots of a multidimensional dose–response surface in a parallel coordinate system, *Biometrics* **46**, 719–735, 1990.
25. H.D. Griffen, Graphic computation of tau as a coefficient of disarray, *J. Am. Statistic. Assoc.* **53**, 441–447, 1958.
26. J.A. Hartigan, *Clustering Algorithms*, John Wiley, New York, 1975.
27. J.J. Helly, Application of parallel coordinates to complex system design and operation, in Proceedings of NCGA 1987, Fairfax, Virginia, Vol. III, pp. 541–546, 1987.
28. H. Hinterberger, Data density: A powerful abstraction to manage and analyze multivariate data, Ph.D. thesis, Informatik-Dissertation No. 4 ETH, Zurich, 1987.
29. P.J. Huber, Projection pursuit, *Ann. Stat.* **13**, 435–475, 1985.
30. A. Inselberg, N-dimensional graphics part I: Lines and hyperplanes, IBM LA Science Center Report No. G320-2711, 1981.
31. A. Inselberg, The plane with parallel coordinates, Special Issue on Computational Geometry of *Vis. Comput.* **1**, 69–97, 1985.
32. A. Inselberg, M. Boz, and B. Dimsdale, Conflict resolution, one-shot problem and air traffic control, *Proc. 1st Canadian Conf. on Computational Geom.*, University of Montreal Press, Montreal, 1989, p. 26.
33. A. Inselberg and B. Dimsdale, Intelligent process control with integrated instrumentation in *Integrated and Intelligent Manufacturing*, C.R. Liu and T.C. Chang, Eds., ASME, New York, 1986, pp. 341–357.
34. A. Inselberg, M. Reif, and T. Chomut, Convexity algorithms in parallel coordinates, *J. ACM* **34**, 765–801, 1987.

35. A. Inselberg and B. Dimsdale, Multidimensional lines I: Representations, IBM LA Scientific Center Report, 1987–2830, 1988.
36. A. Inselberg and B. Dimsdale, Multidimensional lines II: Distance and proximity, IBM LA Scientific Center Report, 1987–2831, 1988.
37. B. Kleiner and J.A. Hartigan, Representing points in many dimensions by trees and castles *J. AM. Stat. Ass.* **76**, 260–269, 1981.
38. R.O. Lejeune, Government-provided complex scenarios for the advanced automated system design competition phase MTR-85W240, MITRE Co, Mc Lean, Virginia, 1985.
39. R.J. Littlefield and W.L. Nicholson, Use of color and motion for the display of multi-dimensional data, Pacific Northwest Laboratory Technical Report No. PNL-SA-10897 also Department of Energy Report No. DE83 0003596, 1982.
40. M.R. Mersereau and A.V. Oppenheim, Digital reconstruction of multidimensional signals from their projections, *Proc. IEEE* **62**, 1319–1337, 1974.
41. J. Rivero, Phase-space in parallel coordinates, IBM LA Scientific Center Report, G320-2742, 1984.
42. S.J. Schwager, Applying statistical graphics to multivariate data, Proceedings of NCGA '86 Anaheim, California, 1986, pp. 704–712.
43. D.W. Scott, Data analysis in three and four dimensions with nonparametric density estimation, in *Statistical Image Processing and Graphics*, E.J. Wegman and D. DePriest, Eds., Marcel Dekker, New York, 1986, pp. 291–305.
44. E.R. Tufte, *The Visual Display of Quantitative Information*, Graphic Press, Cheshire, Connecticut, 1983.
45. J.W. Tukey and P.A. Tukey, Graphical display for data sets in three or more dimensions, in Ref. 9, 1980.
46. F.W. Young, D.P. Kent, and W.F. Kuhfeld, VISUALS: Dynamic hyperdimensional graphics software for visual exploratory data analysis presented at NCGA '86 Anaheim, California, 1986.
47. C.M. Wang and H.W. Gugel, High performance graphics for multivariate data, GM Laboratories, Technical Report No. GMR-5387, Warren, Michigan, 1986.
48. E. Wegman, Hyperdimensional data analysis using parallel coordinates, *J. Am. Stat. Assoc.* **411**(85), 664–675, 1990.

TOOLS FOR INTERACTIVE PROTOTYPING OF TWO-DIMENSIONAL AND THREE-DIMENSIONAL USER INTERFACES

ALESSANDRO GIACALONE AND ARIE KAUFMAN

1. Introduction

This chapter describes a project, currently under way at Stony Brook, concerned with tools to assist in building *visualization environments* for applications primarily in the applied sciences. Active applications are in material sciences, mechanical engineering, economy, oceanography, medicine, applied mathematics, and computer science. The ultimate goal of the project is to develop an integrated environment for interactive prototyping of two-dimensional and three-dimensional user interfaces. The environment consists of an integrated set of tools that support the design, prototyping, reconfiguration, testing, and debugging of graphical user interfaces, without involving any traditional notion of "programming." Instead, the tools allow one to specify interactively and directly both the static and the dynamic parts of a customized user interface by means of a set of graphical "editors."

As an integral part of this project, we are designing a paradigm for a natural and inherent three-dimensional user interface with a configuration that includes a conventional two-dimensional worksurface and a true three-dimensional input device, like a "kite" (six-dimensional sensor) or a glove. The user is furnished with a small-scale artificial three-dimensional reality, which facilitates orientation and visceral communication with the three

ALESSANDRO GIACALONE and ARIE KAUFMAN • Department of Computer Science, State University of New York at Stony Brook, Stony Brook, New York 11794-4400.

dimensions. This serves as the front-end of a three-dimensional user interface.

Building a sophisticated user interface is and will remain a difficult and laborious task even with advanced tools. A distinction between *builders* and *users* can be made among the expected users of our tools. While drawing a clear border line is impossible, the distinction is determined by factors such as degree of expertise and continuity of involvement in developing user interfaces. At present, our work is mainly concerned with builder tools. However, a major objective is to provide layers of interactive tools that can be employed at different levels of expertise (by a user-interface specialist or an end user). Builder tools can be used for developing both "end user" interfaces and higher-level tools that can in turn be utilized by naive builders or users. Replacing "programming" by interactive editing should also bridge the current gap between builders and users, and favor dialogue between the two. We envision naive computer users (e.g., engineers or scientists) generating and molding sophisticated user interfaces tailored to their applications in a couple of man-months, rather than in the several man-years currently required by a computer expert.

Our project constitutes an integration and an extension of two projects that have been evolving independently: the *XY-WINS* environment[1] and the *Cube* system.[2] XY-WINS is an environment of tools for visually designing, implementing, and debugging two-dimensional graphical user interfaces. It includes an interactive editor for a target worksurface, an interactive debugging system, and a concurrent language for defining dialogues. The user interface architecture and the basic framework within which a user interface is developed are inherited from the XY-WINS system. The Cube system and its underlying *Cube architecture* are versatile three-dimensional systems for voxel-based graphics with an inherent three-dimensional user interface. The Cube system provides a platform for three-dimensional user interfaces. Integrating this three-dimensional platform with the XY-WINS results in an *XYZ-WINS* environment. The XY-WINS and Cube projects are also major components of a larger interdisciplinary project concerned with the development of visualization environments for application to applied sciences,[3,4] currently in progress at the Institute for Visualization Environment (IVE) at Stony Brook.

The project has several different but strongly interrelated aspects and is evolving in several directions simultaneously. For example, the development of three-dimensional graphics and three-dimensional interaction capabilities poses problems that are to some extent independent of those posed by the development of tools for dialogue management. Also, many of the tools have first been developed in a two-dimensional graphic context supported by XY-WINS and are now being extended to three-dimensional graphics.

The remainder of this introduction reviews some related work in the context of user interface tools. The conceptual model of user interface supported by the environment is the topic of Section 2. Sections 3 and 4 address tools for interactively configuring the outer levels of the man/machine dialogue, that is, the visual behavior of the system and the abstract modeling of various input devices. In sections 5–7 the Cube project and three-dimensional interaction are discussed. Sections 8–9 cover dialogue management, i.e., the specification of the input behavior and the synchronization of input and output behaviors. In particular, we discuss an approach and tools for visual *debugging* of the dialogue in user interfaces. The final section contains a discussion and an overview of future work.

Interest in user interfaces is demonstrated by the now large number of conferences, workshops, and publications devoted entirely to the subject. A great deal of research has dealt with methodologies and high-level tools for structured development of user interfaces. Ergonomics and design principles have been extensively documented (e.g., Refs. 5 and 6), and libraries and tool-kits have been provided to isolate the user interface implementor at least from the machine-dependent details (e.g., Sun NeWS and X).

An effort has been under way to characterize the notion of *user interface management system* (UIMS [7]). Although the term is not always used consistently, a UIMS is typically viewed as an integrated set of tools that (a) manage the physical I/O devices (and low-level software) automatically and (b) support an abstract specification of the dialogue between the user and the application. Major experiments in this direction include the work of Olsen,[8] Green,[9] Hill,[10] Myers,[11] Barth,[12] and Sibert.[13]

Several approaches have been proposed for dialogue specification and management. Interesting surveys appear in Refs. 14–16. Grammar-based definitions are utilized in the work of Olsen,[17] based on context-free grammars, and in Ref. 18, in which interaction is specified as a tree structure where leaves are "physical" devices. Jacobs[19] presents a language for defining dialogues based on the notion of object and coroutine. Concurrent languages based on communicating processes are presented in Refs. 10, 20, and 1. Some of the more recent efforts have addressed the issue of interactively configuring a user interface.[11, 21–23]

2. *The Structure of the User Interface*

We view a user interface as consisting of two major components: a Graphical Abstract Machine (GRAM) and a High-Level Interaction Component (HIC). This structure and its relationship with the rest of the system is summarized in Fig. 1.

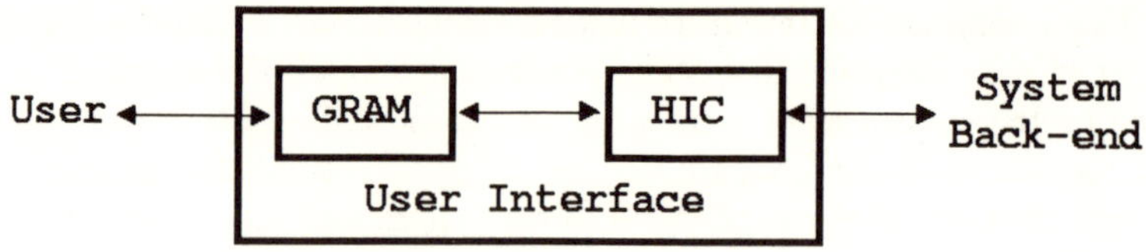

FIGURE 1. Structure of the user interface.

The GRAM is the module that interacts with, and in a broad sense includes, the physical I/O devices and the low-level graphics software. It presents to the HIC an abstract view of the interaction with the user. The HIC manages the dialogue by coordinating the activity of the user through input devices, the production of visual operations through the GRAM, and the interaction with the back-end (or semantic component) of the system.

The GRAM is the locus that incorporates all the technology for producing and manipulating physical representations of pictures, based on abstract specifications. In particular, it includes all the algorithms for rendering and incrementally displaying and updating three-dimensional objects (hidden surface elimination, ray tracing, etc.). Pictures and their manipulations are specified by the HIC in terms of abstract data structures and manipulations of these data structures, respectively.

The task of the GRAM is also to translate physical input events produced by users into *tokens*. A token is an abstract entity that (a) identifies a picture and (b) carries with it symbolic information about its origin (how it was produced) and other attributes that contribute to defining the input event (e.g., a direction or some quantity). Tokens constitute the atomic input events, that is, the lowest level at which input is perceived by the HIC. However, tokens may represent both elementary actions, such as depressing a mouse button, and more complex ones, such as grabbing a three-dimensional object with a glovelike device.

The input model is based on tokens received by the HIC through input *channels*. This basic model has several key advantages. It provides a unified view of input events produced by different devices and naturally extends to support multiple devices that may operate in parallel. It also provides a uniform mechanism for handling both simple and compound input events. Several efficient implementations are possible.

The HIC component implements the dialogue between the abstract view of the user provided by the GRAM and the application. Thus, its task is essentially to "parse" the streams of tokens produced by the GRAM and detect the sequences that require a reaction of some sort on the part of the system. Another task is to manipulate abstract representations of pictures in order to produce visualizations, which may or may not be related to events generated in the back-end of the system. A third task of the HIC is then to *synchronize* visualizations with the input behavior of the user in order to produce various visual effects (such as dragging objects around).

3. *Configuring a Graphical Abstract Machine*

In the XY-WINS environment, the AGIP system[1, 24] supports the generation of the GRAM. AGIP includes a parametric GRAM and a preprocessor. A GRAM for a particular application is generated by defining (a) the pictorial and textual constructs that will be used in the target screen, (b) the tokens that will be produced by the GRAM, and (c) the association between physical input events and pictures used to produce the tokens. Pictures and tokens can be defined by using a high-level textual language or, preferably, constructed interactively and directly by means of the APE (A Picture Editor) tool described below. Pictures may be elementary or structured, that is, composed of other pictures. Elementary pictures are called *Primitive Interaction Components* (PIC). PICs are the atomic elements in terms of which the user–computer dialogue can be expressed. In particular, they cannot be further decomposed except by writing lower-level code. The

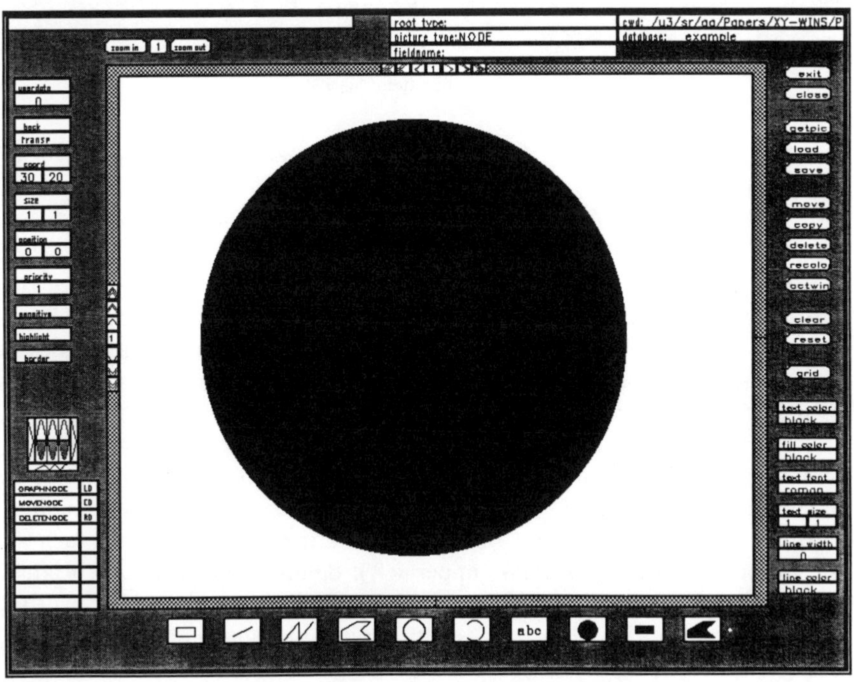

FIGURE 2. APE's display while defining a PIC.

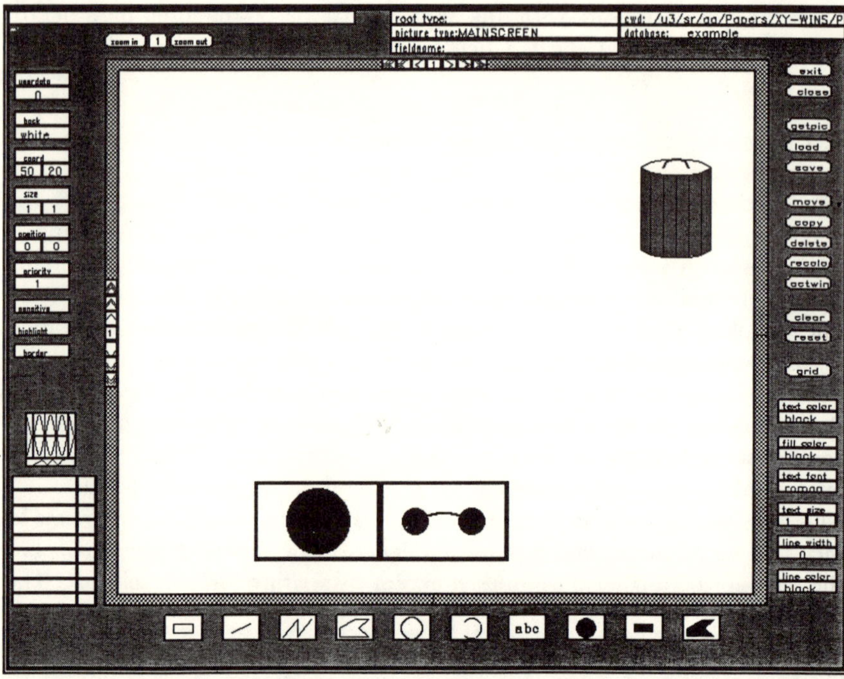

FIGURE 3. APE's display while defining a structured picture.

AGIP system provides this low-level access to picture definition, which is useful when pictures need to be created dynamically.

APE is an interactive system that allows one to define the pictorial and textual constructs of a user interface by drawing and composing pictures. Figure 2 shows what APE's display looks like while defining a picture (in this case a PIC). The picture represents a graph node for a graph editor. Figure 3 shows the display while defining a structured picture, in this case the screen of a graph editor containing a menu (bottom) and a garbage can used to delete objects.

Positioned around the drawing area in the center of the display are various buttons and tools for: identifying working directory and working database of pictures (top right); drawing and defining various graphical attributes (bottom center and right); and defining other attributes, such as the internal coordinate system (upper left); defining the tokens associated with the picture (lower left). The drawing area can be reshaped at will in order to visualize the picture in its actual dimensions and proportions. It is also possible to "pan" around and "zoom" in and out to simulate conditions that may arise in the end-user screen.

4. From Simple to Complex Input Events

The tokens associated with a picture are defined by means of the column of buttons at the lower left of the screen. A major assumption in our treatment of the user's input behavior is that any input event references a picture. While the assumption is obviously accurate in certain simple cases (e.g., "picking" a menu button), it may be less evident in other cases (e.g., pointing in a given direction by means of a three-dimensional glovelike device). Nevertheless, we believe it is valid since even actions that do not specifically identify a picture, such as cursor movement or direction identification, can be viewed as referring at least to some background picture.

In the current version of the APE (and AGIP) system, each token definition associates a physical input event with an input channel and is of the form <*channel name, physical event*>. Physical input events in a particular configuration are denoted by names. Currently, the AGIP prototype operates on a Sun 4 workstation with a keyboard and a three-button mouse. The events known to the system correspond to depressing or releasing each mouse button, to having the cursor enter or exit a picture, and to moving the mouse. A given picture may be sensitive to zero or more events. In our example in Fig. 2, three tokens are defined. One specifies that, when the left mouse button is depressed ("Left Down") on an instance of the picture, the GRAM produces a token and transmits it over input channel GRAPHNODE. The other tokens associate input channels MOVENODE and DELETENODE with the "Center Down" and, respectively, "Right Down" physical events.

The scheme supported by AGIP and APE, based on a fixed set of elementary input events, can be made much more powerful by allowing (expert) designers to (a) define *complex* input events that consist of several elementary events but are mapped onto primitive ones, and (b) tailor the set of acknowledged events for specific applications. While some (simple) notions of complex events have been used repeatedly in various user interfaces (e.g., "double clicking" a mouse button), we are addressing this issue through the systematic development of a *language* that supports the definition of arbitrarily complex events.

Consider, for example, a three-dimensional glovelike input device. The user, wearing the lightweight glove, may point in a certain direction by closing all fingers except for the index finger; in this case, we may want to define *pointing* as a "primitive" event. A glove device usually transmits position, orientation, and shape information of the arm, hand, and fingers and may employ simultaneously more than a dozen sensors. Thus, for example, requirements for our language are that it should express compositions of elementary events under spatial and temporal constraints. Furthermore, it should be possible to further combine events to support an arbitrarily abstract notion of a "physical" event. For example, in the case of the glove, we are developing a "library of gestures" that produce primitive events.

5. *Interacting with the Third Dimension*

The technology described in the previous sections is being extended to the generation of three-dimensional user interfaces, that is, user interfaces that support a direct interaction with three-dimensional objects using true three-dimensional input devices (e.g., a *kite* - a six-dimensional sensor, a glove, and/or a space-ball), but on a conventional two-dimensional work-surface. We thus extend the XY-WINS environment, upgraded and augmented by a new three-dimensional environment, to an XYZ-WINS environment that supports both two-dimensional and three-dimensional interfaces, employing the three-dimensional technology being developed with the Cube project described in Section 6.

The work toward the three-dimensional user interface is concerned with the GRAM component. Three-dimensional pictures still appear to the HIC as abstract data structures with input capabilities. As mentioned, most of the concepts and techniques used for the interactive configuration of two-dimensional interfaces extend naturally to three-dimensional interfaces. An issue that is specific to the three-dimensional work is the natural and inherent three-dimensional interaction using the type of configuration described above (true three-dimensional input devices and a two-dimensional "physical" surface). Interaction with three-dimensional objects is still a developing field. Certain principles and metaphors applied successfully in two-dimensional graphics need to be explored for their corresponding three-dimensional extensions. For example, new principles must be developed for three-dimensional locating, pointing, and movement visualization. In addition, many human factors in such interaction are not yet well understood.

The problem of three-dimensional interaction can be subdivided into two major aspects, which involve (a) the physical devices, that is, true three-dimensional output devices and true three-dimensional input, and (b) the virtual devices and visual metaphors. The first aspect is concerned with determining the best modality or modalities for three-dimensional interaction. True three-dimensional displays, which generate space-filling three-dimensional images (e.g., stereopsis displays, arrays of lenses, holograms, and vibrating mirrors). These systems are currently precluded from routine use. Consequently, true three-dimensional displays are presently excluded from the configuaration of our three-dimensional user interface.

True three-dimensional input devices, however, promise much broader and more effective use. Many approaches have been attempted in the development and manufacturing of three-dimensional input devices, including multiaxis and three-axis joysticks, number wheels, three-axis trackballs, Polhemus electromagnetic 3SPACE sensors,[2, 25] gloves,[26] and more.[27] These devices are relatively inexpensive and usually provide three-dimensional location as well as three-dimensional orientation input and often three-dimensional gesture. We anticipate that a true three-dimensional input device will be an integral part of future three-dimensional workstations, the same way a mouse is an indispensable part of contemporary workstations. We are, therefore, utilizing a true three-dimensional input device in the configuration of our three-dimensional user interface.

Some work has been done in mapping two-dimensional input from a tablet,[28] two-dimensional control devices,[29] or a mouse (termed "triad" mouse)[25, 30] to three-dimensional space. There are also some preliminary solutions to the problems of representing three-dimensional cursors or control points on two-dimensional displays,[25, 30] interactively specifying three-dimensional transformation parameters,[30, 31] and selecting the visualization format and appearance for three dimensions.[32] Unfortunately, these techniques are unnatural, indirect, and do not convey a three-dimensional reality.

6. *The Three-Dimensional Cube System*

The Cube system[2] has been designed and developed at SUNY Stony Brook on color Sun 3 and 4 workstations using SunView. The true three-dimensional input device attached to the workstation is an electromagnetic 3SPACE Polhemus Isotrak device, called the *kite* (or the *bat*), which provides three-dimensional positioning and three-dimensional orientation of a light-weight hand-held sensor. The mouse and the keyboard of the workstation are utilized for command and pick selection, while the kite is used for directly positioning and orienting the *jacks* (three-dimensional cursors).

The Cube system supports a high-level interactive visualization environment for three-dimensional volumetric graphics and imaging. It is based upon the Cube hardware architecture,[33] which is a multiprocessor centered around a large three-dimensional cubic memory, called the *Cubic Frame Buffer* (*CFB*).

The experimental environment of the Cube system provides, in software, a front-end interactive environment, that is, a three-dimensional GRAM. It is supported by six modules: a *scene composer*, a "*room*" *manager*, a *scene presenter*, a *renderer*, a *filing*, and a three-dimensional *user interface* module.

The scene composer is responsible for interactively placing and editing the geometric objects that may be mixed with sampled images. It scan-converts three-dimensional geometric models into their three-dimensional discrete representation within the CFB.[34–36] The room manager is the three-dimensional counterpart of a two-dimensional window manipulator and window manager. Its primitives are unstructured three-dimensional subarrays of the CFB and are of three types: *rooms* (three-dimensional windows), *jacks* (three-dimensional cursors), and *figurines* (three-dimensional icons). The room manager is based upon a *voxblt* engine[37] (an extended two-dimensional *bitblt* engine). The scene presenter is an assortment of tools used to mold both the topology and color of the image of interest for easy visceral exploration and to generate orthographic, parallel, and perspective projections, as well as arbitrary slices. The renderer shades the projected images, employing a fast *congradient* shading technique.[38] The filing module controls the database storage and retrieval of CFB voxel maps, projections and depth buffers, projected images, and display lists or logfiles of the various modules.

Initial experience with the three-dimensional user interface is described in more detail in the next section and in Chapter 11 of this book. In addition to displaying the three-dimensional graphics and controlling the three-dimensional interaction by monitoring the three-dimensional devices, the interface maintains the jack feedback and produces the appropriate input token to the back-end.

7. *A Case Study of Three-Dimensional User Interface*

Figure 4 depicts a snapshot of the Sun screen of the experimental three-dimensional worksurface of Cube. A separate *command panel*, with buttons, sliders, and the like is allocated for interacting with each of the system modules (see bottom-right corner of Fig. 4 for two such panels). A separate *viewing canvas* is designated for depicting a three-dimensional arbitrary parallel or perspective view (top-right corner of Fig. 4), a view of one or all of the six orthographic projections (center of Fig. 4) and/or the three-dimensional "world" view for interactive selection of the eye point, the direction of projection, the projection surface, and the light sources. Where the machine power allows it (e.g., as with a Silicon Graphics 4D) the "world" view can be merged with all the viewing canvases, and the preferred view is selected by dynamically panning and spinning the entire combined cube space.

A viewing canvas is used as both a presentation viewport for shaded projections and a composition and editing canvas for interacting with the existing images or for synthesizing new ones using calligraphic silhouette drafting. This *combination look* of drawings superimposed on shaded images capitalizes on the advantages of both types of looks (Ref. 32). For example, a perspective shaded projection can be overlaid with perspective line-drawing frames that define the rooms of interest. The latter serves as hooks for selecting the rooms and interactively defining their transformations and manipulations.

The interactive session can be limited to a single (e.g., perspective) view or can be echoed in several or all other views. This is a flexible setup allowing one, for example, to shift a two-dimensional cursor in a $+z$ orthographic projection when such a motion is difficult to monitor in a perspective view or, preferably, to shift a jack in a perspective view and have all the two-dimensional cursors in the orthographic projections follow suit. Figure 4 shows a room in the perspective canvas, which also is echoed in the six orthographic views. The jack (at the bottom-right corner of the perspective canvas) has three projections, called the jack *shadows* on the three tiled walls. The jack counterparts in the orthographic projections are presented and behave like common two-dimensional cursors. An alternative to this multiple view would be a mechanism that dynamically spins the entire cube space, facilitating the viewing of back-facing surfaces or complex scenes.

An inherent three-dimensional drafting environment is provided to aid in the selection of the three-dimensional rooms, in the interactive specification of three-dimensional transformations, and in the interactive composition of geometric objects. It relies on interactive constraints assisting the user in maneuvering in three dimensions. Specifically, the following constraints have been implemented in our experimental system: jack gravity and jack constraint to arbitrary points, to lines, and to planes; and jack constraints to move only within planes parallel to one of the primary planes or to move only along lines parallel to one of the primary axes.

8. *The High-Level Interaction Component*

Recall that in our model of the user interface, the HIC implements the dialogue between the abstract view of the user provided by the GRAM component and the back-end of the system. We further assume that the interaction with the (abstract) user is definable in terms of input tokens and visual operations supported by the GRAM as primitive operations.

We have developed an experimental high-level concurrent language (AISLE[1]) that supports the specification of the HIC component as a system of independent processes that execute concurrently and communicate

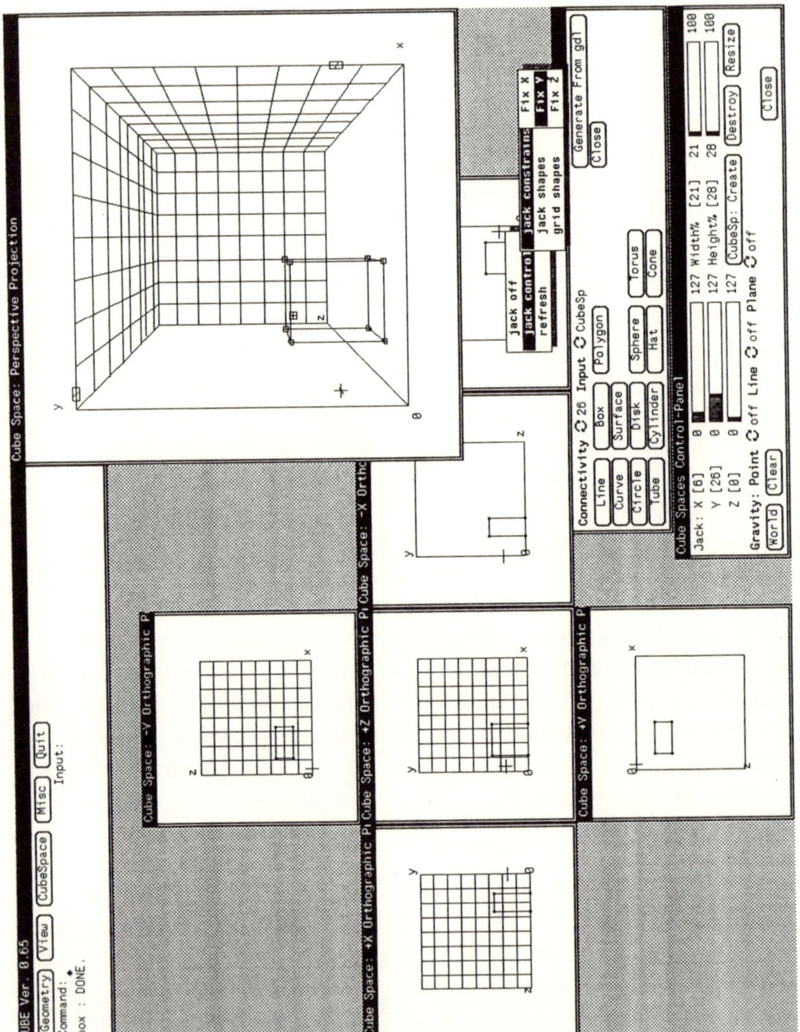

FIGURE 4. The worksurface of the cube experimental three-dimensional user interface.

through *channels*. For a given application, a particular set of channels is used by processes to receive tokens from the GRAM component. AISLE is strongly influenced by Milner's *Calculus of Communicating Systems*[39] and by the OCCAM language.[40] A few researchers have proposed concurrent languages for application at various levels in a user interface (e.g., Refs. 10, 20, 41). An approach to dialogue management based on concurrent processes has several advantages: modularity of the specification, synchronization between user's actions and visual operations, and parallelism for handling features such as multiple I/O devices operating in parallel.

The AISLE language includes a small set of concurrent constructs embedded in a "host" language. The concurrent part of the language includes constructs for creating and terminating processes, creating communication channels, synchronized communication over channels, alternative mutually exclusive executions of a process, and concurrent execution of a set of processes. Concurrent constructs are used to define the control structure of the dialogue, typically, to decide which functionalities of the user interface should be activated and to synchronize the reception of tokens from the GRAM with the activation of visual operations. The host language is used to specify the "internal" execution of each process, which includes visual operations and interaction with the back-end of the system.

As important aspect of AISLE is that it has a two-dimensional syntax, in the sense that its expressions are two-dimensional compositions of pictorial symbols. The syntax derives from a previous effort, documented in Refs. 42 and 43, to produce an *ideographic* syntax for concurrent languages. Consider, for example, the expression in Fig. 5. The two juxtaposed boxes specify an "exclusive or" of possible executions. The process defined by the expression waits for input (inward arrows) either on channel MOVENODE or on channel DELETENODE. In both cases the input value is bound to a variable *node*. Such value may be produced by objects on the screen, i.e., by the GRAM, or by some other process. In the first alternative of the "or" the process simply activates another process called *Movenode*; in the second alternative it executes some internal code and then activates an *Editor* process.

A HIC specified through AISLE can be "compiled" into C code or can be preprocessed and then run in interpreted mode. The latter mode of

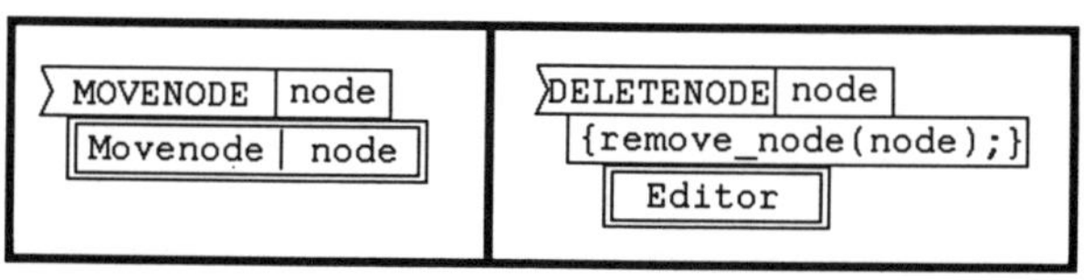

FIGURE 5. An example of AISLE graphical syntax.

execution is the key for an interactive user-interface *debugging* system (discussed in the next section), which in turn provides the key for our approach to interactive dialogue definition.

9. *High-Level Debugging of User Interfaces*

An important aspect of system development is *source level* debugging. Once high-level descriptive tools are available, tools should be provided to debug definitions at the same level. AVID is a highly graphical, interactive system that allows one to execute in debugging mode user interfaces based on a GRAM generated through the AGIP/APE tools and a HIC specified through the AISLE language. AVID allows us to visualize simultaneously the end-user screen, the state of the internal image that corresponds to the user screen, and the state of the HIC in the form of a graphical AISLE expression. Figure 6 shows what the screen looks like during the executions of a simple user interface.

As shown in Fig. 6, the screen is subdivided into three windows: the end-user window (top right), the internal image window (top left), and the

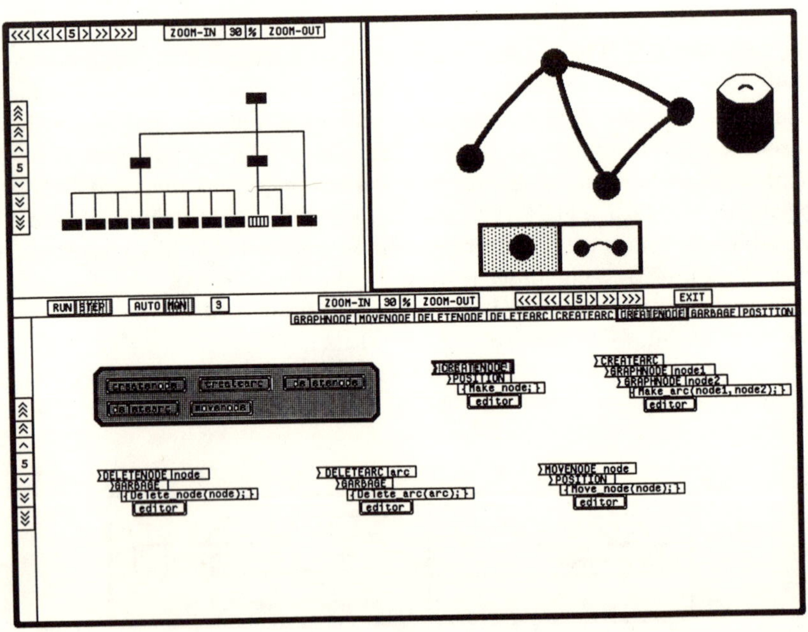

FIGURE 6. A snapshot of the AVID interface debugging system.

HIC window (lower half). The end-user window contains the end-user screen. The internal image window displays the state of the internal image that corresponds to the state of the end-user screen. All changes of the screen are shown by highlighting the affected (sub-)tree. When an input event is generated on a picture in the user screen (and some token is thus generated), the corresponding node of the internal image is highlighted.

The attributes of the picture represented by each node in the tree can be inspected by simply picking the node and opening a specification window. Moreover, the system provides a number of capabilities to deal with internal images that become very large. These include panning, zooming, and the ability of "closing" trees and representing them with icons. All these operations are allowed when the user interface is in a *stable* state, that is when (1) it is being executed in *step mode* (clarified below), and (2) is "frozen" and can be reactivated only by an input from the user.

In the HIC window, the system displays the state of the HIC as an AISLE expression. Executions are viewed as state transitions. The execution can be run in different modes, controlled by the panel appearing across the middle of the display. The two basic modes are *run* and *step*. In *run* mode the current state of the HIC is shown only when the system is stable, so that intermediate events are not visualized. In *step* mode the operations which are currently being executed are highlighted. The traced operations are (a) input from, or output to the GRAM; (b) communication between processes; (c) execution of portions of internal code by processes, viewed at this level as atomic events. While in *step* mode, the execution can be frozen after the occurrence of each of the above events (*manual step* mode), or can proceed automatically after a delay (*automatic step* mode). In *manual step* mode the execution is reactivated by selecting the *restart* button.

In complex systems the display space is normally not adequate to allocate both the user interface of the interactive system being executed and the state of the HIC. To cope with this problem, the HIC window can be run on a separate workstation. This allows one to (a) run the user interface in normal usage conditions, and (b) take advantage of the full display space to visualize the state of the HIC. The execution is driven by the interaction with the end-user screen.

Of course, the user interface of the AVID system has been implemented through XY-WINS. An interesting effect is that AVID could potentially be used to visualize another instance of itself as it is being applied to a third system.

10. Discussion and Future Developments

To summarize, the tools that have been developed so far allow one to: (a) specify interactively, through the APE system and its three-dimensional extensions, the "static" part of a user interface (pictures and their input capabilities); (b) define, through the AISLE language, the control structure of the dialogue. Limitations are that: (a) a significant amount of complexity and implementation effort still lie in the specification of the set of visual operations in terms of which the *output behavior* of the system is expressed; (b) the control structure is defined essentially by programming in a high-level concurrent language. Interactive specification of the dynamic output behavior of the system and specification of the dialogue by direct demonstration are two major themes around which our current work is evolving. Although there is clearly a strong connection between the two aspects, there is also a clear distinction of concerns. Operations on pictures can be specified independently and can then be provided as primitives for the definition of the dialogue. In accord with this scheme, we are developing tools that support the independent specification of operations on pictures and user–system dialogue.

At present, we are implementing an experimental interactive system that expands the capabilities of the APE system and enables a user-interface designer to specify interactively both pictures and *visual operations* on pictures. The new system allows one to define *classes* of pictures. A class of pictures can be thought of as a picture that has (a) a structure, defined in terms of component picture classes, (b) parameters that need to be given values to create instances of the class (e.g., width, color), (c) possible constraints defined on parameters of the class, and (d) a set of visual operations that are applicable to instances of the class. A major feature of the class editor is that it will allow one to define visual operations by operating directly on the pictures themselves through a combination of demonstrations and interactive selection and specification of parameters.

At present, the AVID system allows one to monitor executions of user interfaces that have already been specified. However, the fact that it visualizes the end-user screen and the internal state of the dialogue simultaneously suggests its potential as a system for dialogue *definition.* We are experimenting with a "mixed-mode" approach to interactive dialogue definition, based on a combination of demonstration and direct definition, and supported by the views provided by the AVID system. The basic idea is that the end-user screen can be used to demonstrate a fragment of user–system interaction. In general, we envision that a dialogue will be defined, interactively, through some combination of demonstrations and direct definitions of AISLE expressions. In particular, definitions by demonstration may be used for fragments of interaction that are natural to specify from a user's viewpoint, but have a simple control structure, consisting, for example, of a strictly sequential execution or a branching between two or more sequential executions. We believe that through demonstration alone it is impossible, or unnatural, to specify complex control structures involving, for example, global contexts. These are better and more quickly defined directly in terms of AISLE expressions.

With regard to the three-dimensional virtual devices and visual metaphors, we are investigating several issues: the virtual devices necessary for three-dimensional interaction (locate, pick, and/or stroke); a comprehensive, yet simple, set of visualization tools with which to visualize three-dimensional images on a two-dimensional surface (multiple views, shaded projections, perspective or parallel viewing, the combination of shaded and silhouette images, and/or "turn tables"); the representation of jacks (three-dimensional cursors) in order to enable natural feedback in three-dimensional space (three/six legs, three cross hairs, cubic jack, jack tilting, and/or shadows on the "walls"); interactive specification of three-dimensional parameters (such as rotation axis and angle) in a compact and "human-oriented" manner (edge-based or face-based rotation, gnomons, device direct tilting, and/or three-dimensional gestures); and the utilities and tools necessary for three-dimensional object manipulation (such as constraints, gravity, rubber-banding, grid, dragging, three-dimensional windowing, translucency, highlighting, or more specific application-oriented "electronic tools" e.g., electronic scissor, electronic drill, electronic chisel, and electronic scalpel).

ACKNOWLEDGMENTS

This work was supported by the National Science Foundation under grants Nos. DCR 8319966, DCR 8603676, MIP 8805130, and IRI 90081009.

References

1. A. Giacalone, XY-WINS: An integrated environment for developing graphical user interfaces, Proceedings of SIGGRAPH Symposium on User interface Software, Banff, Alberta, Canada, October 1988.
2. A. Kaufman, The CUBE workstation–A three-dimensional voxel-based graphics environment, *Visual Comput.* **4**(4), 210–221, 1988.
3. A. Giacalone, J. Heller, A. Kaufman, M. Kifer, P.M. Lewis, P. Mishra, I.V. Ramakrishnan, S.A. Smolka, and D.S. Warren, VERITAS: Visualizations environment research in the applied sciences, Proceedings of the SPIE Conference, Vol. 1083, pp. 127–134, January 1989.
4. A. Giacalone, A. Kaufman, S.A. Smolka, and D.S. Warren, Visualization tools for

the applied sciences, in *Advances in Human Factors/Ergonomics*, Vol. 12A, M.J. Smith and G. Salvendy Eds., pp. 451–459, Elsevier, New York, 1989.
5. T. Bleser and J. Foley, Towards specifying and evaluating the human factors of user–computer interfaces, Proceedings of the Conference on Human Factors in Computer Systems, Gaithersburg, Maryland, March 1982, pp. 309–314.
6. M. Green, A methodology for the specification of graphical user interface, *Comput. Graph.* **15**(3), 99–108, 1981.
7. G. Pfaff, Ed., *User Interface Management Systems*, Springer-Verlag, Berlin, 1984. Proceedings of the Seeheim Workshop on User Interface Management Systems, Seeheim, November 1983.
8. D.R. Olsen Jr. and E.P. Dempsey, SYNGRAPH: A graphical user interface generator, *Comput. Graph.* **17**(3), 43–50, 1983.
9. M. Green, The University of Alberta user interface management system, *Comput. Graph.* **19**(3), 205–213, July 1985.
10. R.D. Hill, Supporting concurrency, communication, and synchronization in human–computer interaction: The Sassafras UIMS, *ACM Transs. Graph.* **5**(3), 179–210, 1986.
11. B.A. Myers, *Creating User Interfaces by Demonstration*, Academic Press, New York, 1988.
12. P.S. Barth, An object-oriented approach to graphical interfaces, *ACM Trans. Graph.* **5**(2), 142–178, 1986.
13. J.L. Sibert, W.D. Hurley, and T.W. Bleser, An object-oriented user interface management system, *Comput. Graph.* **20**(4), 259–268, 1986.
14. M. Green, A survey of three dialogue modes, *ACM Trans. Graph.* **5**(3), 244–275, 1986.
15. M. Green, Report on dialogue specification tools, in *User Interface Management Systems*, G.E. Pfaff, Ed., Springer-Verlag, New York, 1985, pp. 9–20.
16. J. Foley, Models and tools for the designers of user–computer interfaces, NATO International Advanced Study Institute, Theoretical Foundations of Computer Graphics and CAD, Italy, July 1987.
17. D.R. Olsen Jr. and E.P. Dempsey, Syntax directed graphical interaction, Proceedings of ACM SIGPLAN '83 Symposium on Programming Language Issues in Software Systems, June 1983.
18. J. van den Bos, Abstract interaction tools: A language for user interface management systems, *ACM Trans. Program. Lang. Syst.* **10**(2), 215–247, 1988.
19. R.J.K. Jacob, A specification language for direct manipulation user interfaces, *ACM Trans. Graph.* **5**(4), 283–317, 1986.
20. L. Cardelli and R. Pike, Squeak: A language for communicating with mice, *Comput. Graph.* **19**(3), 199–204, 1985.
21. P.C.S. Wong and E.R. Reid, FLAIR—User interface dialogue design tool, *Comput. Graph.* **16**(3), 87–98, 1982.
22. L. Cardelli, Building user interfaces by direct manipulation, ACM SIGGRAPH Symposium on User Interface Software, Banff, Canada, October 1988.
23. P.E. Haeberli, ConMan: A visual programming language for interactive graphics, *Comput. Graph.* **22**(4), 103–112, 1988.
24. A. Giacalone, AGIP: A graphical interaction package, Poster Presentation, EUROGRAPHICS '87, Amsterdam, The Netherlands, August 24–28, 1987.
25. C.B. Phillips and N.I. Badler, Jack: A toolkit for manipulating articulated figures, ACM Symposium on User Interface Software, Banff, Alberta, Canada, October 1988, pp. 221–229.
26. S.S. Fisher, M. Mc Greevy, J. Humphries, and W. Robinett, Virtual environment display system, Proceedings ACM Workshop on Interactive 3D Graphics, Chapel Hill, North Carolina, October 1986, pp. 77–87.
27. J.D. Foley, Interfaces for advanced computing, *Sc. Am.* October, 126–135, 1987.
28. I.E. Sutherland, Three-dimensional data input by tablet, *Proc. IEEE* **62**(4), 453–461, 1974.
29. M. Chen, S.J. Mountford, and A. Sellen, A study in interactive three-dimensional using two-dimensional control devices, *Comput. Graph.* **22**(4), 121–130, 1988.

30. G.M. Nielson and D.R.J. Olsen, Direct manipulation techniques for three dimensional objects using two-dimensional locator devices, Proceedings of the ACM Workshop on Interactive Three-dimensional Graphics, Chapel Hill, North Carolina, October 1986, pp. 175-182.
31. E.A. Bier, Skitters and Jacks: Interactive three-dimensional tools, Proceedings of the ACM Workshop on Interactive Three-Dimensional Graphics, Chapel Hill, North Carolina, October 1986, pp. 183–186.
32. A.R. Forrest, User interfaces for three-dimensional geometric modelling, Proceedings of the ACM Workshop on Interactive Three-Dimensional Graphics, Chapel Hill, North Carolina, October 1986, pp. 237–250.
33. A. Kaufman and R. Bakalash, Memory and processing architecture for three-dimensional voxel-based imagery, *IEEE Comput. Graph. Appl.* **8**(6), 10–23, 1988. Also translated into Japanese, *Nikkei Comput. Graph.* **3**(30), 148–160, 1989.
34. A. Kaufman and E. Shimony, Three-dimensional scan-conversion algorithms for voxel-based graphics, Proceedings of the ACM Workshop on Interactive Three-Dimensional Graphics, Chapel Hill, North Carolina, October 1986, pp. 45–76.
35. A. Kaufman, Efficient algorithms for three-dimensional scan-conversion of parametric curves, surfaces, and volumes, *Comput. Graph.* **21**(4), 171–179, 1987.
36. A. Kaufman, Efficient algorithms for three-dimensional scan-converting polygons, *Comput. Graph.* **12**(2), 213–219, 1988.
37. A. Kaufman, The voxblt engine: A Voxel frame buffer processor, in *Advances in Graphics Hardware III*, A.A.M. Kuijk, Ed., Springer-Verlag, Berlin, 1991.
38. D. Cohen, A. Kaufman, R. Bakalash, and S. Bergman, Real-time discrete shading, *Visual Comput.* **6**(1), 16–27, 1990.
39. R. Milner, *A Calculus of Communicating Systems*, Lecture Notes in Computer Science, Vol 92, Springer-Verlag, Berlin, 1980.
40. INMOS Limited, *Occam Programming Manual*, C.A.R. Hoare, Ed., Prentice-Hall, Englewood Cliffs, New Jersey, 1984.
41. H.G. Borufka, H.W. Kuhlmann, and P.J.W. ten Hagen, Dialogue cells: A method for defining interactions, *IEEE Comput. Graph. Appl.* **2**(5), 25–33, 1982.
42. A. Giacalone, M.C. Rinard, and T.W. Doeppner Jr., IDEOSY: An interactive and ideographic program description system, Proceedings of the ACM SIGSOFT/SIGPLAN Symposium on Practical Software Development Environments, Pittsburgh, April 1984, pp. 15–20.
43. A. Giacalone and S.A. Smolka, Integrated environments for formally well-founded design and simulation of concurrent systems: A nonprocedural approach, *IEEE Trans. Software Eng.*, 707–802, 1988.

TOWARD A THREE-DIMENSIONAL USER INTERFACE

ARIE KAUFMAN AND RONI YAGEL

1. Introduction

The term *user interface* refers to the combined environment of data presentation using output devices and different interaction methodologies using input devices. The crucial role that the quality of the user interface plays in the utilization of the system can be appreciated by examining aspects such as responsiveness, capabilities, graphics (colors, two dimensional or three dimensional), ease of use, ease of learning, naturalness, and accuracy.

The importance of user interfaces has motivated the development of countless systems, technologies, and techniques. However, the vast majority of these are exclusively two dimensional, and even the interaction and manipulation of three-dimensional graphics is two-dimensional in nature, utilizing two-dimensional output displays, two-dimensional techniques, and a "flat" input medium (e.g., a mouse), a "linear" medium (e.g., a dial), or even a "discrete" medium (e.g., a keyboard). Examining the state-of-the-art three-dimensional user interfaces indicates that further research is required in order to develop interaction techniques and methodologies that are sufficiently powerful to overcome the flaws inherent in creating realistic three-dimensional interactive tools in a two-dimensional medium.

We describe here the progress made towards a three-dimensional interface as part of the Cube workstation and its environment. It employs special techniques for three-dimensional interaction and direct object manipulation in three dimensions, utilizing a two-dimensional display and either a six-dimensional input device for positioning and orientation, or a gesture input device.

ARIE KAUFMAN and RONI YAGEL • Department of Computer Science, State University of New York at Stony Brook, Stony Brook, New York 11794-4400.

2. *The Cube Workstation*

The Cube workstation supports a high-level interactive visualization environment for three-dimensional voxel-based graphics and imaging. It is based on the Cube hardware architecture,[1] which is a multiprocessor system centered around a large three-dimensional cubic memory, the *Cubic Frame Buffer* (CFB). The CFB is composed of an evenly spaced grid of basic memory elements called *voxels* (volume elements). The three Cube processors input, manipulate, view, and render the images stored in the CFB.

The Cube workstation environment supports different volumetric projection types, input of sampled voxel maps from a database, geometric model input, transformations, manipulation, interaction, arbitrary image slicing and sectioning, color segmentation, translucency, shading, and measurements for a variety of applications. The experimental environment of the Cube workstation provides, in software, a higher-level interactive environment, supported by six components: a *scene composer*, a "*room*" *manager*, a *scene presenter*, a *renderer*, a *filing module*, and a *three-dimensional user interface.*

The scene composer, assisted by the three-dimensional user interface, is responsible for interactively positioning and editing geometric objects. Upon completion of the composition stage, the scene composer scan-converts the three-dimensional geometric models (lines, polygons, parametric or quadratic objects) into their three-dimensional discrete representation within the CFB,[2] which optionally can be intermixed with sampled images.

The room manager is the three-dimensional counterpart of a two-dimensional window manager.[3] It is based on a *voxblt* engine[4] (an extended two-dimensional *bitblt* engine), which performs voxel-map operations, including transformations, overlapping, loading, storing, painting, and voxelwise operations. Its primitives are unstructured three-dimensional sub-arrays of the CFB of three types: *rooms* (three-dimensional windows), *jacks* (three-dimensional cursors), and *figurines* (three-dimensional icons). The room manager, assisted by the filing module, also manages the input of three-dimensional sampled images reconstructed from data obtained from scanning devices (CT, MRI, Confocal Microscope, etc.) or retrieved from an image database.

The scene presenter is an assortment of tools used to mold both the topology and the color of the image of interest for easy visceral exploration and to generate a variety of projections (orthographic, parallel, and perspective), as well as arbitrary slices. The renderer shades the projected images employing a discrete *cogradient shading technique*[5] and provides some translucency and aliasing reduction support. The filing module controls the storage and retrieval to and from the database of CFB voxel maps, projection and depth buffers, projected images, and display lists or logfiles. The three-dimensional user interface is the focus of this paper and is described in the following sections.

The Cube workstation has been designed and developed at the State University of New York at Stony Brook and has been implemented on color Sun workstations using SunView. Currently, the Cube workstation three-dimensional user interface is being ported to X-11 on the Sun, to a Silicon Graphics IRIS workstation using X and GL, and to an HP 400 Turbo VRX using X and Starbase (with great improvement in the display quality and performance for the latter two). All these portings preserve standard user interface capabilities, while taking advantage of special machine features.

3. *Three-Dimensional User Interface—Problem Statement*

Four major issues must be addressed when attempting to design and develop a user interface: which output device(s) will be used, how will the data be visually presented, which input device(s) are provided, and what interaction methodologies will be employed. In this section we describe these four subcomponents of user interfaces and point out some problems arising in three-dimensional environments.

In order to display a three-dimensional image, one can employ a true three-dimensional output device, which generates space-filling true three-dimensional images, such as stereoscopic display,[6] array of lenses,[7] holograms,[8] or vibrating mirrors.[9] These systems, however, have many deficiencies, including low resolution, poor depth/color resolution, limited throughput, user inconvenience, and high price and are thus used only in specific applications. The alternative used by most researches and industry is a conventional two-dimensional screen enhanced with depth cues in order to facilitate the visualization of the three-dimensional scene on a flat medium. The presentation of a three-dimensional scene should convey to the viewer a good understanding of the spatial structure of the scene and should support the additional tools necessary for further scene exploration.

Addressing the subject of three-dimensional data presentation we should keep in mind that the main problem confronting computer graphicists in the last decade is the generation of a realistic two-dimensional image from a model of the spatial scene. The solution to this problem is strongly dependent on the internal representation of the scene. In most CAD applications a polygon-based model will suffice. More demanding scientific visualization problems will require a surface-based model while some volumetric biomedical applications will require a voxel-based representation. Research efforts in the past decade have produced algorithms and hardware architectures that efficiently implement, for each data representation scheme, different image generation techniques such as hidden surface removal, ray tracing, translucency, dynamic transformations, and shading techniques for modeling diverse lighting phenomena. We should, based on the internal data representation and available computing power, decide which image generation techniques will be made available to the user for data visualization.

After dealing with the output devices and the presentation scheme we confront the problem of choosing the appropriate input devices for three-dimensional interaction. Traditional one-dimensional or two-dimensional input devices such as keyboards, mice, and joysticks, can still be used in three-dimensional interaction, augmented with a mapping function that maps a one-dimensional or two-dimensional physical device reading into three-dimensional space. Such an approach has been adopted by several systems.[10–12] Unfortunately, these techniques are unnatural, indirect, and do not convey a three-dimensional reality. They suffer from one or more of the following deficiencies. It is rather cumbersome to obtain and control the third degree of freedom with one-dimensional or two-dimensional input devices. The input techniques are indirect, that is, the user manipulates some control entities or devices that in turn are mapped onto the actual object. The output and feedback are flat, commonly using only the top, front, and side orthographic views. When perspective or parallel projections are used, ray casting is employed in which feedback and the selected objects are specified at the intersection of the eye projector and the view surface. One cannot, for example, directly select a back-facing surface or step into opaque objects.

A preferable strategy is one that employs a true three-dimensional input device, which is an inexpensive, useful, and naturally handled device that provides three-dimensional location, often three-dimensional orientation, and is some cases three-dimensional gesture. Examples of such devices are multiaxis and three-axis joysticks,[13] number wheels,[14] trackballs,[7] Polhemus electromagnetic 3SPACE sensors,[15–17] gloves,[18–20] and others.[19]

Once the I/O devices and the desired data presentation are determined, the environment needed for the actual design of the interaction methodology is in place. We now briefly present several topics and principles that govern the design of a three-dimensional interaction scheme.

The main aspect determining the usability of a user interface is its syntax. The syntax can be of "low level" nature, such as a verbal command language,[21] or it can have moderate power by employing the scheme: push-command-button, specify-parameters, redisplay. According to this scheme, the user directly interacts with command initiation facilities (e.g., menus, buttons) and parameter specification tools (e.g., sliders, dials), rather than with the objects themselves. The most powerful and natural syntax is *direct object manipulation*, in which the user directly interacts with the objects comprising the three-dimensional scene, utilizing true three-dimensional interaction techniques and devices.

Another factor governing interface quality is the set of virtual devices available to the user. In designing a three-dimensional user interface, one needs to determine which of the traditional two-dimensional virtual device should be extended to three dimensions and what other special three-dimensional virtual devices are necessary. For instance, one might need an "orientation" device in addition to the (three-dimensional counterpart of the) "locater" device.

A problem unique to three-dimensional interaction is that of space orientation, that is, how to give the user a sense of his/her precise location in space and support the comprehension of his/her spatial relationship with other objects. We must decide whether the best intuitive, real-time motion feedback is a cursor moving on the flat screen and ray cast into the three-dimensional space, or a true three-dimensional cursor (jack) moving within the three-dimensional scene; which depth cues need to be supported; and what constraints should be imposed on the locater movement (e.g., not allowing the user to go into an opaque object). The effectiveness of the location and orientation feedback can easily be appreciated by trying to perform some accurate pick and locate tasks.

The user interface can be enhanced by adding a diverse set of three-dimensional utility tools. For example, gravity and object highlighting can greatly ease the "pick" operation. Three-dimensional grid and discrete locator movement can improve "locate" operations. Improved scene comprehension can be gained through additional tools that enable the user to manipulate multiple or changing views of the same image, to choose different depth cues, add light sources, change the shading model, edit colors or material, and the like. These aspects of the user interface are subject to implementer ingenuity, which can improve the user interface quality to a large extent by considering the specific needs of the actual application.

4. Overview of the Cube Three-Dimensional User Interface

Based on our observations concerning the main design issues of the three-dimensional user interfaces, we approach the description of the Cube user interface by first describing the I/O devices used and briefly explaining the general layout of the environment. Detailed interaction paradigms and additional techniques are described in the next section (see also Chapter 10).

The output device for a Cube user interface is a color two-dimensional screen of a Sun 3, Sun 4, Silicon Graphics IRIS 4D, or an HP Turbo VRX workstation. In addition to the traditional keyboard and mouse, the HP machine is equipped with a dial box and a button box. The three-dimensional input device attached to these workstations is an electromagnetic 3SPACE Polhemus Isotrak device, called a *kite* (or a *bat*), which provides three-dimensional positioning and three-dimensional orientation of a lightweight handheld sensor. The mouse, button box, and the keyboard are utilized for function selection and text input while the kite directly positions and orients the jack. The dial box is used for valuator control as well as positioning and as a parameter specification device for geometric transformations. On machines lacking a kite, Cube employs the mouse in two modes: a traditional mouse mode and a kite mode. In the latter case the two-dimensional mouse coordinates are mapped to three-dimensional coordinates (i.e., as a triad-mouse)[10–12] transforming the mouse into a restricted three-dimensional input device. When complex interaction is required or desired, a triad mouse can be a useful feature even when a kite is available. For example, the triad mouse can rotate the world while the kite's positioin and orientation are used for object manipulation.

Research is under way to incorporate more sophisticated input devices such as the glove,[18] which is capable of providing three-dimensional location, orientation, and manual postures and gestures. A rich set of natural gestures that are suitable for interaction with a three-dimensional environment is currently being compiled into a "gesture library."

Figures 1 and 2, referred to in the following description of the presentation scheme, are snapshots of a Sun screen showing the three-dimensional user interface worksurface. The interface is composed of two kinds of windows: *command windows* and *viewing windows*. A command window, consisting of a set of buttons, valuators, and other facilities, is designed for interacting with one of the system components. Figure 1 shows several command windows: the main control panel (at the top of the screen labeled "CUBE") from which the Cube modules are activated, the viewport control (labeled Cube Spaces Control Panel), which manages the multiple projection windows, and the Geometry Processor Control Panel. Figure 2 shows also the command window that controls the setup of the kite device driver (labeled "Kite Control Panel").

A viewing window serves as both a presentation viewport for shaded projections and a composition and editing window for interacting with the existing images or synthesizing new ones using calligraphic silhouette drafting. One of the best depth perception cues that a viewing window could support is dynamic scene rotation while experiencing *motion parallax*. Not many machines, however, can support such real-time three-dimensional animation and certainly not when dealing with voxel-based data which requires special purpose hardware for real-time viewing. An alternative way

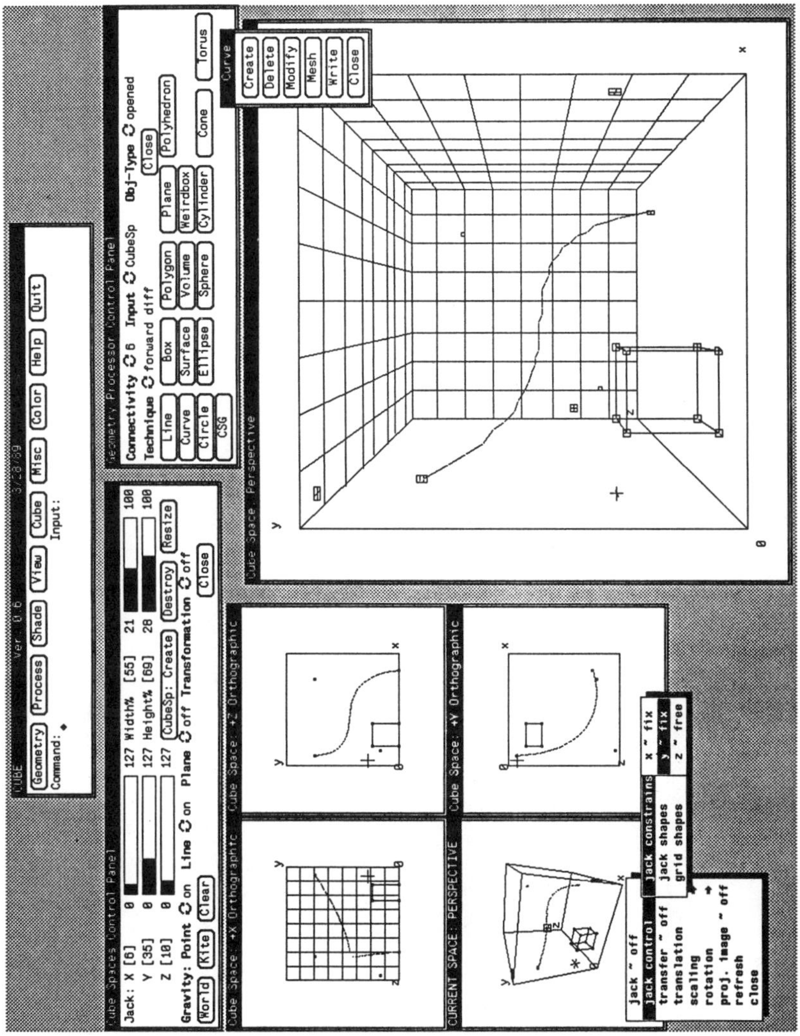

FIGURE 1. A snapshot of the three-dimensional user interface worksurface.

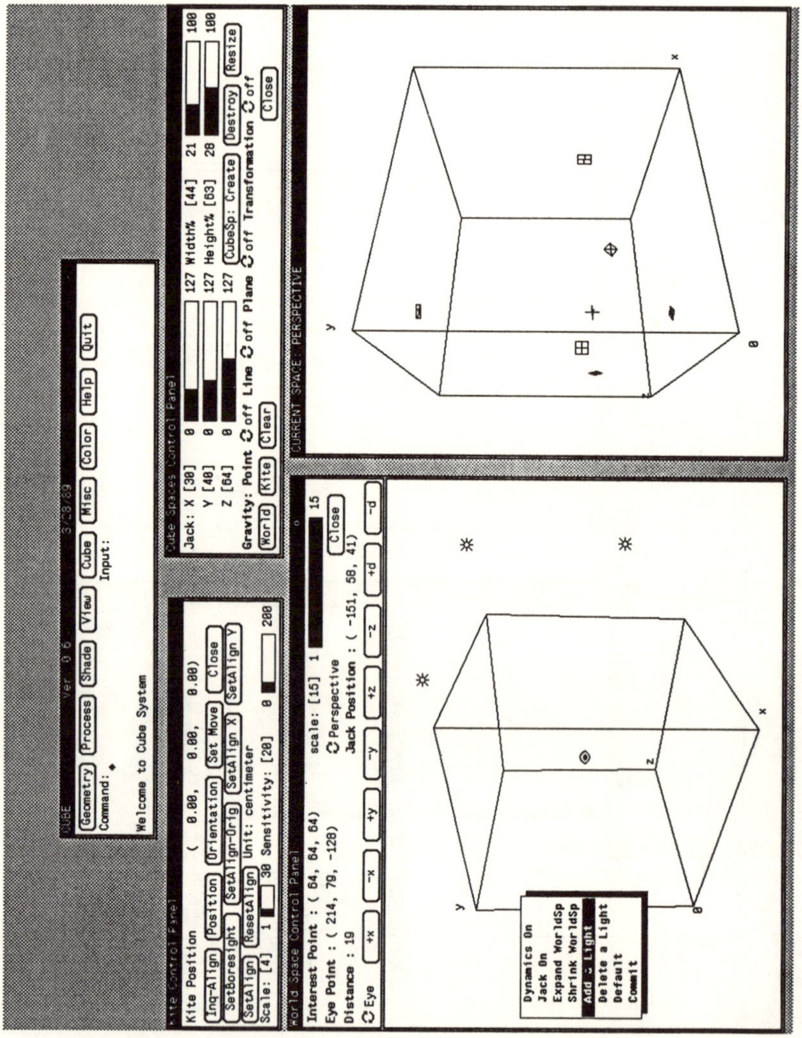

FIGURE 2. A snapshot of the worksurface with the world space.

to approximate dynamic scene rotation is to allow the user to have multiple viewports into the image space, which actually provides a set of static "snapshots" of the dynamic motion. Figure 1 shows two different perspective views and three orthographic views. The viewing windows employ a combination look in which drawings are superimposed on shaded images in order to capitalize on the advantages of both types of looks (cf. Ref. 21). For example, a perspective-shaded projection of a medical image can be overlaid with perspective line-drawing frames defining the rooms (regions) of interest. These serve as hooks for selecting the rooms and interactively defining their transformations and manipulations. Furthermore, new geometric objects can be composed interactively by drafting their silhouettes in the same window.

An additional window, called the *World Space* (bottom-left corner of Fig. 2), is used to dynamically specify the viewing parameters for the current viewing window. The user may choose any arbitrary parallel or perspective view for the current viewing window (e.g., bottom-right corner of Figs. 1 and 2) and a view of any number of the six orthographic projections (left side of Fig. 1). The user can dynamically set the eye point (the small "eye" icon at the center of the World Space Panel in Fig. 2), the direction of projection, the projection surface, the light sources (the small light bulb icons in Fig. 2) and the like. Where the machine power allows it, the world view can be merged with the viewing windows, and the preferred view is selected by dynamically panning and spinning the entire cube space.

5. *Three-Dimensional User Interaction*

Utilizing a gamut of menus, tools, constraints, and feedback (extended versions of those used in conventional two-dimensional graphics editing), the user can place the jack in certain selected three-dimensional positions and three-dimensional orientations in order to specify a room, an object, or a transformation using dynamic rubber-banding, dragging, rotation, and the like. Several techniques that address the problem of user orientation in three-dimensional space have been implemented. First, a surrounding perspective frame representing the CFB boundaries is displayed, and the user is confined to movements within this frame. The jack is represented by three full crosshairs or by a small six-leg toy jack enhanced with shadows on the front and/or back walls of the surrounding frame. The jack and its shadows are shown in the large perspective cube space at the bottom-right corner of Fig. 1. The jack dynamically feeds back the kite location and orientation. This is visualized by the shrinkage or enlargement of the six-leg jack as the kite is moved farther or closer and by the changes in the jack shape and orientation as the kite is being tilted. In order to further enhance user positioning some of the walls can be tiled, giving the effect of a three-dimensional grid (see bottom-right and center-left parts of Fig. 1).

Another way to enhance interaction is provided by the multiple viewport philosophy. The interactive session can be limited to a single (e.g., perspective) view, or it can be echoed in several views, such as six orthographic, parallel, or other perspective views. The multiple viewport allows one, for example, to achieve cursor motion and delicate positioning in a specific orthographic projection when such a motion is difficult to monitor in a perspective view, or to manipulate an object in a perspective view and observe all operations as they are reflected in all the orthographic projections. Figure 1 shows a room in the large perspective window, which is also echoed in the other views. The jack (at the bottom-left corner of the large perspective window) has three back shadows on the tiled walls while the jack counterparts in the orthographic projections are presented and behave like common two-dimensional cursors. An alternative to this multiple view scheme is a combined view coupled with a mechanism that dynamically spins the entire cube space, facilitating the viewing of back-facing surfaces or complex scenes.

To aid in three-dimensional positioning, interactive specification of three-dimensional transformations, and interactive composition of geometric objects, an inherent three-dimensional drafting environment is provided. It consists of utility tools that rely on interactive constraints,[22] thereby assisting the user in maneuvering in three dimensions. When drafting an object, the sytem automatically defines some control points (denoted by small cubes), control lines, and control planes. When drawing a cylinder, for example, the center, the radius, a third point on the base plane, and the cylinder height are defined as either control points or as control lines, while the bases are defined as control planes. All these control entities can later be used as anchors for transformations and object editing. For example, the user may edit a Bezier curve by picking one of the curve's four control points and changing its location. Gravity can be turned on or off for each explicit entity type by toggling the cycles of the Cube Spaces Control Panel (Fig. 1). The system implicitly activates only the appropriate gravities; for example, only point gravity is in effect when the user is expected to pick a point and is turned off once the user makes the choice.

The system relies also on constraint mechanisms (see the menus at the bottom of Fig. 1) to decompose the three-dimensional movement into two-dimensional (or one-dimensional) movements by disabling any one (or two) dimensions and allowing the jack to move only in planes (or lines) parallel to one of the primary planes (or axes). Accurate feedback of one- and two-dimensional movements is achieved by using the appropriate orthographic viewports. If a high level of accurancy is required or when the numeric coordinate value is known, the interactive kite can be totally disabled and a set of one-dimensional valuators can be used instead (e.g., the three sliders of the Cube Spaces Control Panel in Fig. 1). Constraints are also used for the specification of transformation parameters where the user utilizes existing entities (e.g., planes, lines) in order to define angles, distances, and the like, employing an extended version of the mechanisms suggested by Nielson and Olsen.[10]

6. *Interaction Example*

As an example of an interaction session in three dimensions, assume a volume visualization application in which a cylinder representing an injection needle is to be superimposed on a displayed perspective projection of a medical organ structure. The user selects the *Cylinder* "light" button from the Geometry Processor Control Panel, using a right-hand-held mouse. Meanwhile, using a left-hand-held kite moved in *actual* three-dimensional space, the user positions the jack on or within the organ in the simulated three-dimensional space projected in perspective on the window. Once the jack has been positioned, its location is selected by a button click.

The cylinder is defined by first selecting the center of the base circle, rubber extending to specify the radius, followed by selecting a third point to specify the base plane. Then, the "rubber" cylinder is extended to the motion of the kite, which is seen as a jack tugging the cylinder until a button is clicked to fix its silhouette. Based on the interactively sculptured shape, the three-dimensional geometry processor scan-converts it into the CFB, intermixed with the voxel representation of the organ. The scene is immediately reprojected and rendered to form a perspective-shaded image combined with a dynamically manipulatable silhouette representation of the needle. Concurrently, several orthographic and arbitrary parallel projections may be displayed. In order to simulate the insertion of the needle, the user may directly select two points on the gravitized silhouette of the cylinder that identify the object and also define the motion vector. The points can be constrained so that the vector is parallel to the cylinder axis, and this constraint serves to guide the insertion in that direction.

Alternatively, a more sophisticated interaction can be envisioned employing, for example, a hand-worn glove and higher-level interaction tools. For the above example, the user can select a virtual tool, an "electronic injection needle," which then can be directly "grabbed" and driven in by the glove in much the same motion that a clinician would make. In addition, the entire primary view may be smoothly turned for better visualization.

7. *Concluding Notes*

We have presented here a paradigm for a natural and inherently three-dimensional user interface consisting of a conventional two-dimensional work surface complemented with a true three-dimensional input device, like a kite or a glove. It has been our intention to furnish the user with a small-scale artificial three-dimensional reality in order to facilitate his/her orientation and visceral communication with the three dimensions. Further research will investigate the best natural three-dimensional modality (or modalities), tools, specification and visualization techniques, virtual devices, and feedback for three-dimensional interaction between a machine and a human being, who resides and functions in three dimensions.

ACKNOWLEDGMENTS

This work was supported by the National Science Foundation under grants DCR 8603603, MIP 8805130, and IRI 9008109.

References

1. A. Kaufman and R. Bakalash, Memory and processing architecture for three-dimensional Voxel-based imagery, *IEEE Comput. Graph. Appl.* **8**(11), 10–23, 1988.
2. A. Kaufman and E. Shimony, Three-dimensional scan-conversion algorithms for voxel-based graphics, Proceedings of the ACM Workshop on Interactive Three Dimensional Graphics, Chapel Hill, North Carolina, October 1986, pp. 45–75.
3. R. Pike, Graphics in overlapping bitmap layers, *Comput. Graph.* **17**(3), 331–356, 1983.
4. A. Kaufman, The voxblt engine: A voxel frame buffer processor, in *Advances in Graphics Hardware III*, A.A.M. kuijk and W. Strasser, Eds., Springer-Verlag, Berlin, 1989.
5. D. Cohen, A. Kaufman, R. Bakalash, and S. Bergman, Real-time discrete shading, *Vis. Comput.* **6**(1), 1990.
6. J.A. Roese and L.E. Mc Cleary, Stereoscopic computer graphics for simulation and modeling, *Comput. Graph.* **13**(2), 41–47, 1979.
7. K.B. Evans, P.P. Tanner, and M. Wein, Tablet based valuators that provide one, two or three degrees of freedom, *Comput. Graph.* **15**(3), 91–97, 1981.
8. L.B. Lesem and P.M. Hirsch, Computer synthesis of holograms for three-dimensional display, *Communic. ACM* **11**(10), 661–674, 1968.
9. H. Fuchs, S.M. Pizer, L.C. Tsai, S.H. Bloomberg, and E.R. Heinz, Adding a true three-dimensional display to raster graphics system, *IEEE Comput. Graph. Appl.* **2**(7), 73–78, 1982.
10. G.M. Nielson and D.R. Jr. Olsen, Direct manipulation techniques for three-dimensional objects using two-dimensional locator devices, Proceedings of the ACM Workshop on Interactive Three-Dimensional Graphics, Chapel Hill, North Carolina, October 1986, pp. 175–182.
11. I.E. Sutherland, Three-dimensional data input by tablet, *Proc. IEEE* **62**(4), 453–461, 1974.
12. C.B. Phillips and N.I. Badler, Jack: A toolkit for manipulating articulated figures, ACM Symposium on User Interface Software, Banff, Alberta, Canada, October 1988, pp. 221–229.
13. E.G. Britton, J.S. Lipscomb, and M.E. Pique, Making nested rotation convenient for the user, *Comput. Graph.* **12**(3), 222–227, 1978.

14. R.W. Thorention, The number wheel: A tablet based valuator for interactive 3D positioning, *Comput. Graph.* **13**(2), 102–107, 1979.
15. N.I. Badler, K.H. Manoochehri, and D. Baraff, Multidimensional input techniques and articulated figure positioning by multiple constraints, Proceedings of the 1986 Workshop on Three-Dimensional Graphics, Chapel Hill, North Carolina, October 1986, pp. 151–169.
16. A. Kaufman, The CUBE workstation—A Three-Dimensional Voxel-based graphics environment, *Vis. Comput.* **4**(4), 210–221, 1988.
17. C. Schmandt, Spatial input display correspondence in a stereoscopic computer graphic work station, *Comput. Graph.* **17**(3), 253–261, 1983.
18. S.S. Fisher, M. Mc Greevy, J. Humphries, and W. Robinett, Virtual environment display system, Proceedings of the ACM Workshop on Interactive Thee-Dimensional, Chapel Hill, North Carolina, October 1986, pp. 77–87.
19. J.D. Foley, Interfaces for advanced computing, *Sci. Am.* October 1987, pp. 126–135.
20. T. Zimmerman, J. Lanier, C. Blanchard, S. Bryson, and Y. Harvill, A hand gesture interface device, Proceedings Human Factors in Computing Systems and Graphic Interface, 1987, pp. 189–191.
21. A.R. Forrest, User interfaces for three-dimensional geometric modeling, Proceedings of the ACM Workshop on Interactive Three-Dimensional Graphics, Chapel Hill, North Carolina, October 1986, pp. 237–250.
22. A. Borning and R. Duisberg, Constraint-based tools for building user interfaces, *ACM Trans. Graph.* **5**(4), 345–374, 1986.

A REAL-TIME DISPLAY FOR TACTILE IMAGES

FILIP FUMA AND JOHN BRADLEY

1. *Introduction*

The tactile array sensor is a key component of the multisensory integration work at the General Robotics and Active Sensory Perception Laboratory (GRASP Lab) at the University of Pennsylvania.

In our work, tactile sensors provide information about the nature of the contact between an actuated tool, such as a robotic gripper, and its physical domain. This information is used for feedback control of the gripper and robot manipulator to which it is attached, and for tactile perception.

The sensor we currently use, LTS-200 from Lord Corp.[1] consists of two separate modules (see Fig. 1):

- A 10×16 tactile array, which measures displacement (or deflection) via optical technology.
- A 6 degree of freedom force/torque module using strain gauge technology. The force/torque module is mounted under the tactile array. This is not an array–its output is scalar.

The interface and signal processing circuitry is located in an external box.

The data obtained from the LTS-200 tactile array describes the amount of deformation of the sensor cover. The sensor cover, made from a synthetic rubber, can deform, up to a point, to comply with the surface of the object under investigation (see Fig. 2). The sensing array samples this surface and data are obtained describing the texture and local shape of the contact area. This tactile sensor can be compared to a range finder, but with very limited span.

The second module of the LTS-200 provides overall force, in the x, y, and z directions, and overall moments, around the x, y, and z directions, which result from pressing the sensor against an object.

FILIP FUMA and JOHN BRADLEY • Computer and Information Science Department, University of Pennsylvania, Philadelphia, Pennsylvania, 19104.

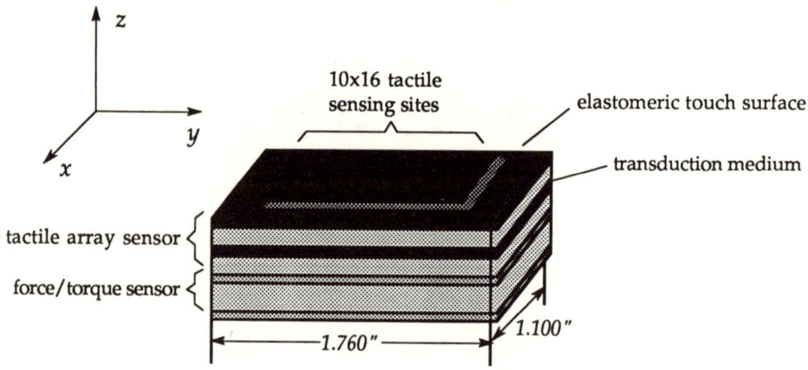

FIGURE 1. LTS-200 Tactile Sensor.

In the course of using the LTS-200, or evaluating new sensors, we felt a need for visualization the tactile information obtained. We looked for a format that provides insight into the tactile image and is suitable for further postprocessing.

2. *Motivation*

Displaying the tactile data in numeric format, organized as an array similar to the arrangement of the actual sites, is not very visually meaningful. We also considered the "three-dimensional mesh" representation. With proper software, the tactile data can be viewed from many angles, and provides some analogy to the actual relief of the explored surface. However, it is our experience that this method of display is not intuitively advantageous, and can become fatiguing with extended viewing.

Given our work with computer version, we decided to represent the tactile data as a gray-scale image. We find this preferable as we are familiar with this format, and, by making it compatible, the tactile image can be

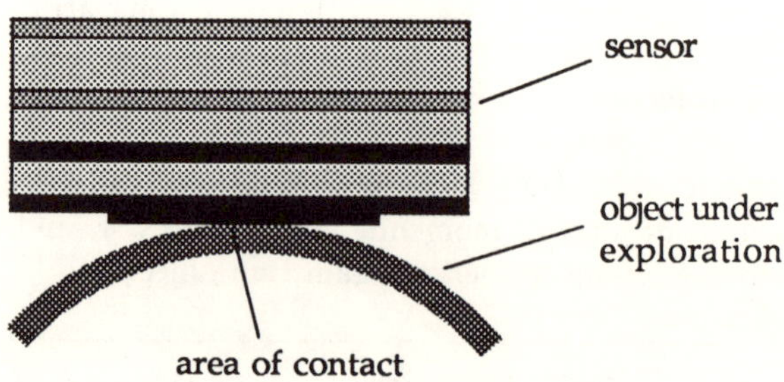

FIGURE 2. Sensor in contact with object.

processed with our predeveloped image processing tools. Such a format requires fewer computations than the three-dimensional mesh, making it easier to display in real time.

Following the anology of the tactile sensor to a range camera, but for proximity use, we chose to think of the tactile sensor as a tactile camera, describing a tactile image, in much the same way that a video camera describes an optical image. This is a useful data-representation metaphor, as it is consistent with the desire to represent tactile data as a gray-scale image. Taking the metaphor further, we have designed the software tools for capturing, retrieving, and displaying tactile data along the lines of a video cassette recorder.

3. Implementation

We have developed an interface[2] that performs the following functions:

- Visualize the tactile output of the sensor array as a gray scale, with choice of color mapping. Forces and moments are displayed numerically, and a subset are displayed as a pitch-yaw graph.
- Record the tactile sensor data for further computer processing.
- Display prerecorded tactile sensor data.

This program is called Tactile Camera Recorder, TCR. It requires a platform consisting of an IBM PS/2 running PC-DOS 3.3, with a color VGA display and mouse. The data are stored on floppy or hard disk, in a binary format suitable for transfer to Unix machines via ftp or kermit, for further analysis.

The interaction between the user and the program was designed to be intuitively simple and friendly. Most actions require "clicking" the mouse on the right iconic object. We have been influenced by the Macintosh interface, yet built an interface that is specific to the application.

3.1. Terminology

TCR operations are, in many respects, analogous to those of a video tape recorder, or a movie camera (with built-in playback unit and editing lab). The Lord sensor does not generate a continuous stream of data. Instead, the controlling program asks it "What is your state?", and a short time afterward the sensor returns a snapshot of its state when the request was made. These snapshots are called *frames*, in keeping with the film analogy. These frames are organized into a *sequence*, the analog of a strip of film. TCR operates on a single sequence, which can be up to 200 frames long.

3.2. *Operation*

TCR has two main modes of operation: Playback Mode and Record Mode.

In *Playback Mode* (Fig. 3) previously recorded sequences from disk can be viewed. It is also possible to inspect measurements at individual sites. *Playback Mode* does not require a Lord sensor attached to the machine.

Record Mode (Fig. 4) requires a Lord sensor to be connected to the computer. "Live" data coming from the sensor can be viewed in real time. The LTS-200 is capable of transmitting this information approximately 4–5 times per second. In Record Mode it is also possible to record sequences, edit them, and save them to disk.

TCR provides two different manners of viewing the tactile image: Detail View and Strip View.

Detail View (Fig. 4) is the default view and features a single, large 10 × 16 grid. Displacement values for each individual site can be read, and selectable cross-sectional profiles along the two axes can be seen.

Strip View (Fig. 3) loses these features, in favor of viewing several frames from a sequence simultaneously. This provides a visual time history of the tactile exploration. Editing sequences with the Recording Controls can be done in this view. A comment which was entered when the sequence was saved is also displayed.

The user interacts with TCR by clicking the mouse on the various objects on the screen.

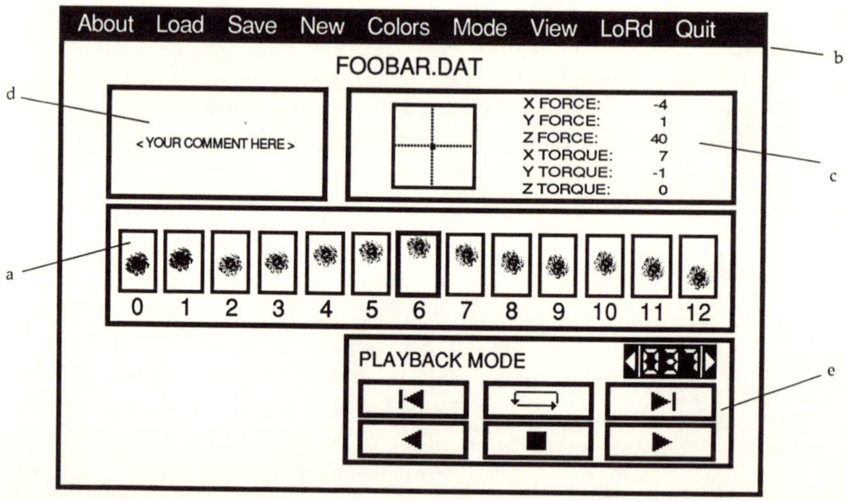

FIGURE 3. TCR *Playback Mode* display: a, Strip window; b, menu bar; c, force/torque window; d, comment window; e, playback mode controls.

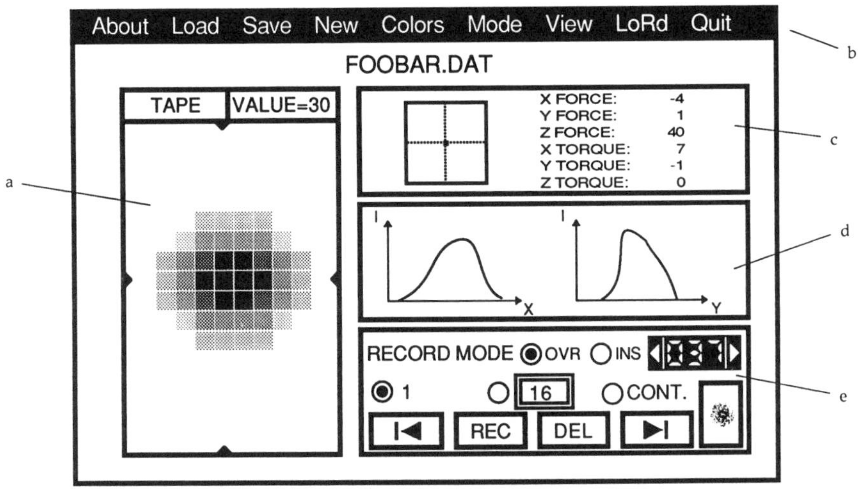

FIGURE 4. TCR *Record Mode* display: a, Detail window; b, menu bar; c, force/torque window; d, cross-section window; e, record mode controls.

The Menu Bar (Fig. 3b or 4b) lists several commands that may be executed at any time. Some of them may be "dimmed," and cannot be selected at that time. (For example, "Save" cannot be selected while in *Playback Mode*, since nothing was recorded.) These commands can be selected by clicking the mouse in the menu bar, and dragging it to the desired command. Alternately, a command can be invoked by typing its first letter (except in the case of "LoRd," where the command-key is "R," the "L" already having been used for "Load").

About. Clicking the mouse on this command brings up an "About" box, which lists the authors and gives the version number of the program.

Load. Loads recorded sequences from disk. Loading a sequence overwrites any sequence that was already in memory.

Save. Only available while in *Record Mode.* Saves the current sequence to disk, for later analysis.

New. Clears the current sequence.

Colors. Cycles between the three color schemes used to represent the amount of sensor displacement. The default is a gray-scale where no displacement corresponds to black, and maximum displacement corresponds to white. A second color scheme is the inverse gray-scale, where white equals no displacement, and black equals maximum displacement. The third color scheme utilizes pseudocolor to represent displacement. The colors follow the standard spectrum, with the "colder" colors (blue, violet) corresponding to small displacement, and the "warmer" colors (yellow, red) corresponding to large displacement.

View. Toggles between the two display modes, *Detail View,* and *Strip View.*

Mode. Toggles between *Playback Mode* and *Record Mode.* Active only when there is a Lord sensor connected to the machine.

Lord. Used to establish communications to an attached sensor. Also used to reset the sensor.

Quit. Used to exit the program.

The *Detail Window* (see Fig. 4a) shows a greatly magnified view of a frame of sensor data. Along the four edges are *Indicators*, which specify two cross sections through the tactile image. The corresponding profiles are displayed in the *Cross Section* window (see Fig. 4d). The indicators can be moved with the arrow keys on the keyboard, or by clicking the mouse in the data area. The value of the site at the intersection of these two lines is shown at the top right of the window, where it says "VALUE = nn." The button at the top left of the window, where it says "TAPE," selects whether the detail window will show the current frame from the recorded sequence, or live action ("LIVE"). Live action can only be selected in *Record Mode.*

The *Force/Torque Sensor* display (Fig. 3d or 4d) presents the force data that the Lord sensor is capable of measuring. The first three values are the forces along the *X*, *Y*, and *Z* axes, respectively, in hundredths of a pound. The last three values are torques around the *X*, *Y*, and *Z* axes. The small window shows three of these values graphically, as a pitch–yaw graph.

Z force range is represented by the color of the small square. The coloration is as follows:

Force (lb)	Color
0.0–0.5	Gray
0.5–8.0	Green
8.0 and above	Red

When taking measurements, best results are obtained when the sensor is operated with normal forces (along the *z* axis) in the "green zone." *X* torque and *Y* torque are indicated by the displacement of the small square from the center point. *X* torque results in vertical displacement, and *Y* torque results in horizontal displacement.

The *Cross Section* window (Fig. 4d) is available only in *Detail View.* This window shows plots of displacement values along the *x* and *y* axes. The axes are selected in the *Detail Window.*

The *Playback Controls* window (Fig. 3e) contains all the controls that handle sequence playback. The large left and right arrows run the sequence forward and backward. The center square stops the playback. The arrows with the vertical lines advance the sequence (in either direction) by a single frame. Holding either of these buttons down continues to advance the sequence until you let go. The top center button specifies whether sequence playback "wraps-around" at the ends, or simply stops.

The frame counter control (at the top-right of the window), performs three separate functions. The two arrows in the counter instantly move to the beginning or end of the sequence. Clicking in the numeric part of the frame counter allows entry of a frame number, and jumps directly to that frame.

The *Recording Controls* window (Fig. 4e) contains all the controls that handle sequence recording and editing. The two buttons with the arrows move forward and backward within the sequence. The "OVR" and "INS" buttons select whether recording new frames OVeRwrites existing frames, or INSerts new frames at the current frame number.

The "1," "16," and "CONT" buttons select how many frames are to be recorded. The "1" button records 1 frame, the "CONT" button records CONTinuously, until stopped, and the "16" button selects any fixed number of frames, between 1 and 199, to be recorded. These buttons only affect the number of frames that will be recorded. Once number of frames to be recorded has been determined, pressing "REC" will start the actual recording. Pressing "REC" again, or any of the controls in this box will stop the recording. Recording will turn off automatically when the selected number of frames has been recorded, unless CONTinuous recording has been selected.

The "DEL" button deletes the current frame. The frame counter operates just like Playback Mode: press the left arrow to move to the beginning of the sequence and press the right arrow to move to the end of the sequence. To move to any random frame within the sequence, click on the numeric part and type in the new frame number.

The *Lord* dialog box (Fig. 5) is used to establish communications with an attached Lord sensor, select the serial port where the sensor is connected, and zero calibrate or reset the sensor. These functions are invoked by clicking the mouse on the appropriate "button." This is a "pop-up" box, it disappears from the screen when its function is completed.

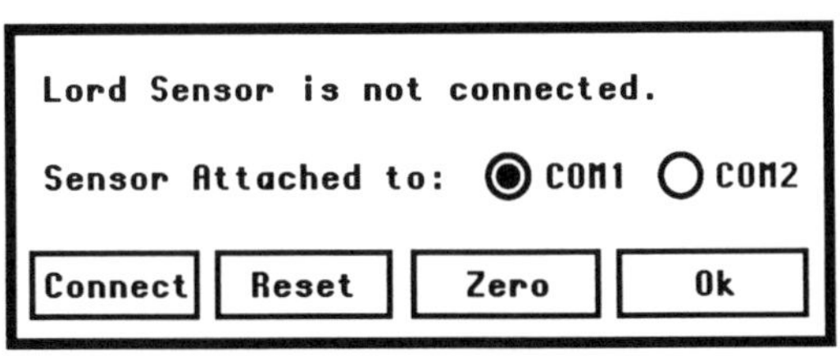

FIGURE 5. TCR *Lord* dialog box.

With the implementation described here, tactile data becomes visually meaningful in a user-friendly environment. The results can then be used for further tactile perception and exploration.

ACKNOWLEDGMENTS

This work was supported in part by the following grants: ONR SB-35923-0, Air Force AFOSR F49620-85-K-0018, ARMY DAAG-29-84-K-0061, NSF-CER DCR-82-19196, AO2, NIH NS-10939-11 as part of the Cerebro Vascular Research Center, NIH 1-R01-NS-23636-01, NSF INT85-14199, NSF DMC85-17315, ARPA N0014-85-K-0807, NATO grant 0224/85, NASA NAG5-1045, as well as by DEC Corporation, IBM Corporation, and LORD Corporation.

References

1. *Installation and Operations Manual for the LTS-200-Sensor*, LORD Corporation; Cary, North Carolina, 1985.
2. F. Fuma, and J. Bradley, A procedure for characterizing tactile sensors, University of Pennsylvania GRASP Laboratory Technical Report No. 171, MS-CIS-88-102, Philadelphia, Pennsylvania, 1988.

A GRAPHICAL INTERFACE FOR A GEOMETRIC MODELING DATABASE

DEEPA KRISHNAN AND TOSIYASU L. KUNII

1. *Introduction*

Geometric modelers form part of application systems involving computer-aided design (CAD), image processing, and animation, many of these fields involving step-by-step design of three-dimensional solid objects with structural modification at each step. For example, in mechanical computer-aided design, components of an assembly are often constructed in this manner. In such a case, it is easier and more convenient to design a shape using features rather than in terms of boundary elements such as vertices, edges, and faces. The term "feature" is used here in a broader sense than in conventional mechanical CAD where it refers to well-formed structural modifications such as holes, steps, or grooves. "Feature" here refers to any topology (collection of boundary elements and adjacency relationships) that is not necessarily that of a well-formed solid. We describe a framework for shape design tasks that allows user-friendly interactions with a geometric modeling database.[1] The framework supports design reusability. It also models the top-down nature of design effectively, and adequately represents the structure, behavior, and constraints associated with an object. The use of this framework makes the system more application-independent and extensible. Since visual interfaces have proven easier to use than conventional command interfaces, we chose graphical means for interaction with this framework. We call it GRIN for GRaphical Interface.

The interface described in this chapter is formulated for a designer. It is essentially special-purpose and employs a mixture of symbolic and graphical modes for increased efficiency. GRIN is based on a semantic model and is better for end-user interaction than an interface based on a conventional low-

DEEPA KRISHNAN and TOSIYASU L. KUNII • Department of Information Science, Faculty of Science, The University of Tokyo, 7-3-1, Hongo, Bunkyo-ku, Tokyo 113, Japan.

level data model. It allows browsing topological and geometrical details of three-dimensional solids; this provides users with a mechanism for exploring the metadata related to an object. Topological queries are answered by viewing the structure of the object at various levels of detail. Thus, formulation of queries that must adhere to complex syntactical rules is eliminated and creation of new topologies and geometries from preexisting ones is also supported by the interface. Windows and motion control using a mouse enable easy interaction. Details of the GRIN environment appear in Section 4; a prototype is being developed on the Personal IRIS workstation.

The paper is organized as follows. the next section surveys existing graphical database interfaces and describes GRIN. Section 3 presents shape representation in GRIN with examples from CAD applications. Section 4 describes the prototype GRIN environment and illustrates its use. The conclusion presents suggestions for further work.

2. *Visual Database Interfaces: A survey*

The function of a visual database interface is to make information more accessible and the database structure more apparent. Existing visual interfaces include those in Refs.2–7. This review presents some of these interfaces, their contributions, and a comparison to GRIN. The latter part of this section briefly introduces the GRIN environment.

Most existing interfaces have been designed to simplify the query formulation task. QBE was one of the earliest unconventional interfaces where a query step involved entering the form of the intended result in a table.[7]SDBMS[4] and TIMBER[8] both have graphical capabilities and use the relational data model. SDBMS eliminates the query formulation step and presents answers to user-queries by browsing. However, complex queries cannot be processed by this method. TIMBER improves over SDBMS by permitting complex queries. All three support the conclusion that a low-level data model has less ability to provide effective user interaction, resulting in efforts to use higher-level semantic models: e.g., GUIDE,[5] LID,[2] ISIS,[3] and GLAD.[6] Most of these were for nonengineering applications. A survey[9] showed that use of conventional database systems for geometric modeling suffers from several drawbacks. One is the need to map object structure into flat relations. Another is the inefficient description of the hierarchical design structure. Also this results in long textual queries and a necessity for the user to be familiar with the database schema and languages in order to be able to use the system. The cause in part is lack of conceptual structure for the database.

By contrast, GRIN is based on a high-level conceptual model describing

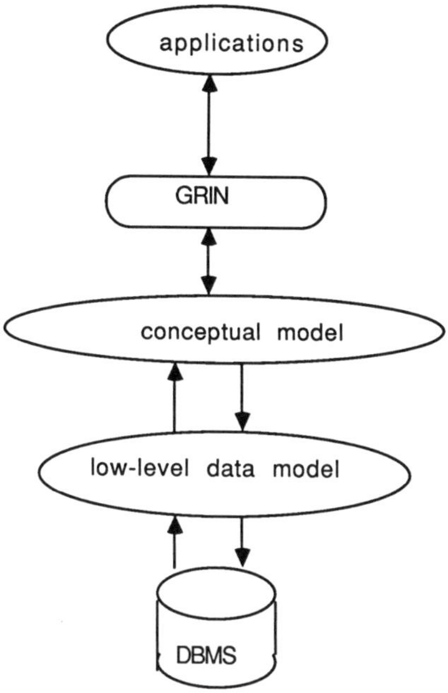

FIGURE 1. The organization of a system that uses the GRIN interface.

shape topology and geometry. Such a model enables the user to comprehend information more easily than in a case where data are presented in the form of relations. In addition, since users do not relate to join clauses and projections, the use of the conceptual model isolates the user from these complex mathematical concepts. Any of the conventional data models (network, relational and hierarchical) can be used to implement the model,* so the interface is low-level model independent. This is an advantage when the database is part of a distributed environment.

GRIN is fundamentally different from earlier visual interfaces[2–4] developed specifically for the relational model. Multiple windows eliminate the need for a facility to answer complex queries as in interfaces like TIMBER.[8] GRIN provides more than just browsing support: it permits creation and editing of shape metadata. The other interfaces discussed are more oriented toward simplifying the query language: they only offer sophisticated browsing facilities. GRIN incorporates the good points of existing interfaces while providing the designer with more specialized facilities.

The overall structure of a system using GRIN is described by Fig. 1. Essentially GRIN interacts with the conceptual model. Hence, we need a

* A similar nonformalized version of this model has been implemented as a relational model by one of the authors.[11]

simple interface for reading DBMS data into GRIN structures, i.e., a connection not shown in the figure between the low-level database and the conceptual models. This function is usually done at the beginning of a design session; thereafter data can be manipulated using GRIN. Such read operations can also be performed during a session as designers use GRIN to create new shapes.

GRIN possesses several novel features. It is a special-purpose interface formulation with the designer in mind, uniquely oriented to engineering databases. The need for such a special-purpose interface is justified by the fact that a user with absolutely no knowledge about a three-dimensional object or the design process cannot comprehend metadata related to shapes, data of use only to a designer. Further, this interface simplifies shape design for users from a wide variety of application fields.

The conceptual framework used to represent shapes in this paper is specific to a geometric database and is supported by operations based on the theory of hypergraphs. These operations facilitate the reusability of designs and creation of new topologies from existing ones. In addition, GRIN maintains a temporary database to support the trial and error design process, and facilitates interaction with the object in this database. Topological queries are answered implicitly, i.e., instead of users formulating complex syntactical queries, they view pictorial answers on the screen as they browse through the database. Windows are provided for simultaneous viewing of topologies at various levels of detail. New topologies and geometries can be created and viewed using this interface. For further details see Section 4, following an introduction of conceptual representations for shapes in the geometric modeling database.

3. *Shape Representation*

Shapes can be designed using features in a hierarchical manner and manipulated using the functions in the interface. The hierarchical top-down design in facilitated by a mechanism* based on hypergraph operations.[1] This new technique permits (1) a broader definition of a "feature" (see Section 1); (2) implementing an evaluation and instancing mechanism which is *independent of the feature used* (most existing feature-evaluation techniques depend on the type of feature used), so new features can be added without addition to, or modification of, code; (3) reusability of designs; and (4) increased system extensibility.

The GRIN conceptual representation for geometric shapes involves a set of object templates defined by a 7-tuple, $\{T, T\text{—OP}, G, G\text{—OP}, R, V, C\}$,

* A formal treatment of this mechanism is given by Krishnan *et al.*[1]

where T denotes the set of topologies (connectivity relations among edges, faces, vertices, and features) available in the database. Here each topology is denoted by a hypergraph, T—OP represents all the topological operations. Each operation modifies object structure. G constitutes the set of all possible geometric information (vertex coordinates, edge equations etc.), while G—OP refers to all possible geometric operations available, i.e., operations that define and manipulate the geometry of an object. R describes all possible relationships between elements in the database, Υ denotes the set of variables and C describes the constraint set. A single object template is given as a 7-tuple,

$$\{t, t\_op, g, g\_op, r, v, c\}$$

where t-$t \subset T$ and is represented as 3-tuple:

$$(H, TE, F\text{-links}) \tag{1}$$

where H is a hypergraph,[1] TE represents the set of tentacles (hyperedges required to attach this object to another object when it is used as a feature),[1] and F-links are pairs (t_i, E_{d_i}). Here t_i is a 3-tuple topology as described in (1) and E_d denotes the base object hyperedges to be deleted while inserting the feature ($t_i \subset T$, $E_d \subset E_P$, P is the set of parents of the feature in the hierarchy). Two types of topological representations are possible. The first is explicit representation. There the pointer to the F-link is NULL and the topology of the object is expressed in terms of hyperedges of H, and connections involve tentacles. This representation is suitable when the instances of the object are frequently used during the design of other objects. The second type of representation is the case when topologies are expressed implicitly in terms of the features. Both kinds of examples are illustrated in Fig. 2. $t\_op - t\_op \subset T_{op}$ and represents the set of operations associated with the object, i.e., object-specific operations and is denoted by $\{\text{top}_1, \text{top}_2, ..., \text{top}_n\}$. Each topological operation involves replacing a portion of a hypergraph by another using the feature-evaluation mechanism.[1] An operation is denoted as follows:

$$\text{top}_i : t_b < -F\text{-link}_i$$

since a new F-link is introduced into a base topology t_b. The F-link is subsequently defined. This is illustrated in Fig. 2b.

$g - g \subset G$ and an explicit representation consists of all geometrical information such as edge and face equations, face normals, edge lengths, and dimensions related to hyperedges such as the angle between two edges.

a

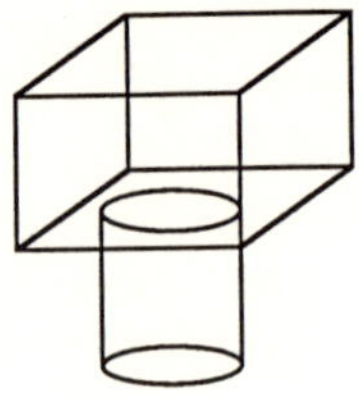

p= { g, g_op, t, t_op, v, r, c}

g: { l_1, l_2, l_3, r, h, a, rpt }

$d_1 = l(e_2) = l(e_4) = l(e_{10}) = l(e_{12})$

$d_2 = l(e_1) = l(e_3) = l(e_9) = l(e_{11})$

$d_3 = l(e_5) = l(e_6) = l(e_7) = l(e_8)$

$r(e_x) = r(e_y)$

coordinates of vertices and other geometry can also be specified explicitly along with their relationships to variables.

g_op: { }

t: { f_1, f_2,...,f_z, l_1, l_2,...,l_z, e_1, e_2,...,e_z, v_1,...,v_z }

$f_1 : (l_1)$

$f_6 : (l_6, l_x)$

$l_1 : (e_1, e_2, e_3, e_4)$

$e_1 : (v_1, v_2)$

tentacles: { fz }

t_op : { o_1 }

o_1: t_{cube} <- f_link$_1$

f_link$_1$ = (t_{round}, v)

v - { length, breadth, height, depth, radius, refpt, axis, vedge}

r - { r_1, r_2 ...,r_8}

r_1 = { length, d_1 }

r_2 = { breadth, l_2 }

c - { c_1, c_2,..., c_5 }

c_1: length > 0

c_6: vedge.type = V-L

FIGURE 2a. An object representation with topology and geometry explicitly specified.

b

p= { g, g_op, t, t_op, v, r, c}

g: { g_1, g_2 }

g_1: CUBOID(o,l_1, l_2, l_3)

g_2: CYCLN(o,r, h, a, rpt)

g_op: { gop_1 }

gop_1 (tweak): CHANGEVERTEX(v, x, y, z)

t - { o }

o: Null <- f_link_1

f_link_1 = (t_{cube}, NULL)

t_{cube} <- f_link_2

f_link_2 = ($t_{cyclnpin}$, NULL)

t_op - { o_1 }

o_1 : t_{cube} <- f_link_x

f_link_x= (t_{round}, v)

FIGURE 2b. Implicit representation of topology and geometry. Note: the rest of the representation remains the same except for the addition of a relationship between the topological operation *o* and the geometric operations g_1 and g_2. The required tentacle is implicitly specified while evaluating the feature. Refer to Fig. 2(a) for diagram.

Dimensions can be defined in reference to other surfaces. Geometry can also be represented implicitly as a sequence of primitive graphical operations.

$g\_op - g\_op \subset G\_op$ and permits the specification of a set of geometric operations associated with the object $\{gop_1, gop_2, ..., gop_n\}$. Each operation alters the geometry of the object in some way and is expressed as a sequence of graphical manipulation operations.

$$r - r \subset R \text{ and } r = r_1 \cup r_2 \cup r_3 \cup r_4 \cup r_5 \cup r_6$$

$$r_1 = \{(g, v) | g \subset G, v \subset V_1\}$$

$$r_2 = \{(t, v) | t \subset T, v \subset V_2\}$$

$$r_3 = \{t\_op, v)|t\_op \subset T\_op, v \subset V_2\}$$

$$r_4 = \{(g\_op, v)|g\_op \subset G\_op, v \subset V_1\}$$

$$r_5 = \{(c, v)|c \subset C, v \subset V\}$$

$$r_6 = \{(t\_op, g\_op)|t\_op \subset T\_op, g\_op \subset G\_op\}$$

$$r_m \cap r_n = \varnothing m, n = 1, 2, ..., 6 m = n$$

V_1, V_2 and V_3 represent the geometrical, topological, and the operation variables respectively.

$v - v \subset V$ and denote variables essential for parametrization. These variables are of four types:

1. Topology variables: variables related to the *F*-links which indicate the number of such features located at that level of the hierarchy.
2. Geometry variables: variables related to dimensions; and
3. Operation variables for example, tentacles.
4. Mating dimensions: these denote the variables that are important during redesign since they are the associated dimensions of two mating components. These variables are of type (2) but are considered separately due to their importance.

$c - c \subset C$ represent constraints, which are of three types:

1. Geometric constraints on dimensions.
2. Connectivity constraints on hyperedges necessary for the integrity of the object.
3. Constraints that govern the mating of components.

Figure 2 shows a representation using this model. The geometry of an object is clearly separated from the topology definition in this model. This would mean that operations (such as a local modification like tweaking) that alter geometry alone can be done independently without traversing the entire connections in the data structure. This is illustrated by the data structures in Fig. 3. In order to change the coordinates of a vertex, the appropriate graph is first retrieved and the list of vertices in the structure (Fig. 3a) is traversed till the required point is found and the alteration in the coordinates is then performed.

Notice that in GRIN the geometry is also accessible from the topological data structure. In addition, the model gives a high-level description of all data related to an object, allowing a user to easily define and manipulate an object. More details about this framework and its associated operations can be found in Ref. 1.

4. *The GRIN Environment*

GRIN provides the user with a mechanism to explore and manipulate the metadata related to a shape. It maintains sufficient information to support a complete conceptual framework. A prototype is being developed in the C programming language for the Personal IRIS workstation.

4.1. *Windows*

GRIN provides the user with a complete set of functions to manipulate windows and explore the feature database. There are three types of GRIN windows: (i) Icon, (ii) Text, (iii) Database.

Icon windows allow pictorial adjustment of graphical variable values. An example appears in Fig. 4, where the "color-edit" window enables the user to make adjustments of the background or objects' color displayed in the current database window. The left button of the mouse is used to adjust color levels in the red, green, and blue color boxes. The window background provides feedback, display of the resultant color. Similar windows are used to set projections and viewing parameters. Such windows do not require any text input from the user and all parameters are set by controlling the motion of the mouse.

Text windows serve three purposes: (1) to indicate errors, (2) output system text, and (3) accept user text input. An example of a text window that accepts the input of a feature name from the user is shown in Fig. 5. The user is provided with the option of altering the input if he clicks the left mouse button on the "alter" rectangle.

Database windows come in two kinds: (1) enable viewing of topological interconnections, and (2) enable viewing and editing of geometry. Examples of window browsing tools with hypergraphs displayed are shown in Figs. 4 and 5. GRIN permits the user to create, close, pop, push, move, resize, and change the color of windows, supporting simultaneous views of topology and geometry of several objects. The small window in Fig. 5 shows some features in the database; it helps users view the set of currently available features.

4.2. *Topological Queries*

Nine main types of low-level topological queries commonly encountered in geometric databases have been listed.[10] These queries concern the connectively relationships between boundary elements such as faces, edges, and vertices. The formulation of such queries can be quite complex in the case of conventional data models.[9] GRIN eliminates the query formulation step altogether and allows the designer to observe the connectivity on the screen. This is easily facilitated since the basic unit of the data structure used is a hyperedge. Thus, the user can pick any feature from the database and the corresponding topological hypergraph is drawn in the appropriate window. This graph can be viewed in a top-down fashion at various levels of detail (see next subsection). Queries have two basic forms:

1. Direct: adjacency information is explicitly maintained in the database; and
2. Inferential: connectivity information is implicit and has to be obtained from contents of hyperedges stored in the database.

Direct queries are easily answered by viewing the topological hypergraph at the appropriate level of detail. Inferential queries are answered by simultaneously viewing topology at the required levels using multiple windows. In the existing version of GRIN, the system does not infer the

a

```
typedef struct blob{
   ident name;
   short tag;
   struct hhlist* connect;
   struct screenlocation* placeblob;
   struct geometry* geom;
   struct blob* prev;
   struct blob* next;
};
```

```
typedef struct flink{
   struct hgraph* feature;
   struct identlist* delete;
   struct flink* prev;
   struct flink* next;
};
```

```
typedef struct hiedge{
   ident name;
   short tag;
   kind typ;
   struct bblist* eblobs;
   struct screenlocation* placedge;
   struct geometry* geom;
   struct hiedge* prev;
   struct hiedge* next;
};
```

```
typedef struct tentacle{
   ident he1;
   ident he2;
   struct tentacle* prev;
   struct tentacle* next;
};
```

```
typedef struct hgraph{
   ident name;
   struct blob* gblobs;
   struct hiedge* gedges;
   struct flink* sgraph;
   struct tentacle* connect;
   struct operation* sequence;
   struct tree* featuretree;
   struct facelist* sfaces;
   struct looplist* sloops;
   struct edgelist* sedges;
   struct vertexlist* svertex;
   struct hgraph* prev;
   struct hgraph* next;
};
```

FIGURE 3a. Topological data structures in GRIN.

connectivity information for these inferential queries automatically since the response time for such a task can be large and depends on the number, and the types of hyperedges stored.

To answer a query such as "what primitive objects compose object A?" the interface maintains a feature tree for each object. This query is answered by a display of the tree on the screen. This eliminates the need to remember a database language syntax to formulate topological queries. Figure 5

b

```
typedef struct tree{
  ident featuretype;
  ident name;
  struct treelist* child;
};

typedef struct object{
  struct hgraph* ptr;
  struct tree* featuretree;
  struct object* prev;
  struct object* next;
};

typedef struct operation{
  ident name;
  struct variable* parameters;
  struct operation* prev;
  struct operation* next;
};

typedef struct variable{
  char type;
  ident name;
  union{
      int i;
      float f;
      controlpoints c;
      vector v;
  } value;
  struct variable* next;
};

typedef struct vertex{
  ident name;
  struct location* place;
};

typedef struct edge{
  ident name;
  char type;
  char dir;
  struct equation* eqn;
  union{
      float length;
      controlpoints curvepoints;
  } kind;
};

typedef struct loop{
  ident name;
  int loopnumber;
  float angle;
  struct equation* eqn;
};

typedef struct face{
  ident name;
  char normaldir;
  vector normal;
  struct equation* eqn;
};

typedef struct geometry{
  union {
      struct vertex* nvertex;
      struct edge* nedge;
      struct loop* nloop;
      struct face* nface;
  } ptr;
};
```

FIGURE 3b. Geometrical data structures in GRIN.

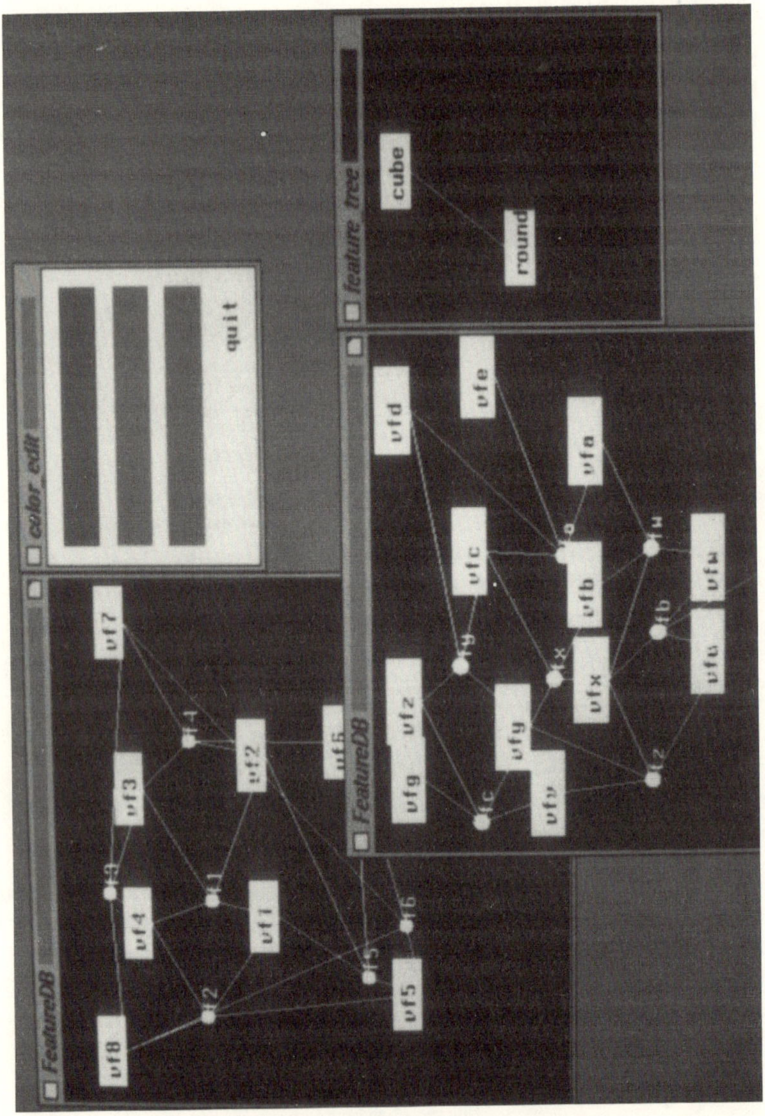

FIGURE 4. Windows in GRIN I.

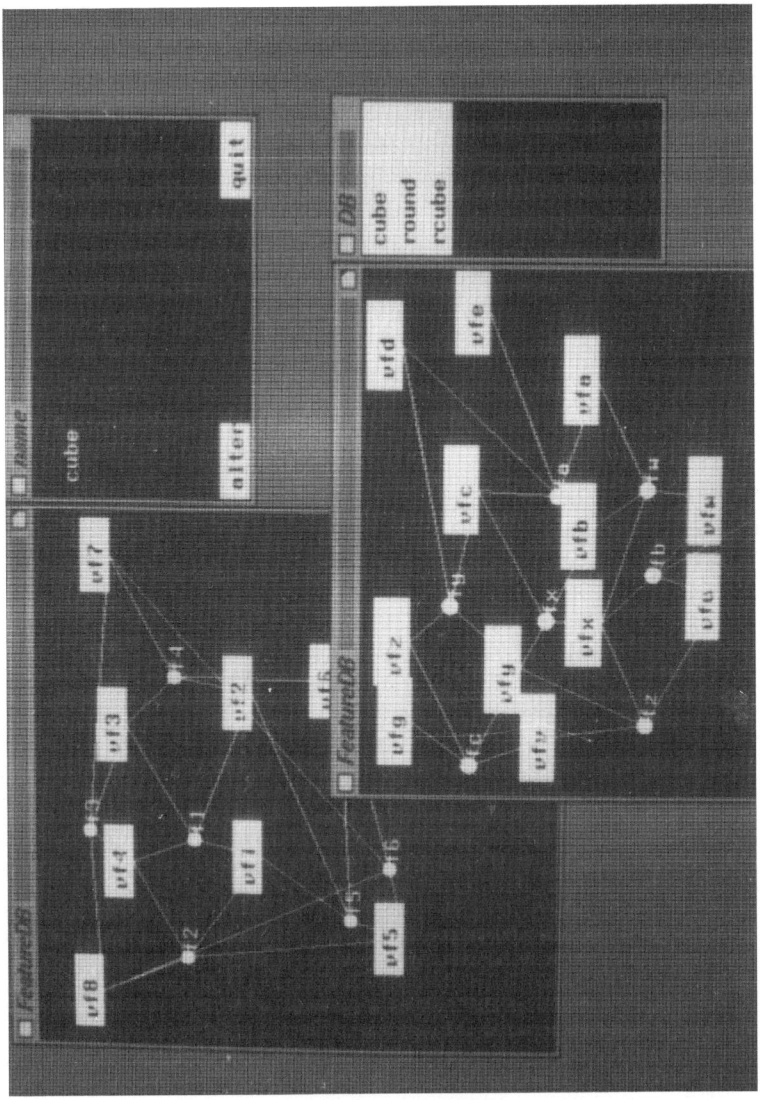

FIGURE 5. Windows in GRIN II.

illustrates a feature tree for an object "rcube": a cube with one rounded side, as in Fig. 2b.

4.3. Browsing

The GRIN browsing facilities enable exploration of metadata. There is a menu "picture" that permits hierarchical viewing of the topological structure. For example, level 1 illustrates the adjacency relationships among the faces while the lowest level illustrates those among the vertices. Level 3 allows the viewing of four types of adjacency relationships as illustrated in Fig. 6. The "picture" menu also enables editing of pictures on the screen to enable the designer to get a better view. For example, some of the hyperedges can be deleted to increase the clarity of displayed information. Figures 4 and 5 show the topological hypergraphs of the features "cube" and "round" at the V–F (vertex–face) level.

A graphic editor GLEE forms part of the system to enable the designer to browse geometric information. For example, consider an object whose geometry is defined implicitly, i.e., in terms of the constituent graphic operations for its features. For example, the geometry of the cuboid with a cylindrical hole in Fig. 7 can be expressed as a sequence of operations,

cuboid(refpoint,length,breadth,height,refaxis)

cylinhole(refpoint,radius,depth,refaxis)

This operation sequence is retrieved from the database and maintained as a part of the GRIN data structure of that feature. The editor then executes these commands in sequence enabling the user to see what the object looks like. Figures 6 and 7 give solids drawn using GLEE's browsing facilities. Explicit geometries can also be displayed by the editor but are more cumbersome since each polygon has to be retrieved and then drawn. The displayed geometry can then be altered if necessary and restored as a part of the description of the object using the functions in GLEE. Some of the commands in the editor are shown in Fig. 7. The commands shown also facilitate the input of geometric information which will be described in the next subsection. The novelty of the interface lies in the fact that the data structures used enable the retrieval of geometric information without traversing the complex topological interconnections.

4.4 Creation of Topologies and Geometries

Topological and geometrical input can be either explicit or implicit; if topology is to be input explicitly, hyperedges and their details are read from

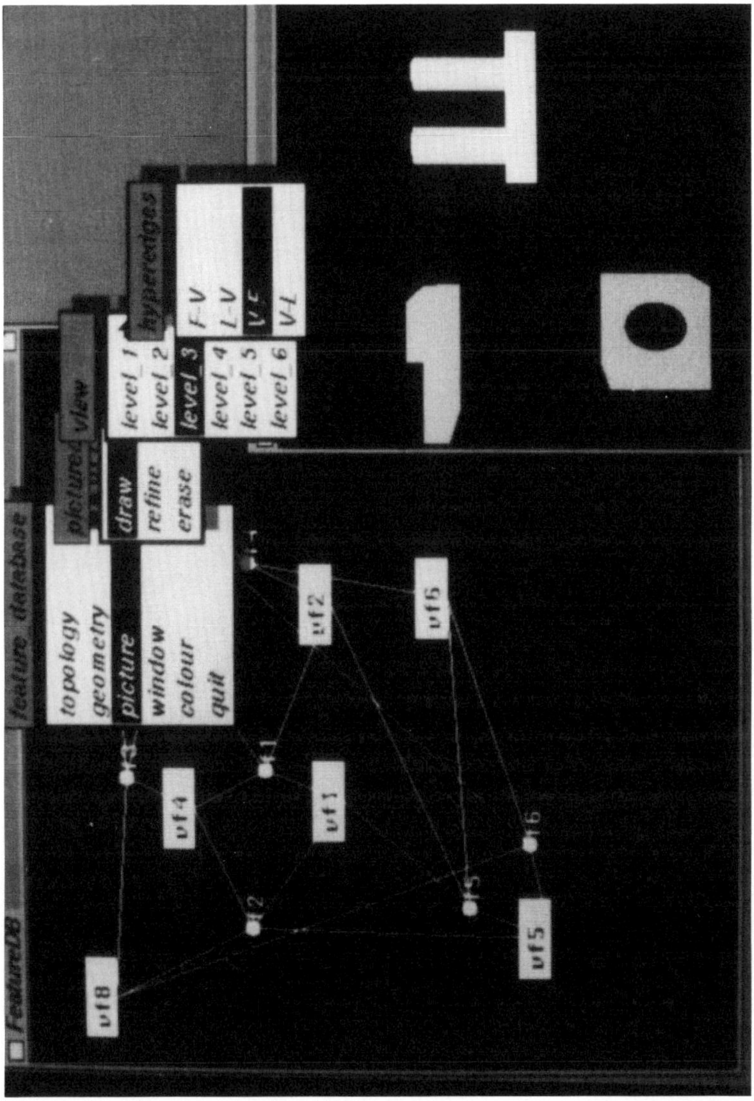

FIGURE 6. Viewing topology hierarchically at various levels.

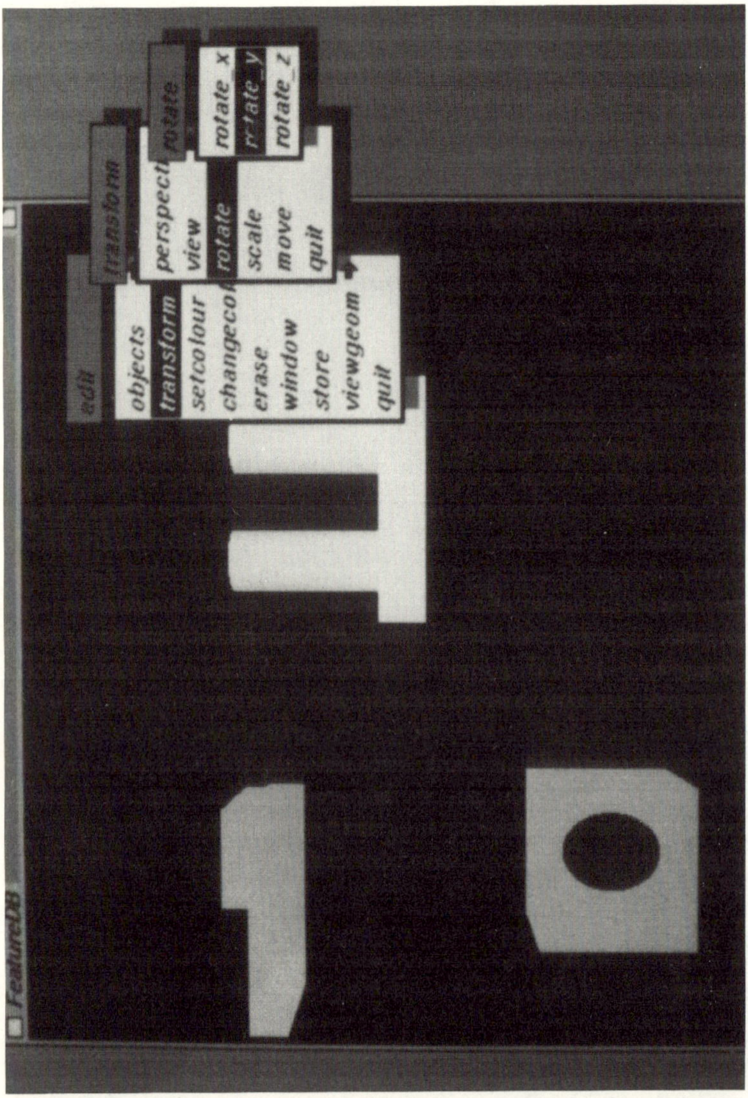

FIGURE 7. GLEE: the graphical editor.

files containing the required information and incorporated into the database. This is convenient in the case of basic features which are to serve as tools in creating complex objects. The commands in the explicit topology menu are shown in Fig. 8. Database features can be created or deleted using these functions. The implicit menu is used to create new topologies from existing ones. GRIN maintains a temporary database which permits the user to create objects but not to evaluate them. The designer starts by defining a base object and then defines the object in a top-down fashion by adding new features at each stage. This top-down design proceeds using the "refine" and "undo" commands in the menu shown in Fig. 9. The hierarchy of features composing an object can be observed using the "displaytree" command in Fig. 9. When the user needs to check the resulting topology, the "evaluate" command is used to compute the boundary of the graph. This function permits evaluation of the entire object or a particular portion of the object (lazy-evaluation) as required. Evaluation of a feature requires the input of a pathname which consists of a sequence of ancestors of the feature seperated by the "/" character. All these ancestors are evaluated before the named feature. The designer can then add this topology to the database using the "addtoDB" function, but this is not mandatory. Since this new topology is to be used at later stages, the designer should be absolutely sure of its correctness before incorporating it in the database. The topological structure can be checked using the facilities described in the previous subsection.

Geometric information can be defined using commands in the implicit and explicit geometry menu. Explicit geometric data is read from files while the graphic editor GLEE is used to input operation sequences. The designer uses the editor's "object" menu to draw, rotate, translate, and scale objects as necessary until their geometric location is correctly determined. A designer can also use a menu that lists the available graphical operations to directly input sequences. The mouse is used to select the appropriate operations from the menu. The "viewgom" command enables the designer to observe the shape obtained by executing an operation sequence associated with an object. An operation sequence can be edited using the "altersequence" command in the implicit geometry menu. The sequence of operations are then recorded as part of the GRIN data structure representing the object using the "storegraph" command. This removes any need for the designer to visualize these sequences and their parameters in their minds.

The data structures used in GRIN are shown in Fig. 3. Figure 3a shows the topological data structures used. It may be observed that the "hgraph" structure maintains separate pointers to lists of faces, loops, edges, and vertices, to permit easy retrieval of geometric information without traversing through the maze of topological interconnections. The structure "object" in

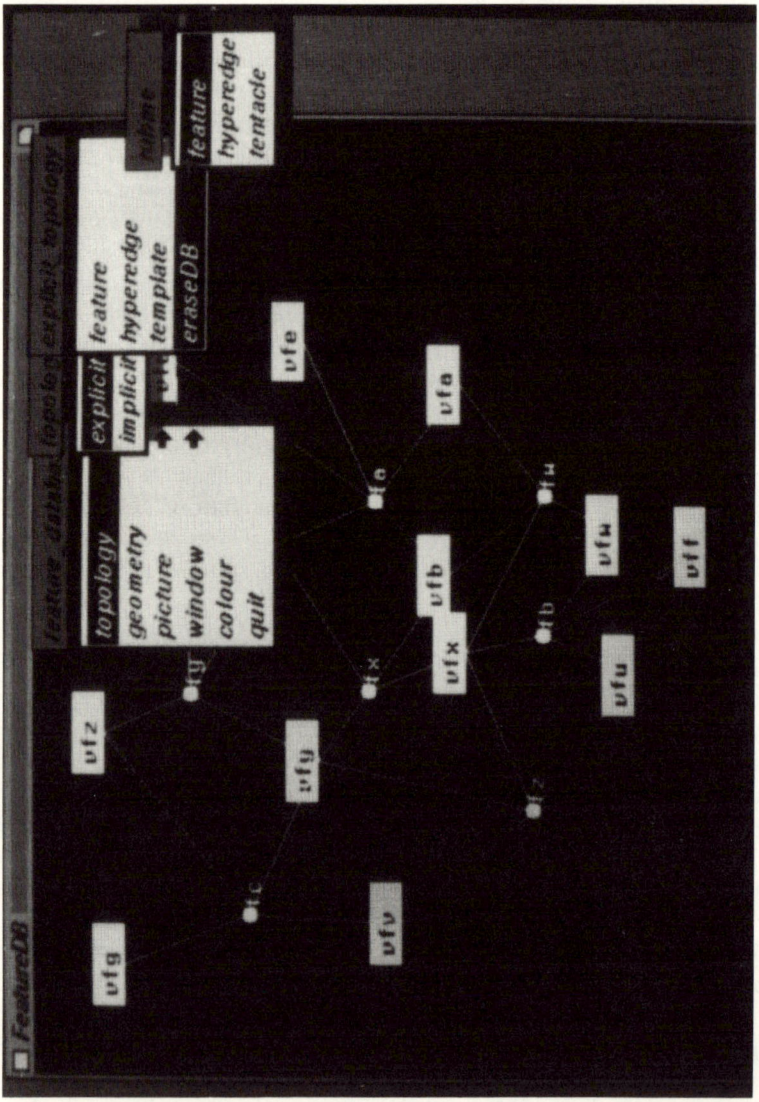

FIGURE 8. Explicit topology menu.

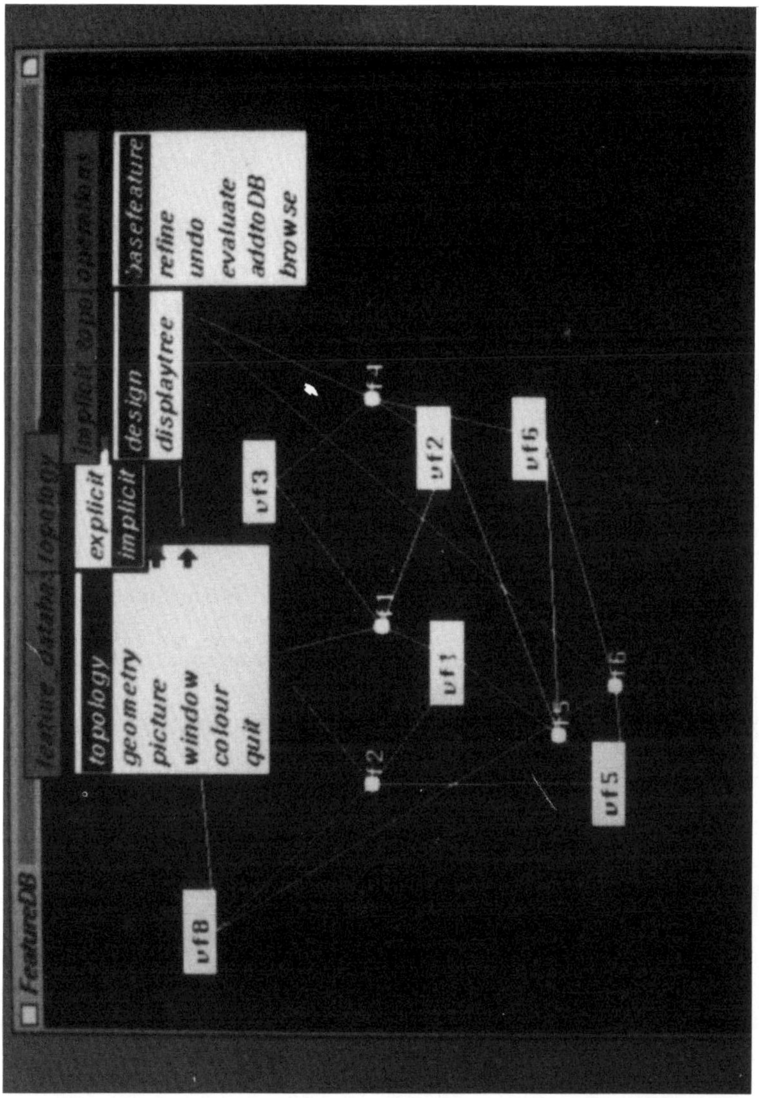

FIGURE 9. Implicit topology menu.

Fig. 3b is used to describe an object in the temporary database. Since constraints have not yet been integrated into the graphical facility, their structure is not described.

Sample Session. At first, a required portion of the database is read into the GRIN data structures on request from the user. GRIN then permits manipulation of these data structures. For example, consider the case where the user wants to create a feature "rcube" representing a cube with one of its edges rounded and incorporate it in the database. In this case, the following steps are executed by the user:

i. Read in the explicit topologies of "cube" and "round" from the database using the "explicit topology" menu. The geometries of both these features which have been previously defined using operation sequences are also retrieved from the database.

ii. Use the "implicit topology" menu to design a new object with an instance of "cube" as its base feature and use the *refine* function to add the feature "round." The feature tree for this object is shown in Fig. 4.

iii. Use the *evaluate* function to evaluate the feature "round" of the object "rcube."

iv. This evaluated object is then added to the database using the *addtoDB* command. Prior to this its topology can be browsed using the "picture" menu. The result of this step is depicted in the small window in Fig. 5.

v. The geometry is then defined using GLEE or simply by combining the operation sequences that were previously retrieved. The resultant sequence will consist of the operations that draw a cube at an appropriate location (cuboid primitive), erase the required corner (eraseline primitive), and then draw the rounded corner (line and curve primitives).

5. *Conclusions and Further Work*

We described the GRIN visual interface to a geometric modeling database, indicated how it is based on a specialized conceptual shape representation and illustrated how the framework represents the design process in top-down terms. Likewise, GRIN describes the structure and operations associated with features, something that is inherently efficient is design situations. The interface is special-purpose and offers a friendly, easy-to-use environment for designers. In addition, it supports design reusability and therefore, a system using GRIN is extensible: no alteration of code is necessary when new features are added to the system. As far as we know, GRIN is one of the few interfaces of this kind for an engineering database.

The present version of GRIN has very few primitives; future work will involve increasing these primitives. Constraints and variables are specified textually but we plan to introduce a graphical specification mechanism as a part of GLEE. Furthermore, graphs are now drawn with the aid of the user who positions them appropriately on the screen. More automated graph drawing facilities are essential. We hope to also extend the "picture" menu commands, and the system, to include zooming and planning. Last but not the least, the specification of tentacles can be complex in the case of certain features. We are considering the introduction of a mechanism which will visually guide a user in the definition of tentacles by providing examples on the screen. This would, however, require that the designer of each feature incorporate the example into its database description.

ACKNOWLEDGMENTS

We would like to thank Silicon Graphics International for making the Personal IRIS workstation available to us. One of the authors was also supported in part by scholarships from from the International Information Science Foundation and from the College Women's Association of Japan and we express our gratitude to them for facilitating this research.

References

1. D. Krishnan and T.L. Kunii, Parametric model: A conceptual framework for geometric modelling database, to appear in the Proceedings of the Second International Conference on Data and Knowledge Systems for Manufacturing and Engineering, Maryland, October, 1989.
2. D. Fogg, Lessons from "Living in a Database" graphical query interface, in Proceedings of SIGMOD, Boston, pp. 100–106, 1984.
3. K.J. Goldman, S.J. Goldman, P.C. Kanellakis, and S.B. Zdonik, ISIS: Interface for a semantic information system, in Proceedings of SIGMOD, Austin, pp. 328–342, 1985.
4. C.F. Herot, Spatial management of data, *ACM Trans. Database Syst.* **5**(4), 493–514, 1980.
5. H.K.T. Wong and I. Kuo, GUIDE: Graphical user interface for database exploration, in Proceedings of the 8th Conference on Very Large Databases, Mexico City, pp. 22–32, 1982.
6. C.T. Wu and D.K. Hsiao, Implementation of a visual database interface using an object-oriented language, in Proceedings of IFIP TC 2/WG 2.6 Working Conference on Visual Database Systems, Tokyo, April, pp. 105–125, 1989.
7. M.M. Zloof, Query-by-example: a database language, *IBM Syst. J.* **4**, 324–343, 1977.
8. M. Stonebraker and J. Kalash, TIMBER: A sophisticated relation browser, in Proceedings of the 8th Conference on Very Large Databases, Mexico City, pp. 7–10, 1982.
9. A. Kemper and M. Wallrath, An analysis of geometric modelling in database systems, *ACM Comput. Surv.* **19**(1), 47–91, 1987.
10. T.C. Woo, A combinatorial analysis of boundary data structure schemata, *IEEE Comput. Graph. Appl.* **5**(3), 19–27, 1985.
11. D. Krishnan and L.M. Patnaik, GEODERM: GEOmetric shape design system using the entity–relationship model, *Comput.-Aided Design* **18**(4), 207–218, 1986.

A DIALECT FOR IMAGING AND GRAPHICS

OLEH J. TRETIAK AND SANJAY BHASIN

1. Introduction

Our group at Drexel University's Image Processing Center is engaged in research and development in Image Data Analysis, Computer Vision, and Image Processing. A typical researcher is expert in some field of Imaging Science, but is not a computer scientist. Nevertheless, a typical project requires substantial computer programming on the part of the researcher. Typically, these programs include computer graphics, program control through a menuing system, as well as imaging calculations specific to the task. Computer graphics involves the generation, representation, manipulation, and evaluation of graphic objects with related nongraphic information, and interactive operations such as drawing or picking. Menuing operations require adherence to a complex protocol of declarations and invocations that are specific to the computer system. Image computations require manipulation of complex data structures in the framework of a Vision Kernel that has a certain class of objects and standard operations. These data structures may or may not be compatible with the Graphics Kernel. Thus a program that creates user interfaces and acts upon objects in the image must cover all these areas.

We find that conventional computer languages, such as C, have great expressive power and permit effective use of a computer, but they expose the application programmer to unnecessary detail. Just as a scientist using FORTRAN does not need to know *in detail* how operations on floating point or complex numbers are performed, so an imaging scientist or application programmer should not be required to know, *in detail*, the operation of the graphics, windowing system, or standard imaging libraries. We propose to extend the C language by endowing it with data types and primitive operations that permit simple, natural expression of image processing

OLEH J. TRETIAK and SANJAY BHASIN • Image Processing Center, Drexel University, Philadelphia, Pennsylvania 19104.

programs. In effect, we are developing a dialect of C that is closely connected to the windowing system and the user's application discipline (we call this concept Babytalk). In this system, DIG (*D*ialect for *I*maging and *G*raphics), image processing and graphics reflect the same semantics to make the generation of the user interface simpler. DIG enables the association of various parts of a application program to separate menus and enhances the user's perception of the order in which operations must be carried out while providing a distinct level of abstraction from the windowing system and details of the image processing objects. To aid in conceptualization, semantics are oriented towards this class of applications and the language has primitives which reflect the nature of objects native to the application area. Conventional computer languages deal with primitive entities such as numbers and arrays. In image processing it is necessary to deal with visual objects, so DIG has such objects as primitives.

Nowadays a microcomputer attached to a TV camera is a practical way of performing interactive data collection tasks. Application programmers who write interactive packages need an environment that is independent of the graphics/display system. We have a technology for developing programs that require substantial program adaption to the multitude of windowing standards such as MacOs, Sunview, X windows, which are powerful but complex. And at the same time we have a large resource of image-processing technologies. There is a need for a language that allows an application programmer access to these and that allows him to concentrate on the application at hand rather than details of the underlying implementation. This work draws on a background of interface design, image processing, and window systems.

In this paper, we first relate our work to other research on interface design and image-processing languages. We then discuss the philosophy behind DIG, and of the features that describe the scope and workings of this language. Finally, we consider possible extensions and augmentations of DIG.

2. *Background*

Architectures for user interfaces based on the concept of direct manipulation without time-sequential user–machine dialogues which impose no syntactic structure on the command-language have been developed.[22] User interface management systems (UIMS) have developed as a result of the need to have separate user interface module that facilitate design, construction, and maintenance of user interfaces.[9] But they separate user interface concerns from those of the application and concentrate on the problem of input dialogue management alone. Toolkits that offer a rich set of compostion mechanisms and a variety of predefined objects that allow easy implementation of complex user interfaces have been extensively used.[12] The interactive portion of DIG draws from this material.

Graphical Kernel System (GKS) provides a complete, standardized, well-constructed set of graphic functions which reflect current thinking.[10] It is oriented towards device-independent output. Programmer's Hierarchical Interactive Graphics System (PHIGS) is a draft standard for an interface between an application program and an underlying graphics system that controlls all the devices.[2] PHIGS is being developed as a more powerful interface that is needed to write highly interactive, three-dimensional applications. It utilizes and builds on many GKS concepts. The data are organized to enable easy transfer of multilevel application models into the graphics system.[1] PHIGS also provides facilities that allow dynamic modification and rendering of the graphics database, thereby providing the ability to create truly interactive applications. These standards develop a fairly good basis for a device-independent interactive graphics system. We draw on them to design the graphics portion of DIG.

IDL (Interface Description Language) is a technology for the generation of tool intercommunication support with roots in compiler generation. It is a nonprocedural specification of data structures which describes the interface between components and supplies a programming language independent specification of the abstract data types used in the implementation of the interface. A full description of IDL can be found in Ref. 14 Generation of software based on IDL description and the tool supporting the generation and discussed in a thesis by Lamb. An IDL specification makes assertions about relations between different elements in the interface data structure and provides hints how to efficiently implement the data structures and to optimize data storage. The specification is processed by a set of tools to generate a package of subprograms that implement the specified abstract data types. This specification also provides valuable documentation for the programmer, tester, and maintainer. It supplies the programming languages with consistent packages for use in implementing the system. The class concept in IDL is similar to that is Smalltalk. IDL was used to define the intermediate representation for Ada™ called Diana.[6] In 1985 members of the SoftLab project at the University of North Carolina at Chapel Hill completed an implementation of an IDL translator and a set of IDL development support tools running on UNIX™.[19]

Image-processing software systems are most frequently implemented as subroutine libraries.[15] Other systems, such as the University of Maryland Image Processing System (UMIPS), provide operating system commands, which may either be invoked interactively or from a program (script file).

Some systems provide a degree of programming through a macro capability for replacing command sequences. PICASSO SHOW 1.6 and PAL are high-level languages for picture processing.[5] Subroutine libraries place great demands on the programmer, since the function parameters usually have no logical order, so that programs are hard to write and to maintain, while command-based systems require substantial operating system overhead and allow limited programming expression.

Smalltalk-80 uses two-dimensional nature of display surface and pointing device to provide multiple, concurrent dialogue streams.[8] It uses windows as a uniform interactive interface. Windows allow a large amount of information to be organized on a display screen and allow a programmer to keep track of several activities at a time. Smalltalk introduced object-oriented programming, the systematic treatment of programming as simulation of a real problem.[4] But none of these allow the description of categorical elements as high-level structures. Need exists for a system in which the user dictates what responses need to be taken on certain conditions and the control mechanism to manage that is produced by the compiler.

We are novel in that we think of image as an object, and not as a file. We also incorporate the notion of objects in an image and integrate these entities with a graphics system. In conventional languages, program flow is specified by loop and conditional statements, whereas in DIG one specifies an action for each case in terms of a menu and program flow is generated by a compiler. We have idiosyncratic or application-specific semantics resulting in a simpler view of both the program and the user interface.

3. *An Overview of DIG*

3.1. *Illustrative Example*

To motivate further discussion, we will describe the hardware environment for DIG, and present an illustrative example of an application. The system diagram in Fig. 1 shows the feedback loop involving hardware and software. Hardware consists of a grey-value display and television camera connected to a computer. User control of the software is with a mouse (or some other sort of pointing tool) and/or a keyboard. The application program draws on DIG utilities such as the Reader, Menu Manager, and Window Manager. The task at hand is to digitize an image, designate a region, and to calculate the average grey level. The operator chooses the SCAN–NEW option, as shown in Fig. 2a, which causes an image from the camera to appear on the screen (see Fig. 2b). (This is an image of the metabolic activity along a section, through a rat brain.) The operator now adjusts the film to center this image on the screen, as shown in Fig. 2c, and signals to the computer that this is the image to be digitized: the computer stores the digital data in memory, and leaves the image on the screen. Subsequently, the operator wishes to designate a region in the brain section, which is done by selecting the ANALYSIS-AUTO QUANT option on the menu (Fig. 2d). The cursor turns into a drawing tool, and the operator draws an outline of the region of interest. When the outline is completed (Fig. 2e) the program computes the average grey value of the portion of the image within the outline, and the interior of the outline is filled with a uniform color (Fig. 2f).

To the user, this sort of interface is attractive for several reasons. The process is controlled through a menu system. This makes the operation natural and easy to learn. Geometric entities, e.g., the outline of the region to be measured, are delineated through natural graphical actions, such as pointing and drawing. The image and graphics are in natural apposition. Computer activity, the measurement of a feature over a region, is communicated to the operator by generating and displaying a natural graphical entity.

DIG includes expressions that make it easy to write this sort of program. Menus are generated with declarative statements that reflect the control structure of the program. Menus appear on the same display as the image and graphics. Primitive expressions in the language are used to invoke interactive actions such as image scanning and selection of objects on the screen. The image analysis functions have a simple, natural syntax.

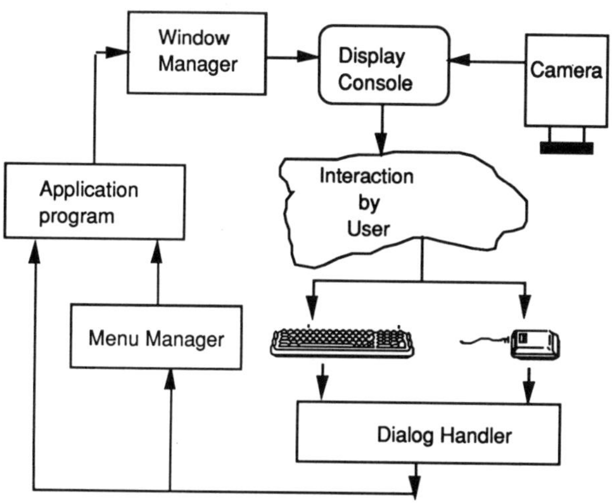

FIGURE 1. System operation diagram.

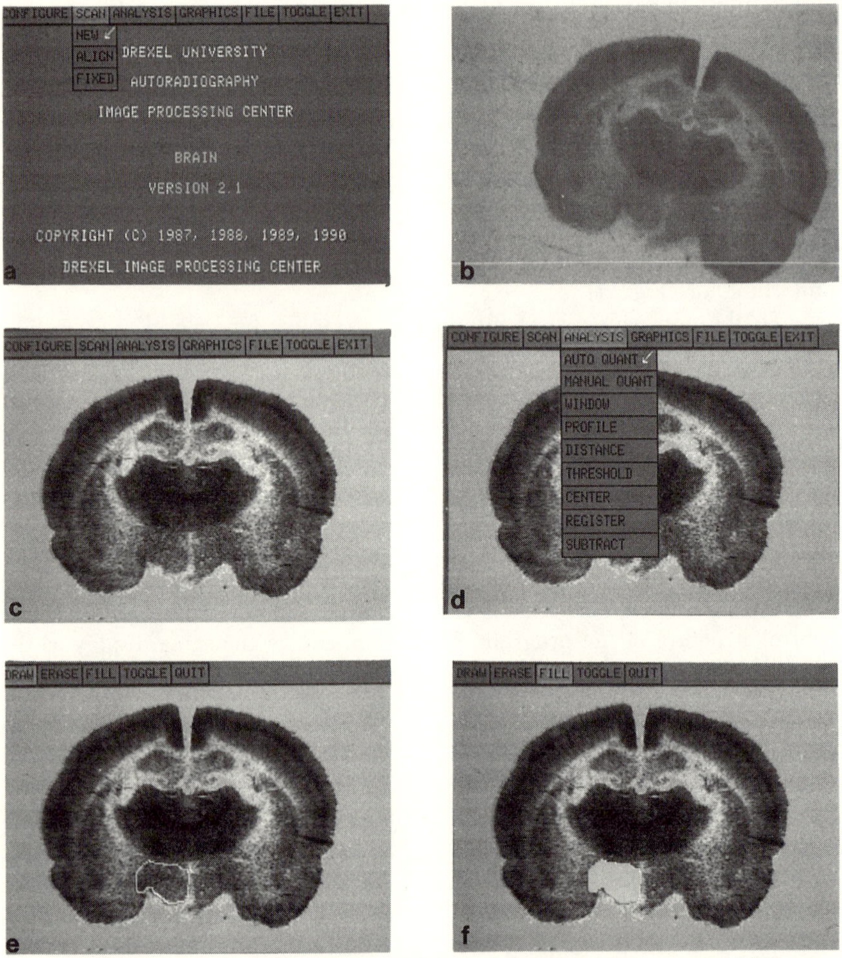

FIGURE 2. (a) The menu with SCAN selected. (b) Live video signal from camera. (c) Capture of centered image. (d) Choosing ANALYSIS option. (e) Outline around region of interest. (f) The area of interest is shaded white to indicate the part of the image it covers.

The design of DIG is based on a several of fundamental notions. One is a scheme for relating images, the experimental data, to geometric entities, the computed attributes of images. Another is a framework for separating interactive "pointing" actions into references to the program control actions (the menuing system) and graphical actions (object selection, delineation, and modification). Finally there is the conceptual scheme for integrating the computed and interactive operations on images and graphics.

3.2. Imaging Semantics

One of the most powerful features of natural language is that words are polysemous—they have the potential of having several meanings. Yet a statement can convey a clear notion even though the individual words are ambiguous. Good language design should imitate this attribute. This allows terse and clear imperative statements. This is achieved when programming constructs reflect the deep structure of the application domain and the programmers general factual knowledge and syntactic inclinations.

Images are our fundamental datatype. They are digitized pictures, documents, or scenes. As such, they take a range of forms and meanings. For example, we may have images produced with x-rays, light, sound, or other modalities. They may be monochrome or in color, they may be still or dynamic. An image may depict objects in three-dimensional space, a section through a three-dimensional body, or some coded distribution such as a diffraction pattern, spectrum, or an alphanumeric document. In view of this variety it is not surprising that image processing methods are many and diverse. We define an image as a function of two continuous variables that takes on a range of continuous values: in mathematical terms, an image is a function from a subset of the Euclidean plane into a real n-dimensional vector space. Of course, images on the computer are digitized: they are specified on a grid of points, and take on discrete values. This is an important matter since it affects the accuracy of various image processing operations, yet our goal is to devise programs that act as if images were objects in real space.

In applications that concern us images are composed of objects at various locations of the image plane: the objects may be internal organs in an x-ray, lines in a spectrum, or the graphemes in a document. Our job is to locate and identify them. *Our second fundamental datatype are geometric objects, which represent subsets of the plane.* Circles, rectangles, and polygons are all examples of geometric objects. Our basic classification of thes objects is

1. Points
2. Lines—straight or curved
3. Regions—that is, point sets with open interiors.

For example, an ordered pair of numbers $p = (x, y)$ is a point, the set

$$\{(x, y) \mid x^2 + y^2 = 1\}$$

is a line (a circle), and

$$\{(x, y) \mid x^2 + y^2 < 1\}$$

is a region (a circular disk). We are interested in geometric objects because they can be used to identify the locations of the more abstract objects that comprise the image.

Perhaps the most fundamental image-processing operations are geometric transformations: mappings of the plane such as translations, scalings, rotations, projective mappings, and affine transformations.[13] This is not an exhaustive list: transformations from polar to rectangular coordinates, and cartographic transformations which are used to map the surface of a sphere to a plane are useful in certain applications. The important property of geometric tranformations is that they can be applied to geometric objects and to images. Geometric transformations are relevant in mathematical morphology,[17] which is a calculus based on plane mapping and point set unions, intersections, and complements.

While geometric transformations are based on the geometry of the plane, topological operations are founded on the plane's topology. They include the evaluation of boundaries of regions (lines), the inverse operation of filling the interiors of boundaries, and the identification of connected subsets of a set.[24] More primitive are set-theoretic operations, such as unions, intersections, and complements.

Geometric transformations act on geometric objects and on support sets of images. Another collection of operations act on the image values, the range set. These include unary operations such as contrast enhancement, and binary operations such as image addition. Much study has been devoted to linear and quasilinear image-to-image functions such as convolutions, Fourier transforms, and edge detection operators.[23] These should be contrasted to *segmentation operations*, ones that produce geometric objects (regions) from images. The most fundamental such operation is thresholding: for example, if $f(x, y)$ is an image, one can define a region by the relation

$$R = \{(x,y) | f(x,y) < t\}$$

Image segmentation is the most important problem in computer vision, and DIG will be of use in research this field.

We have discussed the primary data types, namely, images and geometric objects, and have mentioned some of the more important functions for transforming images to images, regions to regions, as well as segmentation operators. The remaining class of operations are *feature extractors*: they apply to images, regions, or image–region combinations, and compute various data, which we call features. Some of them are simple and rapid, such as the computation of the average value of an image value. Others have considerable technical complexity, such as the moments of an image or region.[25] The most important features are ones that have direct physical meaning, such as the area, length, or sensitometric attributes of an object. We cannot end this list without mentioning the brightness value histogram, a complex feature that is easy to compute and one that has found great application in image processing.

We see that imaging provides a rich set of objects well based on mathematical foundations. The design of the imaging kernel is founded on images and geometric objects as main object categories. Geometric transformations of both geometric objects and images and arithmetic operations on image range types are the main functions. To this basis we add functions that process both images and geometric objects, and functionals that compute features. The set of functions and feature types is so large that it is unreasonable to implement all but the most important. The key design issue is to define the basic objects in such a way that the most frequently used operations are efficient. Of course, one need not restrict oneself to only one representation for a given object type. The specific representations are hidden from the applications programmer: it is the task of the language designer and translator to choose the best representation, and to convert among representations whenever necessary.

3.3. Interactive Interface

Interactive image processing involves the building and display of control structures, actual images, and objects computed from the images. DIG provides tools that allow a programmer to build application programs such as a custom graphics editor. The built-in constructs provide a level of abstraction that allows the programmer to concentrate on the application and code at an algorithmic level, and the imaging tools reflect a human's natural conceptual organizations of visual objects.

The salient features of DIG include the functionality to create and manipulate windows with associated image and graphics data, to manipulate the colors, images, and graphics on an output device, and to obtain input from available devices using a canonical, programmatic interface which embodies a high degree of device independence. The usefulness of the Babytalk methodology of using natural semantics lies in expressing a user interface with ease and demonstrates the simplicity and power of such a system of dialects. There is a clear dichotomy of on-screen and off-screen operations. The off-screen part comprises of a rich set of objects based on mathematical foundations. The on-screen part consists of the graphical counterpart of the imaging objects which are treated as passive data and menus which are active control constructs. In addition, there is support for interactive operations on images such as contrast enhancement, adding pseudocolor to the image, and modifying regions on a display. Such interactive operations may be used to develop presentation graphics, or to input image and geometric data into the computer. The running process has a point-and-click interface, a pointing device such as a mouse is the controller. The mouse action has a different meaning based on the context it is in. In the menu bar, mouse operations allow the user to move around in the application program: the selection of an item in the menu invokes an associated action. In the data windows, the mouse is used to select objects and draw outlines on the image. The iconic interface is intuitive and thus requires little or no learning time.

The graphical interface consists of three types of displays. There is a single control element through which the user communicates with a DIG process, namely, a menu bar on the screen that sets up a context for interpretations of the user actions. The menu may be switched under program control, but only one menu bar is visible at one time. The second type of display consists of graphical widgets called alert boxes, which are used for getting response to queries (e.g., "Do you want to quit the program?"), and to allow the user to chose an item from a list (e.g., to pick from a list of files presented on the scree). The third type of display consists of imaging windows (data windows) that display images and image-related graphics. Traditionally, one window has been associated with a single process, but DIG allows multiple data windows under control of a single DIG program.

The menu is a decision object that is decoupled from other windows and is declared and specified separately from them. It allows you to initiate operations. The essential part of program design is reduced to describing a certain set of tableaux in which user operates, and each tableau is associated with an object called *bar menu.* The set of bar menus has the structure of a general directed graph so that it is possible to write a program in which bar menus lie on cycles. Each bar menu, however, is an acyclic graph (more specifically, a rooted tree of height not greater than 2), so that a each part of an application has a heirarchical division. This means that that there can be no recursion within a bar menu, that intramenu options are isolated and localized to one menu. Each item in the menu is either a submenu (a pop-up menu) or, in the case of a leaf, a block of code in the implementation language. Only one menu is current or active at a time and has full control of all parts of display. Flow of control is reflected in activation being passed from menu to menu and the new menu replacing the old one. As a result, one can keep the same image and graphical objects but change menus with ease. A menu set is represented in Fig. 3. This example depicts two bar menus: one containing six items, and a second with three items. All items in the first menu are primary (height 1 in the tree). The second menu, invoked through the leaf "TRANSFORM," has one

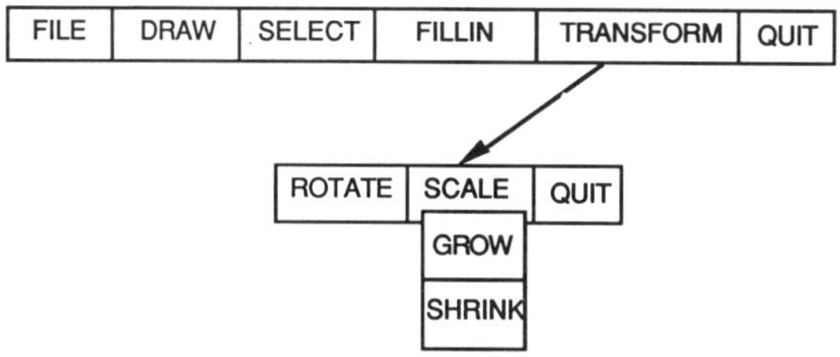

FIGURE 3. Example of a menu bar generated by the code in Table 2.

primary item "ROTATE" and a node "SCALE" which produces a popup menu with two leaves, "GROW" and "SHRINK."

Each data window is independent and different data sets can be displayed and manipulated in them. A data window contains grey scale image plus graphical objects and text which are overlayed. They share coordinate systems; writing/erasing of graphics/images are mutually independent within each window and also have no effect on other data windows. The number of image bit planes and graphics planes depends on the hardware.

DIG graphics provides support for the Image Processing Kernel datatypes. The objects exist in the imaging library and have graphics counterparts used for their rendering. Device independence is achieved by supporting the description of device interfaces within the language rather than having these defined in the libraries performing the rendering of the objects.

Semantics in DIG simplify the handling of input devices, such as the mouse or frame grabber. Primitive operations are provided that query/interact with the device, e.g., Scan returns a grey-level image to the caller without the fuss of opening the device and reading all the data byte-by-byte. In addition, it provides the data in a format readily usable by all the routines dealing with regions so that the user never has to deal with the internal structure of any object and can do all his coding at an algorithmic level. For the output device, the screen, there is a Show primitive that posts objects, and a Remove primitive that deletes graphic objects. The interface allows certain bidirectional communication between the program and the data window: Select allows one to choose objects by pointing to them on the screen or passing reference to the object from the program to the data window. Transformation primitives allow the dragging (translation), scaling, and rotation to be mirrored onto the Vision Kernel copy of the object. They also are bidirectional, allowing similar actions to be initiated by both the program and the user. All high-level display functions include clipping, i.e., they make sure that the object is within the edges of the window.

In summary, the salient features of DIG are the funtionality to create and manipulate windows with associated image and graphics data, to manipulate the colors, images, and graphics on an output device, and to obtain input from available devices using a canonical, programmatic interface, which embodies a high degree of device independence. The environment consists of an event handler, a bit-mapped display, and a pointing device (mouse or table). The key to the portability of the system across hardware platforms is the independence of DIG from the underlying windowing system. The graphical user interface consists of menus that change under a pointing-device control. The usefulness of the Babytalk methodology of using natural semantics lies in expressing a user interface with ease and demonstrates the simplicity and power of such a system of dialects.

4. *Implementation*

In implementing a language, one is faced by conflicting demands between simplicity and expressiveness, between power and performance. Foremost is the need for reliability: the intended operations should be carried out correctly, the design of the language should discourage constructs that are prone to programmer error. The key technique used to achieve reliability is the use of objects that hide the implementation, initialization, and transformation of data structures from the programmer.

4.1. *The Imaging Kernel*

A central feature of the Vision Kernel is the scheme used for dealing with point coordinates. There is a clear duality: points in the plane are represented by pairs of real numbers, while image pixel coordinates are pairs of integers. A feature of image processing hardware is that most scanning equipment produces rectangular pixels: typically, spatial increments between horizontal and vertical pixels are different. To deal with this, the abstract datatype "point" is implemented in two ways: *realpoint* and *pixel.* The *realpoint* data structure consisting of two floating-point numbers, while the *pixel* consists of two integer fields. The latter structure can be used to access image arrays. User programs should operate with *realpoints*: these are the appropriate coordinates for reporting distances, perimeters, and areas. In many experimental situations, locations of objects in an image can be measured with subpixel accurancy, and real coordinates provide a mechanism for dealing with such measurements. With proper calibration, *realpoints* can be in physical units such as meters, micrometers, or angles.

The imaging kernel software is a subroutine system with three levels: user (U), derived (D), and primitive (P). The applications programmer is expected to operate at the U level: this level provides the simplest function invocation and the highest degree of reliability. The DIG scanner translates the U level constructs into D level code, which consists of C language operations and function invocations. The D level system supports the functionality of DIG, but is complex and therefore requires considerable programming astuteness. P level structures and operations are utilized to develop the D level structures: there are relatively few such structures, and are designed for efficiency and portability.

To illustrate this layering, we will describe some of the methods used for representing regions. Recall that a region represents a point set on the plane. At the U level, a region consists of the following (1) a "magic number," (2) a pointer to a D level region structure, and (3) a transformation. The meaning of this is "point p is in region X if $T^{-1}(p)$ is in the D level region object." For example, if the complex region object is a circle of radius 2 with center at (1,1), and the transformation represents a shift of (2, 3) and followed by a twofold scaling, then the region object denotes a circle centered at (3, 4) with radius equal to 4. The magic number is a unique code that signifies that this object has the abstract datatype "region." This scheme permits a certain efficiency and accuracy in region operations. For example, a function that creates a new region by shifting an existing region by a given amount may be invoked (at the U level) by

$$R2 = delta \gg R1$$

where $R1$ and $R2$ are regions and *delta* is a real vector. This causes $R2$ to contain a pointer to the same D region object denoted in $R1$, while the transformation part of $R2$ is computed by composing the transformation in $R1$ with a shift of amount delta. This function can also be invoked with the assignment operator

$$R1 \gg = \text{delta}$$

which causes the transformation portion of $R1$ to be modified. Another operation in this family is

$$A = B\&C$$

which produces a region that contains all points common to regions B and C (set intersection). A region is deleted with the statement

$$A = 0$$

which causes identifier A to denote an empty region.

There are several derived data structures that can be used to represent: one such is the interval-list, which represents a region in terms of pixel coordinates. It consists of a header, *Interval_list*, and a doubly-linked list of *Interval_bins*, each containing pointers to bins and an array of *Intervals*. The *Interval* is a triplet specifying a horizontal run of points: it contains three integers y, $x1$, and $x2$, with the notion that all pixels with the given y coordinates, and with x values in the range $x1 \leqslant x < x2$ are members ofthe specified region. Operations on the interval-list is through "primitive" functions:

new_I()—creates an empty interval-list
augment_I()—adds an interval to an interval-list
open_I()—initiates access to an interval-list
next_I()—obtains the next interval from an opened interval-list
close_I()—terminates creation or traversal of an interval-list
delete_I()—annihilates an interval-list

An interval-list can be in one of three states: closed, open for creation, or open for traversal. When an interval-list is open it is accessed through a *I_list_tool* structure: this is done to provide multiple access to an interval-list. Memory for interval-lists is obtained from the operating system: it is requested in the *new_I* and *augment_I* functions, and returned to the operating system in the *delete_I* call.

The implementation of an interval-list contains a number of safety features. The header structure contain a "magic number" field. Functions that access interval-lists (such as *open_I*) check this field, and produce fatal errors if it is not valid. Another field monitors the creation/traversal state of the interval-list and a count of U level structures that refer to it.

There are several data structures for regions. The *Regionbit* codes arbitrary pixel sets through a one-bit-per-pixel scheme (bitplane), and the *Box* type represents rectangular regions through the coordinates of the upper left and lower right pixels. The *Circle*, *Ellipse*, and *Rectangle* datatypes represent regions in terms of real coordinates, rather than pixels. Each is identified by an appropriate magic number, so that a function that obtains a region-type structure as a parameter can identify it through the magic number and invoke the appropriate code. These operations may involve "casts" between region objects. For example, the intersection of two *Circles* will typically yield an interval-list, since the resulting region is not a circle.

The second major class of geometric objects are lines. As in regions, a U level line consists of a magic number, a pointer to a derived line object, and a transformation. The most versatile line object is a *Border*: this is a sequence of chaincodes which denote line segments between adjacent pixels, as shown in Fig. 4. The scheme for storing, creating, and traversing borders

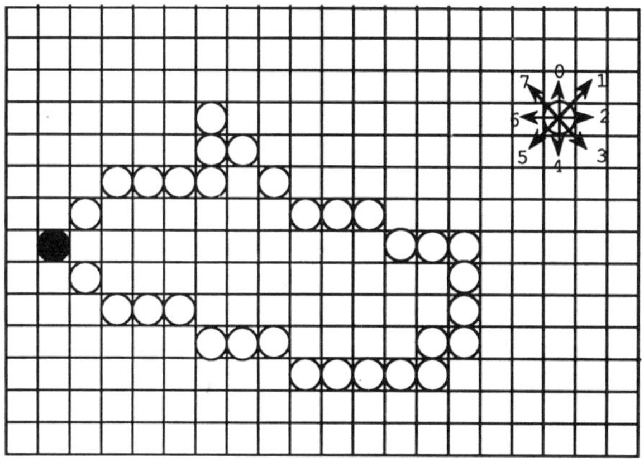

FIGURE 4. Using the reference coordinates shown and the black pixel as the starting point, the chain code for the line is (1,1,2,2,2,0,0,3,3,3,2,2,3.3,3,4,4,4,6,4,6,6,6,6,7,6,6,7,6,6,7).

mimics that for interval-lists. Another type of line is a *Broken_line*: this consists of a sequence of line segments whose end points are real vectors. There are two types of lines: line segments and loops. A loop is a line whose start and end points are identical. A special type of loop is a Jordan curve: this is a loop that does not cross itself. The interior of a loop is a region, and the interior of a Jordan curve is a connected region. The calculus of line operations supported by DIG is fairly poor: lines may be transformed, and one may evaluate lengths of lines, loops may be converted to regions, and one can compute functionals of images along lines. The most frequent use of lines is to delineate boundaries of regions: these lines are Jordan curves. Such lines are used in the graphics interface of DIG to display region outlines.

The principle followed in implementing the U level image object is the same as for lines and regions: a image consists of a magic number, a derived image structure, and a transformation. In the practice of image processing and computer vision one can anticipate an enormous variety of image structures. To illustrate this, we will describe a few possibilities. The image obtained from a monochrome camera is a two-dimensional array of integers denoting digitized values of brightness in a scene. A color camera produces a two-dimensional array of three-dimensional vectors denoting the *R*, *G*, *B* values of the camera signal. Consider the following examples: the FFT of an image is typically represented by an array of complex numbers indexed by rectangular coordinates; other transforms, such as the Hough or Radon transforms, produce arrays indexed by polar coordinates. Image transformations such as shifts, rotations, and scalings of the underlying geometric space produce well-defined but ideosyncratic effects on transform images.

There is a potential for a rich but bewildering collection of datatypes and operations, but in accordance with the "dialect" philosophy, we restrict ourselves to a powerful yet limited subset of possiblities. We feel that just as different languages are associated with different cultures, brevity of communication is lost if one attempts to encompass all possibilities. We have therefore chosen to use the following definition at the U level: *An image derived object consists of a magic number, a support set (region), a two-dimensional array, and a gray value denotation object.* The denotation object indicates the interpretation for the values stored in the array (signed/unsigned, color/monochrome, etc.) Since gray value denotation is in itself an object, it can be a procedural function, a tabular function (as is the usual practice), or the indication of the range type used (signed or unsigned numbers, binary numbers with arbitrary radix points, etc.). If contrast enhancement is required for purposes of display, this may be accomplished by changing the grey value denotation. To illustrate this, consider the following example:

> A spectroscopic instrument produces a print in which the data are restricted to a circular region. Within this region, the physically relevant data are proportional to the optical density of the print. A scanning device that produces 8 bit numbers over a rectangular array is used to digitize the print. The digitized values are proportional to the reflectivity of the print. If N is the brightness value (grey value) from the print then the optical desity is given by $D = \log_{10}(225/N)$. The derived object for this image consists of a *Box* representing the array dimensions, two-dimensional array of 8-bit numbers (grey values), a region object denoting the circle in which the data are stored, and an array of 255 floating point numbers (look-up table) containing the map from grey values to density values.

This general scheme permits a simple method for dealing with rectangular images, and encourages representation of grey values with small integers, yet allows one to represent images of arbitrary shape.

Images are created through input/output operations, such as scanning or reading a file, or through image operations. For example,

$$I = J - K$$

computes a new image I by subtracting the values of K from those of J. The support set of the new image is the intersection of those of the two operands. As another example,

$$I^* = a$$

where a is a floating-point number recomputes I by multiplying each pixel value by a. This operation affects the denotation object only. Another operation is

$$I = delta \gg J$$

which shifts image J by vector amount delta.

Many operations on images are invoked through function calls. For example,

$$I = Edge(J)$$

computes an image by applying an edge filter (e.g., the Sobel operator). Convolutions and Fourier transforms are performed in this way. Feature extraction is again a function:

$$F = Feature(I, R, stat)$$

where F is a feature datatype (e.g., a structure consisting of the mean and standard deviation), I is an image, and R is a region. *Feature* is a generic method that takes a keyword third parameter. The feature is computed over the intersection of region R and the support set of the image.

Transformations of the plane are a crucial part of DIG semantics. These are defined as functions from the plane to the plane. The basic definition is: If T is a transformation, and p is any point in the plane, then $T(p)$ is a point in the plane. Tranformations have the natural semigroup property of composition, i.e., for any two transformations T_1 and T_2 there is a third transformation T_3 such that for all points p

$$T_1(T_2(p)) = T_3(p)$$

and we write $T_3 = T_1\ T_2$. Transformations induce mappings on point sets,

$$T(S) = \{p | p = T(p') \text{ for some } p' \text{ in } S\}$$

and on images,

$$(T[I])(p) = I(T(p))$$

whenever $T(p)$ is in the support set of I. DIG supports the class of affine transformations, which can be represented by the equations

$$x' = a_{11}x + a_{12}y + a_{13}$$

$$y' = a_{21}x + a_{22}y + a_{23}$$

where (x, y) is a point in the plane, (x', y') is the image of (x, y), and a_{ij} are real numbers. This class includes shifts, rotations, reflections about the axes, independent scalings along the two axes, and is closed (the composition of two affine transformations is an affine transformation). The representation of an affine transformation in terms of the a_{ij} coefficients is awkward: we normally think of transformations as being composed of simpler transformations such as shifts, rotations, etc. We therefore implement an algebra with operations such as

$$T2 = T1 \gg p \ (preshift)$$

where $T2$ is an affine transformation and p is a vector. This denotes the transformation obtained by first shifting a distance p and then transforming with T, i.e.,

$$T_2(p') = T_1(p + p').$$

In general, shifts do not commute with other transformations, so that

$$T \gg p \neq p \gg T$$

The operator notation for transformation composition is given in Table 1. Unary operators have higher precendence than binary, and order of

TABLE 1
Operator Notation for Transformations[a]

Notation	Action
$S * T$	Function composition
$\gg p$	Shift by p
$T \gg p$	$T * (\gg p)$ preshift
$p \gg T$	$(\gg p) * T$
$\hat{}\, a$	Rotate by a
$T \hat{}\, a$	$T * (\hat{}\, a)$
$a \hat{}\, T$	$(\hat{}\, a) * T$
$* a$	Scale by a
$T * a$	$T * (* a)$
$a * T$	$(* a) * T$
$* p$	Scale by components of p
S/T	$S * T^{-1}$
$1/T$	T^{-1}
$S + T$	$[S + T](p) = S(p) + T(p)$

[a] In this table S and T are transformations, a is a scalar, and p is a vector.

evaluation is right to left except where parentheses are used. With this notation, a transformation that rotates by angle a about point p is produced by the expression

$$T = p \gg \hat{} a \gg (-p)$$

The expression $T = a$ (a is a scalar) is a shorthand for $T = * a$, so that an identity transform is produced by the statement $T = 1$. There are assignment forms for all operators except for/, and the convention is

$$Tf = S \equiv T = SfT$$

This scheme is at variance with the interpretation of assignment operators in C language, but postoperations are more natural and common than preoperations.

The design of the objects in the vision kernel was made with the goal of efficient computation and memory utilization. The basic fact of computations on images and regions is that these objects require substantial amounts of memory and operations on these objects require many computer cycles. We use the principle of virtual objects, as contrasted with actual objects. To explain this principle, consider the predicate function member (p, R) which is true if point p is an element of set (region) R. If $S = T(R)$,where T is a transformation, then

$$\text{member}(p, S) \int \text{member}(T^{-1}(p), R)$$

Assume that region R in the following code fragment is represented with a bulky defined object such as a *Regionbit* structure (bitplane), and consider the following code fragment:

```
S = T(R); /*transform R by T*/
if(member(p, S)) ...
```

If the transformation was to involve the calculation of a new *Regionbit* structure this would consume substantial time and memory. Parenthetically, this typically produces an approximate rendition of the transformed region. But because the U level region R consists of a bitplane and a transformation, the first statement causes S to consist of a pointer to the bitplane R and a transformation reflecting the effect of T. The *member* function in the second statement maps p with this transformation, and performs the test on the bitplane. This is efficient if only a few such tests are required. However, if many such tests are to be performed, it may be more efficient to compute an actual rather than virtual region object. This may be done by

$$S = \text{Eval}(T(R))$$

In summary, the scheme allows efficient and accurate implementation of image computations, and leaves control in the programer's hand, allowing intelligent program optimization. The "virtual object" notion is applied with both regions and images. For example,

$I = (*.5)(I);$ /**reduce image by a factor of* 2*/

$x = I[p];$ /**find the grey value at p**/

would require massive calculation to find the actual image array in the first statement. With a virtual image object, this code segment requires the calculation of a transformation in the first statement, and the transformation of a point in the second.

4.2. Interactive Interface

The major thrust of this work is to provide an environment to develop interactive programs by simply dictating the responses to certain user actions. Menu is a decision object object that is decoupled from windows that display images, graphics, and text. Menuing requires a protocol of communication to a windowing system, gathering input events produced by the operator and causing the appropriate action on the required object in the correct window. The associations between menus specified by the programmer are converted to a directed graph during parsing. Menus can be of two types: popup or bar. The former are transitory and do not change the current menu bar, whereas the latter replace the current menu bar. The DIG compiler produces the control mechanism necessary to manage the user interaction. In bar menus, the option QUIT is added on by the system so that the parent (or calling menu) node can be activated and the present menu can be erased. The code segment in Table 2 sets up the menu in Fig. 3 and calls to the appropriate functions when those menu options are selected. The "top level" bar menu (lines 5–12, and 23) consists of items FILE, DRAW, SELECT, FILLIN, TRANSFORM, and QUIT (the laser is inserted by the system). TRANSFORM invokes a new bar menu (lines 13–16) with items ROTATE, SCALE and QUIT. The SCALE item is a popup (lines 17–21) with two items, GROW and SHRINK. The action associated with each menu is a program between braces after the menu item name. For example, the action for FILE is to call function *getfile*(), which reads an image into the data window, while DRAW invokes the *Draw* primitive, which returns a line-type object to a program (the outline of a region). The SELECT menu option invokes a primitive *Pick* function, which identifies one of the line objects on the screen. The object can be operated on in either the FILLIN or TRANSFORM options. The TRANSFORM option produces a

TABLE 2
Code Fragment for Creating Menus

```
Region pixmap;
Line item, outline;
position menu at x = 240 y = 240;
make_window size x = 100 y = 100 at x = 240 y = 300;
barmenu in Helvetica Bold Oblique
begin
        FILE                { getfile(); }
        DRAW                { outline = Draw(); }
        SELECT              { item = Pick(); }
        FILLIN              { pixmap = (Region) item;
                                          Show(pixmap); }
        TRANSFORM
        barmenu in Times Roman
        begin
                            ROTATE          { item = 5 ^ item; }
                            SCALE
                            popmenu in Courier
                            begin
                                    GROW            { item *= 2; }
                                    SHRINK          { item *= 0,5; }
                            end
        end
end
```

new bar menu (erasing the previous menu) and allows two choices: ROTATE rotates the item by a fixed amount, and SCALE produces a popup menu with two alternatives, GROW and REDUCE.

Though objects in the vision kernel are independent of the graphics system, there is a correspondence between the image processing world (internal representation) and graphics representation (external display). These objects are represented by geometric models and are independent of resolution. The display gives visual feedback to the operator, who can verify the accuracy of the operations and draw inferences relevant to the underlying experiment. The graphics system renders all classes of objects and supports presentation graphics which include annotations and displays such as histograms of grey values falling in a certain region of the image.

Graphical interactions comprise display of static and moving objects. The primitive Show posts all classes of objects. Interactive primitives that transform the object track the operator's movements and respond accordingly. They need to erase the object from its present position, compute the new position and shape of the object, and display the result of this computation. Having planes for graphics display independent from those for the image makes this action simpler as then we can erase the whole graphics plane and post all freshly computed graphics objects at their new position. For lack or real time performance in the state of the art monitors to display large arrays of 8-bit values we do not implement an interactive image-shifting primitive.

Interactive Graphics Support Environment (IGSE) is set of subroutines or tools that allow you to program at a higher level than the native windowing system by judicious layering of subroutines. The interactions between the various aspects of the system can be visualized as shown in Fig. 5. A user process communicates with all managers that handle inquiries and actions in that category. Input handler uses coordinates from mouse to decide which window and in turn which object inside the window is currently selected. The display subsystem generates the low-level calls that are device dependent. We have a two-level manager consisting of a *Dialog Handler* that returns a coordinate and a *Menu Manager* that filters out interaction with menu bar and passes other events to the user code. One may need to control the *Dialog Handler* directly in some cases and bypass the heirarchical structure imposed by the window manager. A situation that warrants the preemption of the menu control is in a draw routine, so that the routine does not quit if the user accidently strays onto the menu bar.

Simple primitives are provided which communicate with the *Dialog*

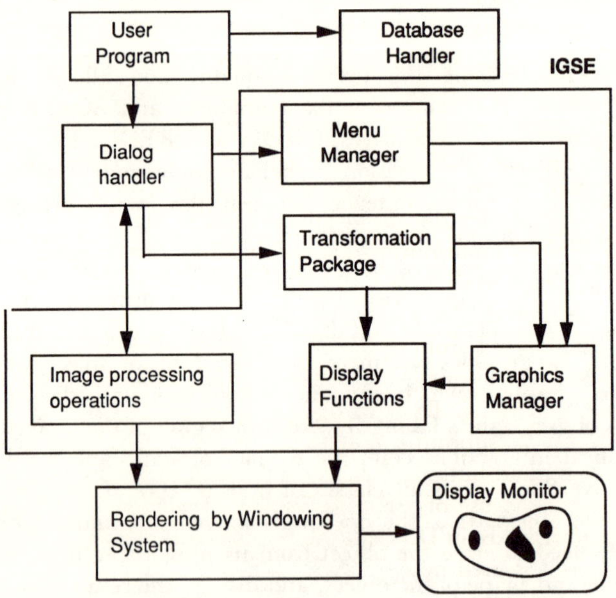

FIGURE 5. Interaction of the separate parts of the system.

Handler and the screen drivers and perform simple activities. The primitive PICKMOUSE returns the coordinates of the mouse. REFRESH posts the image and the graphics objects indicated by the call.

Fancy primitives are the ones that perform a series of activities and return an object or silently update all necessary structures with no effect on program flow. They include Draw, which allows the operator to sketch an outline made up of a bunch of points in a window and then makes a single unit out of those points so that the operator can choose the object using the primitive SELECT and then modify it using interactive transformation (under operator's control) or noninteractive transformation primitives (under program's control).

The interactive transformations described earlier involve the update of the corresponding structure in the vision kernel and also require real-time update of the display. All of this is behind the scenes and does not have to be done explicitly by the programmer.

The basic problem in portability is the diversity of windowing environments. We utilize layering and modularity to decrease the complexity for the user. For the sake of simplicity we need to restrict choices; thus, there is a restricted graphics model supported by system. Layering gives generality and simplicity but slow execution speed. In a usual layered system we layer on one system, but here we are layering on two things: objects in image (borders, regions) and windowing (display system). A transformation matrix must be applied to determine the real coordinates and image pixels, and between real coordinates and display representation. Mapping from the pixel space to display space (in digital coordinates) compensates for the aspect ratio of the display device.

The alert boxes are widgets and need to be tended to by the operator before he/she can go on to something else as they supersede the menu bar. They are useful in providing a level of security (protecting people from themselves) so that they do not accidentally erase hours of hard work. The boxes have limited input/output capability; they are capable of displaying text alone and can receive only PICK inputs. No modification or drawing of graphics is allowed. The alert box returns the option picked as an object. If it is a file selected from a list of filenames, the object can be passed to an OPENFILE routine. If it is the response to a question, the answer is provided as a string. Because of the diversity of use, the alert box is supported through subroutine calls. A future version will allow one to describe the text and buttons to be put up and have a single object returned no matter what the use of the box. But before we are able to provide the semantic of alert boxes in DIG, all the problems in unification of the various ways in which it is used need to be solved.

The program translator shown in Fig. 6 depicts the interface between

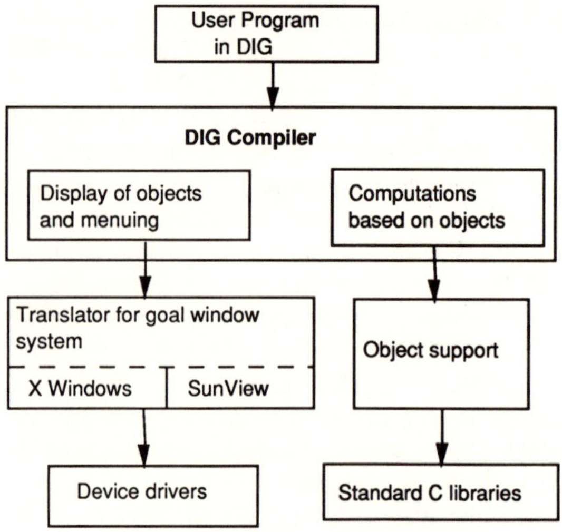

FIGURE 6. Translation process for DIG.

the graphics library and the underlying windowing system. The interface consists of a code generator that produces calls in the specific system being used. By changing the supporting libraries one can produce code to run on different environments as the inclusion of device-dependent code is delayed till the library resolution or linking stage.

With a language as opposed to a set of subroutine calls we can support new objects and do strong typing (a feature not in C). This leads us to the issue of dependencies on the numerical and display hardware. As the DIG compiler is written in C under the UNIX environment, most of it can port to other UNIX machines with recompiling and relinking. The dependencies on the display hardware are, however, not easily portable. The device level libraries will have to be rewritten as the scanner technology and graphics hardware evolves. Also the manipulation of the color map of the monitor and the number of graphics planes available differ from machine to machine. On top of this, we have the handling of the cursor and the menuing which should coexist with the graphics and image display, i.e., share the same real estate resources. But there is hope that the need for custom adaptation will also soon be reduced. X Windows is fast becoming a widely accepted platform for graphics much in the same way that C is for nongraphical code. With the graphical calls based in Xlib, we will get a large degree of device independence.

5. *Discussion and Conclusions*

5.1. *Paradigm and Instance*

In this chapter we present the design of a language that supports both imaging and interactive graphics. The design and implementation of such a language poses a number of problems. First, for a language to be effective, we need to have an extensive set of conventional operations, and these are supported by conventional languages. We respond to this need by embedding the language in a standard compiler and operating system—C and UNIX,™ in our implementation: this makes our language a dialect of C. Second, we strive for natural expression for the intended application area. We do this by defining objects that have a natural meaning in the application domain, and operators that have intuitive sense in the intended application context. In this paper, the objects are images, geometric entities, and interactive graphics and program controls constructs. The domains of computer vision and interactive graphics are very large, and expressing a substantial portion of these domains is an overwhelming task. A language that would express a large choice of image processing and computer graphics operations as primitives would be difficult to implement, maintain, and learn. We choose a collection of operations from this domain: this is a second sense in which our language is a dialect. DIG is intended for research in image segmentation, and for the construction of interactive software for quantitative data acquisition of images. The design we present is an instance of paradigm for designing other such dialects.

5.2. *Language Translation*

We considered several alternatives for constructing a translator for this language. They were (1) subroutine libraries, (2) macro preprocessor, (3) definitions in an object oriented language, and (4) a translator consisting of a scanner/parser that produces object code in C. We feel that a syntactic processor, i.e., a language in itself, is the best choice. Subroutine libraries in a language such as C are a necessary component for this sort of project: the D-level structures are best built in this way. However, the syntax for invoking subroutines is quite awkward, and poses limitations on efficient implementation and natural expression. Consider the example in Section 4.1, where the statement

$$T = p \gg {}^{\wedge}a \gg (-p)$$

is used to construct a rotation by angle *a* about point *p*. If we have functions *Indent*( ), *Shift* (*trans*, *vector*), and *Rotate* (*trans*, *scalar*) that, respectively, construct an identity transformation, produce a shift of a transformation, and produce a rotation of a transformation, then the above statement could be programmed by

$$\{\ vect\ x;$$
$$x = Negate(p);\ \ T = Ident();$$
$$T = Shift(T, x);\ \ T = Rotate(T, a);$$
$$T = Shift(T, p);\ \ \}$$

Packaging this into a subroutine will reduce the amount of code, but one has to remember more subroutine names and calling sequences. The operation can be implemented more campactly by

$$\{\ vect\ x = Negate(p);$$
$$T = Shift(Rotate(Shift(Ident(), x), a), p);\ \ \}$$

but nobody will applaud this construct. (A macro preprocessing suffers from similar limitations.) The naturalness gained with binary infix operators is clear.

Another advantage to using a compiler is the possibility of having various destination domains. Presently, the display is handled through software (calls to the host window system). With the increasing need for faster display, one sees the emerge of hardware to handle image-processing functions with built-in display routines. These cards have their own instruction set, much like that of a coprocessor. With a compiler we can generate code in the card's instruction set. Extending this further, multiprocessor machines can be supported. Code can be produced in a parallelized fashion to speed up image processing functions, such as convolutions.

A possible extension to the system is logical connections between data window, whereby drawing in one window is reflected in others as well and already drawn outlines could be copied over. An example of this would be the propagation of the outline of a structure in an image to serial sections. To refine the outline, one could run an adaptive routine that takes the original rough outline and finds the structure in the new image more accurately.

Infix operators and operator overloading can be achieved with an object-oriented language such as C++.[20] The reason we have not pursued this path is, primarily, that we have had difficulties with the versions of C++ available to us. We feel that a translator constructed with LEX/YACC is a more flexible in that it allows more freedom in the design of the syntax.

References

1. S.S. ABI-EZZI AND A.J. BUNSHAFT, An implementer's view of PHIGS, *IEEE Comput. Graph. Appl.* **6**(2), 12–23, 1986.
2. D. SHUEY, D. BAILEY, AND T.P. MORRISSEY, PHIGS: A standard, dynamic, interactive graphics interface, *IEEE Comput. Graph. Appl.* **6**(8), 50–57, 1986.
3. P.S. BARTH, An object-oriented approach to graphical interfaces, *ACM Trans. Graph.* **5**(2), 142–172, 1986.
4. B.J. MAC LENNAN, *Principles of Programming Languages: Design, Evaluation, and Implementation*, Holt, Rinehart and Winston, New York, 1987.
5. M.J.B. DUFF AND S. LEVIALDI, *Languages and Architectures for Image Processing*, Academic Press, London, 1981.
6. A. EVANS, JR. AND K.J. BUTLER, (Eds.), *Diana*—An intermediate Language for Ada, Revised Version, Springer-Verlag, New York, 1983.
7. P.H. FELLER, Relationship between IDL and structure editor generation technology, *SIGPLAN Not.* **22**(11), 1987.
8. A. GOLDBERG AND D. ROBINSON, *Smalltalk-80: The Language and its Implementation*, Addison-Wesley, Reading, Massachusetts, 1983.
9. M. GREEN, The University of Alberta User Interface Management System, *SIGGRAPH '85 Proceedings*, 205–213, 1985.
10. G. ENDERLE, K. KANSKY, AND G. PFAFF, *Computer Graphics Programming GKS—The Graphics Standard*, Springer-Verlag, Berlin, 1984.
11. D.A. LAMB, Sharing intermediate representations: The interface description language ph. D. dissertation, Computer Science Department, Carnegie-Mellon University, May 1983.
12. M.A. LINTON, J.M. VLISSIDES, AND P.R. CALDER, Composing user interfaces with interviews, *IEEE Comput.* **22**(2), 8–22, 1989.
13. P.S. MODENOV AND A.S. PARKHOMENKO, *Geometric Transformations*, Vols. 1 and 2, Academic Press, New York, 1965.
14. J.R. NESTOR, W.A. WULF, AND D.A. LAMB, IDL—Interface description language—Formal description, Technical Report, Software Engineering Institute, Pittsburg, Pennsylvania, February, 1986.
15. W.M. NEWMAN AND R.F. SPROULL, An Approach to Graphics Systems Design, *Tutorial: Computer Graphics*, IEEE Computer Society, New York, 2nd ed., April 1982.
16. R.W. SCHEIFLER AND J. GETTYS, The X window system, *ACM Trans. Graph.* **5**(2), 79–109, 1986.
17. J. SERRA, *Image Analysis and Mathematical Morphology*, Academic Press, London, 1982.
18. K. SHANNON, T. MARONEY, AND R. SNODGRASS, Using IDL with C, SoftLab Document No. 6, Computer Science Department, University of North Carolina at Chapel Hill, May 1985.
19. R. SNODGRASS, Ed., IDL manual entries (Version 2.0), SoftLab Document No. 15. Computer Science Department, University of North Carolina at Chapel Hill, December 1985.
20. B. STROUSTRUP, *The C++ Programming Language*, Addison-Wesley, Reading, Massachusetts, 1986.
21. W.B. WARRREN, J. KICKENSON, AND R. SNODGRASS, A tutorial introduction to using IDL, *SIGPLAN Not.* **22**(11), 1987.
22. A. WILLIAMS, An architecture for user interface R&D, *IEEE Comput. Graph. Appl.* **6**(7), 39–50, 1986.
23. A. ROSENFIELD AND A.C. KAK, *Digital Picture Processing*, Vol. 1, Academic Press, New York, 2nd ed., 1982.
24. A. ROSENFIELD AND A.C. KAK, *Digital Picture Processing*, Vol. 2, Academic Press, New York, 2nd ed., 1982.
25. M.K. HU, Visual pattern recognition by moment invariants, *I.R.E. Trans. Inf. Theory*, 1962.

INDEX

CARDIFF
UWCC LIBRARY